Applied Mathematics in Engineering and Computer Science

Applied Mathematics in Engineering and Computer Science

Edited by **Frank West**

Willford Press

New York

Published by Willford Press,
118-35 Queens Blvd., Suite 400,
Forest Hills, NY 11375, USA
www.willfordpress.com

Applied Mathematics in Engineering and Computer Science
Edited by Frank West

International Standard Book Number: 978-1-68285-186-9 (Hardback)

Printed in China.

Contents

Preface

As a branch of mathematics, applied mathematics has applications in a variety of disciplines ranging from science and engineering to business and economics. This innovative and comprehensive book integrates the well-developed theory and practical applications of applied mathematics especially in the fields of engineering and computer science. Differential equations, algebra, computational mathematics, etc. are some of the areas that have been elucidated in the text. It includes contributions of experts and scientists which will provide innovative insights into this field and will be extremely beneficial for research scholars.

Various studies have approached the subject by analyzing it with a single perspective, but the present book provides diverse methodologies and techniques to address this field. This book contains theories and applications needed for understanding the subject from different perspectives. The aim is to keep the readers informed about the progress in the field; therefore, the contributions were carefully examined to compile novel researches by specialists from across the globe.

Indeed, the job of the editor is the most crucial and challenging in compiling all chapters into a single book. In the end, I would extend my sincere thanks to the chapter authors for their profound work. I am also thankful for the support provided by my family and colleagues during the compilation of this book.

Editor

A three-layer supply chain integrated production-inventory model under permissible delay in payments in uncertain environments

Dipak Kumar Jana[1]*, Kalipada Maity[2] and Tapan Kumar Roy[3]

*Correspondence:
dipakjana@gmail.com
[1] Department of Engineering Science, Haldia Institute of Technology, Haldia Purba Midna Pur 721657, West Bengal, India
Full list of author information is available at the end of the article

Abstract

In this paper, an integrated production-inventory model is presented for a supplier, manufacturer, and retailer supply chain under conditionally permissible delay in payments in uncertain environments. The supplier produces the item at a certain rate, which is a decision variable, and purchases the item to the manufacturer. The manufacturer has also purchased and produced the item in a finite rate. The manufacturer sells the product to the retailer and also gives the delay in payment to the retailer. The retailer purchases the item from the manufacture to sell it to the customers. Ideal costs of supplier, manufacturer, and retailer have been taken into account. An integrated model has been developed and solved analytically in crisp and uncertain environments, and finally, corresponding individual profits are calculated numerically and graphically.

Keywords: Production inventory system; Uncertain variables; Three-layer supplier chain; Delay in payment

Introduction

Supply chain management has taken a very important and critical role for any company in the increasing globalization and competition in the market. A supply chain model (SCM) is a network of suppliers, producers, distributors, and customers which synchronizes a series of interrelated business process in order to have (1) optimal procurement of raw materials from nature, (2) transportation of raw materials into a warehouse, (3) production of goods in the production center, and (4) distribution of these finished goods to retailers for sale to the customers. With a recent paradigm shift to the supply chain (SC), the ultimate success of a firm may depend on its ability to link supply chain members seamlessly.

One of the earliest efforts to create an integrated SCM has been developed by Oliver and Webber [1], Cohen and Baghanan [2], and Cachon and Zipkin [3]. They developed a production, distribution, and inventory (PDI) planning system that integrated three supply chain segments comprising supply, storage/location, and customer demand planning. The core of the PDI system was a network model and diagram that increased the decision maker's insights into supply chain connectivity. The model however was confined to a single-period and single-objective problem. Viswanathan and Piplani [4] were concerned an integrated inventory model through common replenishment in the SC. Khouja [5] was the first to consider a three-stage supply chain with one or more firms at each stage. Agarwal et al. [6] have developed a dynamic balancing of inventory model in supply chain management. Rau et al. [7] developed an integrated SCM of a deteriorating item with shortages. Lee [8] added a new dimension to the single vendor-single buyer problem by setting the number of raw material shipments received by the vendor per cycle to be a decision variable. Ben-Daya et al. [9] have developed an integrated production-inventory model with raw material replenishment considerations in a three-layer supply chain. Sana [10] has integrated a production-inventory model of imperfect quality products in a three-layer supply chain. Recently, Pal et al. [11] have developed a three-layer supply chain model with production-inventory model for reworkable items. All of the abovementioned SCMs are considered with constant, known demand and production rates in a crisp environment.

Different types of uncertainty such as fuzziness, randomness, and roughness are common factors in SCM. In many cases, it is found that some inventory parameters involve fuzzy uncertainty. For example, inventory-related costs such as holding cost and setup cost, demand, and selling price depend on several factors such as bank interest, stock amount, and market situation which are uncertain in a fuzzy sense. To be more specific, inventory holding cost is sometimes represented by a fuzzy number, and it depends on the storage amount which may be imprecise and range within an interval due to several factors such as scarcity of storage space, market fluctuation, human estimation, and/or thought process. The following papers have been developed in these environments.

Wang and Shu [12] developed a fuzzy decision methodology that provides an alternative framework to handle SC uncertainties and to determine SC inventory strategies, while there is a lack of certainty in data or even a lack of available historical data. Fuzzy set theory is used to model SC uncertainty. A fuzzy SC model based on possibility theory is developed to evaluate SC performances. Based on the proposed fuzzy SC model, a genetic algorithm approach is developed to determine the order-up-to levels of stock-keeping units in the SC to minimize SC inventory cost subject to the restriction of fulfilling the target fill rate of the finished product. The proposed model allows decision makers to express their risk attitudes and to analyze the trade-off between customer service level and inventory investment in the SC, so that better SC inventory strategies can be made.

Das et al. [13] have presented a joint performance of an SC with two warehouse facilities in a fuzzy environment. A realistic two-warehouse and multi-collection-production-inventory model with constant/stock-dependent demand, defective production system, and fuzzy budget constraint has been formulated and solved in an SC context. Later Chen et al. [14] developed a multi-criteria fuzzy optimization for locating warehouses and distribution centers in a supply chain network.

Peidro et al. [15] developed a fuzzy linear programming model for tactical supply chain planning in a multi-echelon, multi-product, multi-level, multi-period supply chain network in a fuzzy environment. In this approach, the demand, process, and supply uncertainties are jointly considered. The aim is to centralize multi-node decisions simultaneously to achieve the best use of the available resources along the time horizon so that customer demands are met at a minimum cost. This proposal is tested using data from a real automobile SC. The fuzzy model provides the decision maker with alternative decision plans with different degrees of satisfaction.

Chu [16] developed the supply chain flexibility that has become increasingly important. This study thus builds a group decision-making structure model of flexibility in supply chain management development. Recently, Jana et al. [17] have developed a fuzzy simulation via contractive mapping genetic algorithm approach to an imprecise production-inventory model under volume flexibility. This study presents a framework to evaluate the supply chain flexibility that comprises two parts: (1) an evaluation hierarchy with flexibility dimensions and related metrics and (2) an evaluation scheme that uses a three-stage process to evaluate the supply chain flexibility. This study then proposes an algorithm to determine the degree of supply chain flexibility using a fuzzy linguistic approach. Evaluations of the degree of supply chain flexibility can identify the need to improve supply chain flexibility and identify specific dimensions of supply chain flexibility as the best directions for improvement. The results of this study are more objective and unbiased for two reasons. First, the results are generated by group decision-making with interactive consensus analysis. Second, the fuzzy linguistic approach used in this study has more advantages to preserve no loss of information than other methods. Additionally, this study presents an example using a case study to illustrate the availability of the proposed methods and compare it with other methods.

Kristianto et al. [18] developed an adaptive fuzzy control application to support a vendor-managed inventory (VMI). This paper also guides the management in allocating inventory by coordinating with suppliers and buyers to ensure minimum inventory levels across a supply chain. Adaptive fuzzy VMI control is the main contribution of this paper.

However, the uncertainty theory was developed by Liu [19], and it can be used to handle subjective imprecise quantity. Much research work has been done on the development of the uncertainty theory and related theoretical work. You [20] proved some convergence theorems of uncertain sequences. Liu [21] has defined uncertain process and Liu [22] has discussed uncertain theory. In this paper, we developed for the first time a three-layer supply chain model under delay in payment in an uncertain environment.

In the traditional economic order quantity (EOQ) model, it often assumed that the retailer must pay off as soon as the items are received. In fact, the supplier offers the retailer a delay period, known as trade credit period, in paying for the purchasing cost, which is a very common business practice. Suppliers often offer trade credit as a marketing strategy to increase sales and reduce on-hand stock level. Once a trade credit has been offered, the amount of the tied up retailer's capital in stock is reduced, and that leads to a reduction in the retailer's holding cost of finance. In addition, during the trade credit period, the retailer can accumulate revenues by selling items and by earning interests. As a matter of fact, retailers, especially of small businesses which tend to have a limited number of financing opportunities, rely on trade credit as a source of short-term funds. In this research field, Goyal [23] was the first to establish an EOQ model with a constant demand rate under the condition of permissible delay in payments. Khanra, Ghosh, and Chaudhuri [24] have developed an EOQ model for a deteriorating item with time-dependent quadratic demand under permissible delay in payment. Also,

Maihami and Abadi [25] have established joint control of inventory and its pricing for non-instantaneously deteriorating items under permissible delay in payments and partial backlogging.

The proposed model considers a three-layer supply chain involving the supplier, manufacturer, and retailer who are responsible in transforming the raw materials into finished product and making them available to satisfy the customer's demand time. Inventory and production decisions are made at the supplier, manufacturer, and retailer levels in uncertain environments. The problem is to coordinate production and inventory decisions across the supply chain so that the total profit of the chain is maximized.

Necessary knowledge about uncertain variables

To better describe subjective imprecise quantity, Liu in [19] proposed an uncertain measure and further developed an uncertainty theory which is an axiomatic system of normality, monotonicity, self-duality, countable subadditivity and product measure.

Definition 1. Let Γ be a non-empty set and L be a σ algebra over Γ. Each element $\Lambda \in L$ is called an event. A set function $\mathcal{M}\{\Lambda\}$ is called an uncertain measure if it satisfies the following four axioms of Liu [19]:

Axiom 1. (Normality) $\mathcal{M}\{\Lambda\} = 1$

Axiom 2. (Monotonicity) $\mathcal{M}\{\Lambda\}+\mathcal{M}\{\Lambda^C\} = 1$, for any event Λ

Axiom 3. (Countable subadditivity) For every countable sequence of events $\Lambda_1, \Lambda_2, \ldots,$ we have

$$\mathcal{M}\left\{\sum_{i=1}^{\infty} \Lambda_i\right\} \leq \sum_{i=1}^{\infty} \mathcal{M}\left\{\Lambda_i\right\}.$$

Definition 2. (Liu [19]) The uncertainty distribution $\Phi : R \to [0, 1]$ of an uncertain variable ξ is defined by

$$\Phi(t) = \mathcal{M}\{\xi \leq t\}.$$

Definition 3. (Liu [19]) Let ξ be an uncertain variable. Then the expected value of ξ is defined by

$$E[\xi] = \int_0^{\infty} \mathcal{M}\{\xi \geq r\}dr - \int_{-\infty}^{0} \mathcal{M}\{\xi \leq r\}dr,$$

provided that at least one of the two integrals is finite.

Theorem 1. (Liu [26]) *Let ξ be an uncertain variable with uncertainty distribution Φ. If the expected value exists, then $E[\xi] = \int_0^1 \Phi^{-1}(\alpha)d\alpha$.*

Lemma 1. *Let $\xi \sim \mathcal{L}(a, b, c)$ be a zigzag uncertain variable. Then its inverse uncertainty distribution $\Phi^{-1}(\alpha) = \frac{1}{2}[(1 - \alpha)a + b + \alpha c]$, and it can be expressed as*

$$E[\xi] = \int_0^1 \frac{1}{2}[(1 - \alpha)a + b + \alpha c]\, d\alpha = \frac{a + 2b + c}{4}. \tag{1}$$

Theorem 2. (Liu [26]) *Let $\xi_1, \xi_2, \ldots, \xi_n$ be independent uncertain variables with uncertainty distributions $\phi_1, \phi_2, \ldots, \phi_n$, respectively. If f is a strictly increasing function, then $\xi = f(\xi_1, \xi_2, \ldots, \xi_n)$ is an uncertain variable with inverse uncertainty distribution*

$$\Phi^{-1}(\alpha) = f\left(\Phi_1^{-1}(\alpha), \Phi_2^{-1}(\alpha), \ldots, \Phi_n^{-1}(\alpha)\right).$$

Theorem 3. (Liu [26]) *Let ξ_1 and ξ_2 be independent uncertain variables with finite expected values. Then for any real numbers a_1 and a_2, we have*

$$E[a_1\xi + a_2\eta] = a_1E[\xi] + a_2E[\eta]. \quad (2)$$

Assumptions and notations

Assumptions

The following assumptions are considered to develop the model:

(i) Models are developed for single item product.
(ii) Lead time is negligible.
(iii) Joint effect of supplier, manufacturer, retailer is consider in a supply chain management.
(iv) Supplier produced the item with constant rate unit per unit time, which is a decision variable.
(v) Total production rate of manufacturer is equal to the demand rate of manufacturer.
(vi) The manufacturer give the opportunity to the retailer conditionally permissible delay in payment.
(vii) Idle cost of suppliers, manufacturer and retailer are also assumed.

Notations

The following notations are considered to develop the model:

- p_s = production rate for the suppliers, which is a decision variable.
- p_m = demand rate or production rate for the manufacturer.
- D_r = constant demand rate for the retailer.
- D_c = constant demand rate of customer.
- C_s = purchase cost of unit item for suppliers.
- C_m = selling price of unit item for suppliers which is also the purchase cost for manufacturer.
- C_r = selling price of unit item for manufacturer which is also the purchase cost for retailers.
- C_{r_1} = selling price for retailers.
- t_s = production time for supplers.
- T_s = cycle length for the suppliers.
- T_r = time duration where order is supplied by the manufacturer, by retailer's cycle length.
- T' = last cycle length of the retailer.
- T = total time for the integrated model.
- h_s = holding cost per unit per unit time for suppliers.
- h_m = holding cost per unit per unit time for manufacturer.
- h_r = holding cost per unit per unit time for retailers.

- A_s = ordering cost for suppliers.
- A_m = ordering cost for manufacturer.
- h_r = ordering cost for retailers.
- $\mathrm{id}_s, \widetilde{\mathrm{id}_s}$ = idle cost per unit time for suppliers in crisp and uncertain environments, respectively.
- $\mathrm{id}_m, \widetilde{\mathrm{id}_m}$ = idle cost per unit time for manufacturer in crisp and uncertain environments, respectively.
- $\mathrm{id}_r, \widetilde{\mathrm{id}_r}$ = idle cost per unit time for retailers in crisp and uncertain environments, respectively.
- n = number of cycle for retailers.
- r = number of cycles where manufacturer stops production.
- M = retailer's trade credit period offered by the manufacturer to the retailers in years, which is the fraction of the years.
- I_p = interest payable to the manufacturer by the retailers.
- $I_e, \widetilde{I_e}$ = interest earned by the retailers in crisp and uncertain environments, respectively.
- ATP = average total profit for the integrated models.
- P_m^* = optimum value of P_m for integrated models.
- ATP* = optimum value of average total profit for the integrated models.

Model description and diagrammatic representation

The integrated inventory model (Figure 1) starts when $t = 0$ and stock is zero. At that time, the suppliers start their production with the rate p_s unit per unit time and purchase at the rate p_m unit per unit time to the manufacturer. When $t = t_s$, suppliers stop their production, and at $t = T_s$, the inventory level of suppliers become zero. The total time of the integrated model is T, so the idle time for suppliers is $T - T_s$. Similarly, the

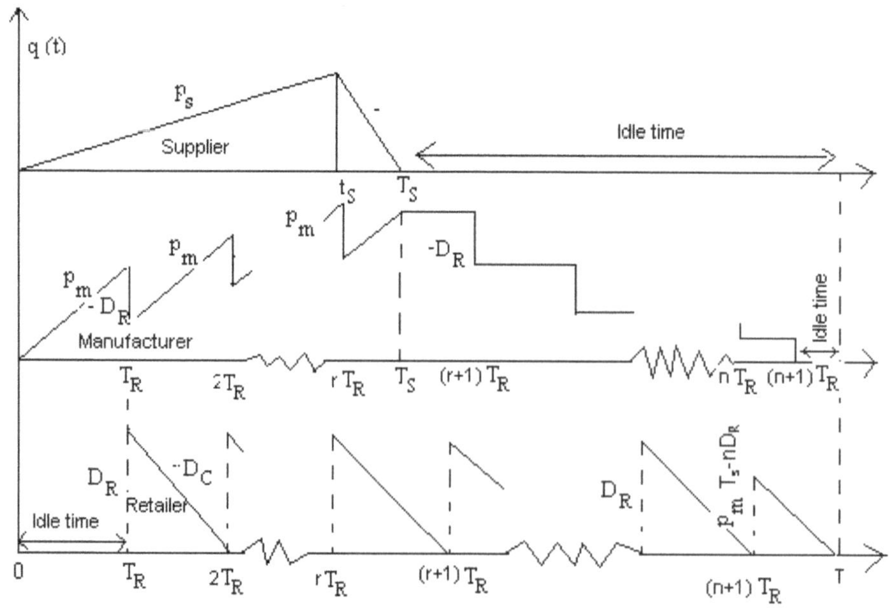

Figure 1 Inventory level for the integrated model.

manufacturers start their production at the same time $t = 0$ with the production rate p_m unit per unit time and purchase this production D_r unit to the retailer in the time gap T_R, which is the bulk pattern. At time $t = T_s (= [\frac{T_s}{T_R}])$, manufacturers stop their production, and at $t = (n+1)T_R$ $(n = [\frac{p_m T_s}{D_r}])$, the stock of manufacturer is zero. Thus, idle period for the manufacturer is $T - (n+1)T_r$. Retailers start selling this product to the customers at time $t = T_r$ and end selling at $T = (n+1)T_r + \frac{p_m T_s - nD_r}{D_c}$. The idle period for retailers is T_r.

Mathematical formulation of the model

Formulation of suppliers' individual average profit

Differential equation for the supplier in Figure 2 in $[0, T]$ is given by

$$\frac{dq_s}{dt} = \begin{cases} p_s - p_m, & 0 \le t \le t_s \\ -p_m, & t_s \le t \le T_s, \\ 0, & T_s \le t \le T \end{cases}$$

with boundary conditions $q_s(t) = 0$ and $t = 0, T_s$. Solving the differential equation with the boundary condition, we have

$$q_s(t) = \begin{cases} (P_s - P_m)t, & 0 \le t \le t_s \\ P_m(T_s - t), & t_s < t \le T_s \\ 0, & T_s < t \le T \end{cases} \tag{3}$$

$$\begin{aligned} H_s &= \text{Holding cost of supplier.} \\ &= h_s \left[\int_0^{t_s} (p_s - p_m) t\,dt + \int_{t_s}^{T_s} p_m (T_s - t) dt \right] \\ &= h_s \left[\frac{p_s t_s^2}{p_m} - p_s t_s^2 \right]. \end{aligned}$$

The total idle cost $= \text{id}_s \left[T_{R+P_s t_s} \left(\frac{1}{D_c} - \frac{1}{p_m} \right) \right]$, purchase cost $= c_s p_m T_s$, selling price $= c_m p_m T_s$, and ordering cost $= A_s$.

$$\begin{aligned} \text{APS} &= \text{Average profit for supplier.} \\ &= \frac{1}{T} [\text{revenue from sale-purchase cost-holding cost-idle cost-ordering cost.}] \\ &= \frac{1}{T} \left[(c_m - c_s) p_s t_s - h_s \left(\frac{p_s^2 t_s^2}{p_m} - p_s t_s^2 \right) - \text{id}_s \left(T_R + p_s t_s \left(\frac{1}{D_c} - \frac{1}{p_m} \right) \right) - A_s \right]. \end{aligned} \tag{4}$$

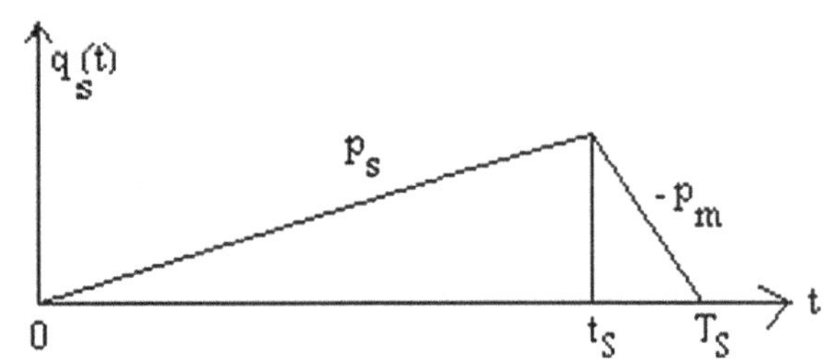

Figure 2 Inventory level of supplier.

Formulation of manufacturer individual average profit

Inventory level of manufacturer in Figure 3 in $[0,T]$ is given by

$$q_m(t) = \begin{cases} p_m t, & 0 \le t \le T_R \\ p_m t - iD_R, & iT_R < t \le (i+1)T_R \;\; i = 1,2,\ldots,(r-1) \\ p_m t - rD_R, & rT_R < t \le T_s \\ p_m T_s - rD_R, & T_s < t \le (r+1)T_R \\ p_m T_s - iD_R, & iT_R < t \le (i+1)T_R \;\; i = r+1, r+2, \ldots, n-1 \\ p_m T_s - nD_R, & nT_R < t \le (n+1)T_R \\ 0, & (n+1)T_R \le t \le T \end{cases} \tag{5}$$

with boundary conditions $q_m(0) = 0$ and $q_m(iT_R + 0) = q_m(iT_R) - D_R$.

$$\begin{aligned} H_m &= \text{Holding cost for manufacturer.} \\ &= h_m\left[\int_0^{T_R} p_m t dt + \sum_1^{r-1}\int_{iT_R}^{(i+1)T_R}(p_m t - iD_R)dt + \int_{rT_R}^{(T_s}(p_m t - rD_R)dt\right. \\ &\left.+\int_{T_s}^{(r+1)T_R}(p_m T_s - rD_R)dt + \sum_{r+1}^{n-1}\int_{iT_R}^{(i+1)T_R}(p_m T_s - iD_R)dt + \int_{nT_R}^{(n+1)T_R}(p_m T_s - nD_R)dt\right] \\ &= h_m\left[np_m T_s T_R - \frac{n^2+n-2r-2}{2}T_R D_R - \frac{p_s^2 t_s^2}{2p_m}\right]. \end{aligned}$$

The total idle cost $= \text{id}_m\left[\frac{p_m T_m - nD_R}{D_c}\right]$, purchase cost $= c_m p_m T_s$, selling price $= c_r p_m T_s$, and ordering cost $= A_m$.

Case 1. (When $M \le T' \le T_R$)

$$\begin{aligned} I_{em} = I_{pr} &= \text{Amount of interest earned by the manufacturer in } [0,T] \text{ from retailer.} \\ &= \text{Amount of interest paid by the retailer to the manufacturer in } [0,T]. \\ &= c_r I_p\left[n\int_M^{T_R}(D_R - D_c t)dt + \int_M^{T'}(p_m T_s - nD_R - D_c t)dt\right] \\ &= \frac{nc_r I_p}{2}\left[T_R D_R + D_c M^2 - 2MD_R\right] + c_r I_p\left[\left(\frac{(p_m T_s - nD_R)^2}{2D_c}\right.\right. \\ &\left.+(p_m T_s - nD_R)M + \frac{D_c M^2}{2}\right] \end{aligned}$$

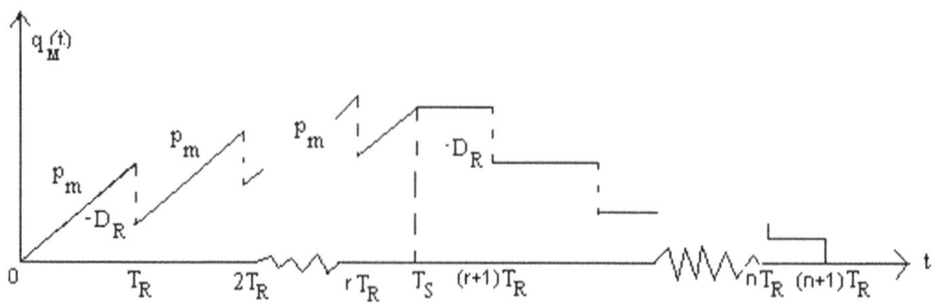

Figure 3 Inventory level of manufacturer.

$$\begin{aligned}
\text{APM}_1 &= \text{Average profit of manufacturer.}\\
&= \frac{1}{T}[\text{revenue from sale-purchase cost-holding cost-idle cost}\\
&\quad + \text{earned interest-ordering cost.}]\\
&= \frac{1}{T}\left[(c_r - c_m)p_m T_s - h_m\left(np_m T_s T_R - \frac{n^2+n-2r-2}{2}T_R D_R - \frac{p_s^2 t_s^2}{2p_m}\right)\right.\\
&\quad - \text{id}_m\left(\frac{p_m T_m - nD_R}{D_c}\right) + \frac{nc_r I_p}{2}\left[T_R D_R + D_c M^2 - 2MD_R\right]\\
&\quad \left. + c_r I_p\left(\frac{(p_m T_s - nD_R)^2}{2D_c} + (p_m T_s - nD_R)M + \frac{D_c M^2}{2}\right) - A_m\right]
\end{aligned} \tag{6}$$

Case 2. (When $T' \leq M \leq T_R$)

$$\begin{aligned}
I_{em} = I_{pr} &= \text{Amount of interest earned by the manufacturer in } [0,T] \text{ from retailer.}\\
&= \text{Amount of interest paid by the retailer to the manufacturer in } [0,T.]\\
&= c_r I_p\left[n\int_M^{T_R}(D_R - D_c t)dt + \int_M^{T'}(p_m T_s - nD_R - D_c t)dt\right]\\
&= \frac{nc_r I_p}{2}\left[T_R D_R + D_c M^2 - 2MD_R\right] + c_r I_p\left[\ (\frac{(p_m T_s - nD_R)^2}{2D_c}\right.\\
&\quad \left. +(p_m T_s - nD_R)M + \frac{D_c M^2}{2}\right]
\end{aligned}$$

$$\begin{aligned}
\text{APR}_2 &= \text{Average profit of retailer.}\\
&= \frac{1}{T}[\text{revenue from sale-purchase cost-holding cost}\\
&\quad + \text{earned interest-payable interest-idle cost-ordering cost}].\\
&= \frac{1}{T}\left[(c_{r_1} - c_r)p_m T_s - \frac{h_r}{2}\left(\frac{p_m^2 T_s^2}{D_c} - 2np_m T_s T_R - (2n+1)T_R D_R\right)\right.\\
&\quad + \frac{nc_{r_1} i_e D_c M^2}{2} + \frac{c_{r_1} i_e}{2}(p_m T_s - nD_R)(2M - T')\\
&\quad \left. - \frac{nc_r I_p}{2}\left[T_R D_R + D_c M^2 - 2MD_R\right] - \text{id}_r T_R - A_r\right]
\end{aligned} \tag{7}$$

Formulation of retailer individual average profit

Inventory level of retailer in Figure 4 in $[0, T]$ is given by

$$q_r(t) = \begin{cases} D_c t, & iT_R \leq t \leq (i+1)T_R \\ p_m T_s - nD_r - D_c t, & (n+1)T_R \leq t \leq T \end{cases} \tag{8}$$

with boundary conditions $q_r((n+1)T_R) = 0$ and $q_r(T) = 0$.

$$\begin{aligned}
H_r &= \text{Holding cost of retailer.}\\
&= nh_r\left[\int_0^{T_R}(D_R - D_c t)dt + \int_0^{T'}(p_m T_s - nD_R - D_c t)dt\right]\\
&= \frac{h_r}{2}\left[\frac{p_m^2 T_s^2}{D_c} - 2np_m T_s T_R - (2n+1)T_R D_R\right].
\end{aligned}$$

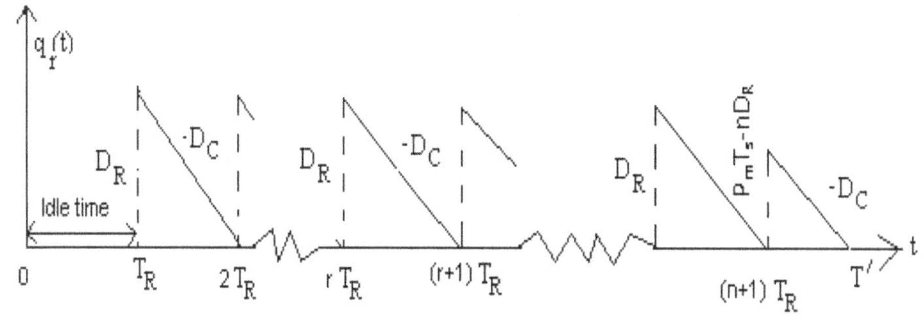

Figure 4 Inventory level of retailer.

The total idle cost $= \mathrm{id}_r T_R$, purchase cost $= c_r p_m T_s$, selling price $= c_{r1} p_m T_s$, and ordering cost $= A_r$.

Case 1. (When $M \le T' \le T_R$)
Interest earned by the retailers for $(n+1)$ cycle is given by

$$I_{er} = \text{Amount of interest earned by the retailer from the bank in } (n{+}1) \text{ cycle.}$$
$$= (n+1)c_{r_1} i_e \left[\int_0^M (M-t)D_c dt\right]$$
$$= \frac{(n+1)c_{r_1} i_e D_c M^2}{2},$$
$$I_{pr} = \text{Amount of interest paid by the retailer to the manufacturer in } [0,T].$$
$$= c_r I_p \left[n\int_M^{T_R} (D_R - D_c t)dt + \int_M^{T'} (p_m T_s - nD_R - D_c t)dt\right]$$
$$= nc_r I_p \left[\frac{T_R D_R + D_c M^2 - 2MD_R}{2}\right] + c_r I_p \left[\frac{(p_m T_s - nD_R)^2}{2D_c}\right.$$
$$\left. +(p_m T_s - nD_R)M + \frac{D_c M^2}{2}\right],$$

$$\mathrm{APR}_1 = \text{Average profit for retailer.}$$
$$= \frac{1}{T}[\text{ revenue from sale-purchase cost-holding cost}$$
$$+ \text{ earned interest-payable interest-idle cost-ordering cost.}]$$
$$= \frac{1}{T}\left[c_{r_1} p_m T_s - c_r p_m T_s - \frac{h_r}{2}\left(\frac{p_m^2 T_s^2}{D_c} - 2np_m T_s T_R - (2n+1)T_R D_R\right)\right. \tag{9}$$
$$+ \frac{(n+1)c_{r_1} i_e D_c M^2}{2} - \frac{nc_r I_p}{2}\left[T_R D_R + D_c M^2 - 2MD_R\right] - \mathrm{id}_r T_R - A_r$$
$$\left. + (p_m T_s - nD_R)M + \frac{D_c M^2}{2}\right) - c_r I_p \left(\frac{(p_m T_s - nD_R)^2}{2D_c}\right]$$

Case 2. (When $T' \le M \le T_R$)
Interest earned by the retailers for $(n+1)$ cycle is given by

$$
\begin{aligned}
I_{re} &= \text{Retailers' earned interest.} \\
&= c_{r_1} i_e \left[n \int_0^M (M-t) D_c dt + \int_0^{T'} (T'-t) D_c dt + (M-T')(p_m T_s - nD_R) \right] \\
&= \frac{n c_{r_1} i_e D_c M^2}{2} + \frac{c_{r_1} i_e}{2} (p_m T_s - nD_R)(2M - T')
\end{aligned}
$$

Interest payable by the retailers for the first n cycle is given by

$$
\begin{aligned}
I_{rp} &= \text{Retailers' payable interest.} \\
&= c_r I_p \left[n \int_M^{T_R} (D_R - D_c t) dt \right] \\
&= \frac{n c_r I_p}{2} \left[T_R D_R + D_c M^2 - 2MD_R \right]
\end{aligned}
\tag{10}
$$

$$
\begin{aligned}
\text{APR}_2 &= \text{Average profit for retailer.} \\
&= \frac{1}{T}[\text{revenue from sale-purchase cost-holding cost} \\
&\quad + \text{earned interest-payable interest-idle cost-ordering cost}] \\
&= \frac{1}{T} \left[(c_{r_1} - c_r) p_m T_s - \frac{h_r}{2} \left(\frac{p_m^2 T_s^2}{D_c} - 2np_m T_s T_R - (2n+1) T_R D_R \right) \right. \\
&\quad + \frac{n c_{r_1} i_e D_c M^2}{2} + \frac{c_{r_1} i_e}{2} (p_m T_s - nD_R)(2M - T') \\
&\quad \left. - \frac{n c_r I_p}{2} \left[T_R D_R + D_c M^2 - 2MD_R \right] - \text{id}_r T_R - A_r \right]
\end{aligned}
\tag{11}
$$

Crisp environment

Case 1. $(M \le T' \le T_R)$

$$
\begin{aligned}
\text{ATP}_1 &= \text{Total average profit for integrated model.} \\
&= \text{APS} + \text{APM}_1 + \text{APR}_1 \\
&= \frac{D_c}{p_m T_s} + D_R \left[(c_m - c_s) p_m T_s - h_s \left(\frac{p_s t_s^2}{p_m} - p_s t_s^2 \right) - \text{id}_s \left(T_R + p_s t_s \frac{1}{D_c} - \frac{1}{p_m} \right) - A_s \right. \\
&\quad + (c_r - c_m) p_m T_s - h_m \left(np_m T_s T_R - \frac{n^2 + n - 2r - 2}{2} T_R D_R - \frac{p_s^2 t_s^2}{2p_m} \right) \\
&\quad - \text{id}_m \left(\frac{p_m T_m - nD_R}{D_c} \right) \\
&\quad - A_m + (c_{r_1} - c_r) p_m T_s - \frac{h_r}{2} \left(\frac{p_m^2 T_s^2}{D_c} - 2np_m T_s T_R - (2n+1) T_R D_R \right) \\
&\quad \left. + \frac{(n+1) c_{r_1} I_e D_c M^2}{2} - \text{id}_r T_R - A_r \right] \\
&\quad - \frac{D_c}{p_m T_s + D_R} \left[A p_s^2 t_s^2 - B p_s t_s + E \right],
\end{aligned}
\tag{12}
$$

where $P_m T_s = P_s t_s$,

$$
A = \frac{h_m - h_s}{2p_m} - \frac{h_r}{2D_c} \tag{13}
$$

$$
B = (c_{r_1} - c_s) + n(h_r - h_m) T_R - \frac{\text{id}_m + \text{id}_s}{D_c} \tag{14}
$$

and

$$E = \left[h_m \frac{n^2+n-2r-2}{2} + h_r \frac{2n+1}{2}\right] T_R D_R + \frac{(n+1)c_{r_1} I_e D_c M^2}{2}$$
$$+ (n\mathrm{id}_m - \mathrm{id}_s - \mathrm{id}_r) T_R + \mathrm{id}_s T_s - (A_s + A_m + A_r) \quad (15)$$

$$\frac{d}{dp_s}(\mathrm{ATP}_1) = 0$$
$$\Rightarrow p_s = \frac{2D_R - B \pm \sqrt{(2t_s - A)^2 - 4(2t_s - A)(E - BD_R)}}{2(2t_s - A)} \quad (16)$$

$$\frac{d^2}{dp_s^2}(\mathrm{ATP}_1) < 0, \qquad \text{if} \quad 2Ap_s + B < 4p_s t_s + Bt_s + 2D_R. \quad (17)$$

Therefore, ATP_1 is concave if

$$2Ap_s + B < 4p_s t_s + Bt_s + 2D_R. \quad (18)$$

Case 2. $(T' \leq M \leq T_R)$

$$\begin{aligned}
\mathrm{ATP}_2 &= \text{Total average profit for integrated model.} \\
&= \mathrm{APS} + \mathrm{APM}_1 + \mathrm{APR}_1 \\
&= \frac{D_c}{p_m T_s + D_R}\left[(c_m - c_s)p_m T_s - h_s\left(\frac{p_s t_s^2}{p_m} - p_s t_s^2\right) - \mathrm{id}_s\left(T_R + p_s t_s \frac{1}{D_c} - \frac{1}{p_m}\right) - A_s\right. \\
&\quad + (c_r - c_m)p_m T_s - h_m\left(np_m T_s T_R - \frac{n^2+n-2r-2}{2} T_R D_R - \frac{p_s^2 t_s^2}{2p_m}\right) \\
&\quad - \mathrm{id}_m\left(\frac{p_m T_m - nD_R}{D_c}\right) \\
&\quad - A_m + (c_{r1} - c_r)p_m T_s - \frac{h_r}{2}\left(\frac{p_m^2 T_s^2}{D_c} - 2np_m T_s T_R - (2n+1)T_R D_R\right) \\
&\quad \left. + \frac{nc_{r_1} I_e D_c M^2}{2} + \frac{c_{r_1} I_e}{2}(p_m T_s - nD_R)(2M - T) - \mathrm{id}_r T_R - A_r\right] \\
&= \frac{D_c}{p_m T_s + D_R}\left[Ap_s^2 t_s^2 - Bp_s t_s + F\right]
\end{aligned} \quad (19)$$

where A and B are given in (16) and (17), respectively and

$$F = \left[h_m \frac{n^2+n-2r-2}{2} + h_r \frac{2n+1}{2}\right] T_R D_R + \frac{(n+1)c_{r_1} I_e D_c M^2}{2}$$
$$+ (n\mathrm{id}_m - \mathrm{id}_s - \mathrm{id}_r) T_R + \mathrm{id}_s T_s - (A_s + A_m + A_r) \quad (20)$$

$$\frac{d}{dp_s}(\mathrm{ATP}_2) = 0$$
$$\Rightarrow p_s = \frac{2D_R - B \pm \sqrt{(2t_s - A)^2 - 4(2t_s - A)(F - BD_R)}}{2(2t_s - A)} \quad (21)$$

$$\frac{d^2}{dp_s^2}(\mathrm{ATP}_1) < 0, \qquad \text{if} \quad 2Ap_s + B < 4p_s t_s + Bt_s + 2D_R. \quad (22)$$

Therefore, ATP_1 is concave if

Table 1 Input data of different parameters for Case 1 and Case 2

Parameters	c_s	c_m	c_r	c_{r1}	h_s	h_m	T_s	p_m	n	r	ρ	id_s	id_m	id_r	I_e	A_s	A_m	A_r	m	i_p	D_C	D_R
Case 1	7	14	25	35	0.15	0.8	10	16	4	4	0.4	1	2	3	1	25	40	45	20	16	55	100
Case 2	10	12	26	25	0.17	0.9	12	25	17	5	0.3	1.5	2.5	3.5	1	30	28	52	24	45	120	130

$$2Ap_s + B < 4p_st_s + Bt_s + 2D_R. \tag{23}$$

Proposed inventory model in uncertain environment

Let us consider $\widetilde{id}_s$, $\widetilde{id}_r$, $\widetilde{id}_m$, and $\widetilde{I}_{re}$ as zigzag uncertain variables where $\tilde{\mathrm{id}}_s = L(\mathrm{id}_{s1}, \mathrm{id}_{s2}, \mathrm{id}_{s3})$, $\tilde{\mathrm{id}}_r = L(\mathrm{id}_{r1}, \mathrm{id}_{m2}, \mathrm{id}_{r3})$, $\tilde{\mathrm{id}}_m = L(\mathrm{id}_{m1}, \mathrm{id}_{m2}, \mathrm{id}_{m3})$, and $\tilde{I}_{re} = L(I_{re1}, I_{re2}, I_{re3})$. Then, the objective is reduce to the following:

- For Case 1 ($M \le T' \le T_R$)

$$\begin{aligned}
\mathrm{A\tilde{T}P}_1 = \frac{D_c}{p_mT_s + D_R}\Bigg[& 0(c_m - c_s)p_mT_s - h_s\left(\frac{p_st_s^2}{p_m} - p_st_s^2\right) - \widetilde{id}_s\left(T_R + p_st_s\left(\frac{1}{D_c} - \frac{1}{p_m}\right)\right) \\
& - A_s + (c_r - c_m)p_mT_s - h_m\left(np_mT_sT_R - \frac{n^2+n-2r-2}{2}T_RD_R - \frac{p_s^2t_s^2}{2p_m}\right) \\
& - A_m + (c_{r_1} - c_r)p_mT_s - \frac{h_r}{2}\left(\frac{p_m^2T_s^2}{D_c} - 2np_mT_sT_R - (2n+1)T_RD_R\right) \\
& - \widetilde{id}_m\left(\frac{p_mT_m - nD_R}{D_c}\right) + \frac{(n+1)c_{r_1}\tilde{I}_{re}D_cM^2}{2} - \widetilde{id}_rT_R - A_r\Bigg].
\end{aligned} \tag{24}$$

- For Case 2 ($T' \le M \le T_R$)

$$\begin{aligned}
\mathrm{A\tilde{T}P}_2 = \frac{D_c}{p_mT_s + D_R}\Bigg[& (c_m - c_s)p_mT_s - h_s\left(\frac{p_st_s^2}{p_m} - p_st_s^2\right) - \widetilde{id}_s\left(T_R + p_st_s\frac{1}{D_c} - \frac{1}{p_m}\right) - A_s \\
& + (c_r - c_m)p_mT_s - h_m\left(np_mT_sT_R - \frac{n^2+n-2r-2}{2}T_RD_R - \frac{p_s^2t_s^2}{2p_m}\right) \\
& - \widetilde{id}_m\left(\frac{p_mT_m - nD_R}{D_c}\right) \\
& - A_m + (c_{r1} - c_r)p_mT_s - \frac{h_r}{2}\left(\frac{p_m^2T_s^2}{D_c} - 2np_mT_sT_R - (2n+1)T_RD_R\right) \\
& + \left\{\frac{nc_{r_1}D_cM^2}{2} + \frac{c_{r1}}{2}(p_mT_s - nD_R)(2M - T')\right\}\tilde{I}_{re} - \widetilde{id}_rT_R - A_r\Bigg].
\end{aligned} \tag{25}$$

Table 2 Optimum results for objective functions and other parameters

Parameter	Case 1	Case 2
ATP*	1,039.45	1,108.93
p_s^*	70.49	79.23
T	2.85	1.59
APS*	246.51	253.56
APM*	468.25	463.18
APR*	321.43	390.47

Table 3 Optimum value changes due to parametric changes

Parameter name	Parametric value	T^*		P_s^*		ATP*	
		Case 1	Case 2	Case 1	Case 2	Case 1	Case 2
c_s	1.5	1.92	1.86	79.18	78.31	10.89	1,529.43
	2.5	1.61	1.76	46.76	78.31	10.14	1,526.25
	3.5	1.12	1.45	18.18	78.31	10.06	1,522.47
c_m	1.5	1.82	1.82	56.14	78.31	10.56	1,528.36
	2.5	1.75	1.65	69.76	78.31	10.46	1,525.64
	3.5	1.10	1.33	64.15	78.31	10.15	1,520.25
c_r	1.5	1.75	1.96	65.18	78.31	10.56	1,528.19
	2.5	1.52	1.69	62.17	78.31	10.56	1,522.58
	3.5	1.06	1.14	58.14	78.31	10.56	1,520.43
h_s	1.5	1.86	1.76	69.18	78.49	10.56	1,536.45
	2.5	1.64	1.54	64.76	78.31	10.14	1,533.25
	3.5	1.25	1.19	60.19	78.04	10.56	1,520.43
h_m	1.5	1.09	1.76	88.76	78.74	10.56	1,530.52
	2.5	1.21	1.52	76.78	78.54	10.18	1,525.19
	3.5	1.13	1.01	55.45	78.17	10.56	1,523.57
h_r	1.5	1.75	1.89	69.76	78.31	10.19	1,520.43
	2.5	1.25	1.15	54.76	78.01	10.56	1,518.29
	3.5	1.12	1.06	48.16	78.00	10.04	1,518.21

The equivalent crisp model

Using Lemma 1 and applying Theorem 2, the expected total average profit is given by the following:

- For Case 1 ($M \le T' \le T_R$)

$$
\begin{aligned}
E[\tilde{\mathrm{ATP}}_1] = \frac{D_c}{p_m T_s + D_R} \Bigg[& 0(c_m - c_s) p_m T_s - h_s \left(\frac{p_s t_s^2}{p_m} - p_s t_s^2 \right) \\
& - E[\tilde{id}_s] \left(T_R + p_s t_s \left(\frac{1}{D_c} - \frac{1}{p_m} \right) \right) \\
& - A_s + (c_r - c_m) p_m T_s - h_m \left(n p_m T_s T_R - \frac{n^2 + n - 2r - 2}{2} T_R D_R - \frac{p_s^2 t_s^2}{2 p_m} \right) \\
& - A_m + (c_{r_1} - c_r) p_m T_s - \frac{h_r}{2} \left(\frac{p_m^2 T_s^2}{D_c} - 2 n p_m T_s T_R - (2n+1) T_R D_R \right) \\
& - E[\tilde{id}_m] \left(\frac{p_m T_m - n D_R}{D_c} \right) + \frac{(n+1) c_{r_1} E[\tilde{I}_{re}] D_c M^2}{2} - E[\tilde{id}_r] T_R - A_r \Bigg].
\end{aligned}
\tag{26}
$$

- For Case 2 ($T' \le M \le T_R$)

Table 4 Input data of different zigzag parameters for Case 1 and Case 2

Parameters in uncertain environments	$\tilde{id}_s$	$\tilde{id}_m$	$\tilde{id}_r$	$\tilde{I}_e$
Case 1	$L(0.8,1.2,1.4)$	$L(1.5,2.0,2.5)$	$L(1.4,2,2.3)$	$L(0.04,0.06,0.08)$
Case 2	$L(1.4,1.8,2.4)$	$L(2,2.3,2.9)$	$L(1.4,2.1,2.5)$	$L(0.06,0.08,0.1)$

Table 5 Optimal values of objective and decision variables

Parameter	Case 1	Case 2
ATP*	1,056.43	1,520.43
p_s^*	69.76	78.31
T	1.75	1.76
APS*	248.71	276.76
APM*	463.43	467.35
APR*	324.43	393.73

$$
\begin{aligned}
E[\mathrm{A\tilde{T}P_2}] = {} & \frac{D_c}{p_m T_s + D_R}\Bigg[(c_m - c_s)p_m T_s - h_s\left(\frac{p_s t_s^2}{p_m} - p_s t_s^2\right) - E[\widetilde{id}_s]\left(T_R + p_s t_s \frac{1}{D_c} - \frac{1}{p_m}\right) - A_s \\
& + (c_r - c_m)p_m T_s - h_m\left(n p_m T_s T_R - \frac{n^2 + n - 2r - 2}{2} T_R D_R - \frac{p_s^2 t_s^2}{2p_m}\right) \\
& - E[\widetilde{id}_m]\left(\frac{p_m T_m - n D_R}{D_c}\right) - A_m + (c_{r1} - c_r)p_m T_s \\
& - \frac{h_r}{2}\left(\frac{p_m^2 T_s^2}{D_c} - 2n p_m T_s T_R - (2n+1) T_R D_R\right) \\
& + \left\{\frac{n c_{r_1} D_c M^2}{2} + \frac{c_{r_1}}{2}(p_m T_s - n D_R)(2M - T')\right\} E[\tilde{I_{re}}] - E[\widetilde{id}_r]\, T_R - A_r\Bigg].
\end{aligned}
\tag{27}
$$

Numerical example

Crisp environment

The input data of different parameters for Case 1 and Case 2 are shown in Table 1, and the expected optimum value of the total profit is given in Table 2.

Sensitivity analysis

The major contribution of the supply chain is mainly in the inclusion of the manufacturer. We consider product reworking of defective items which are reworked just after regular production, with a different holding cost for good and defective items in the three-layer supply chain. An integrated production-inventory model is

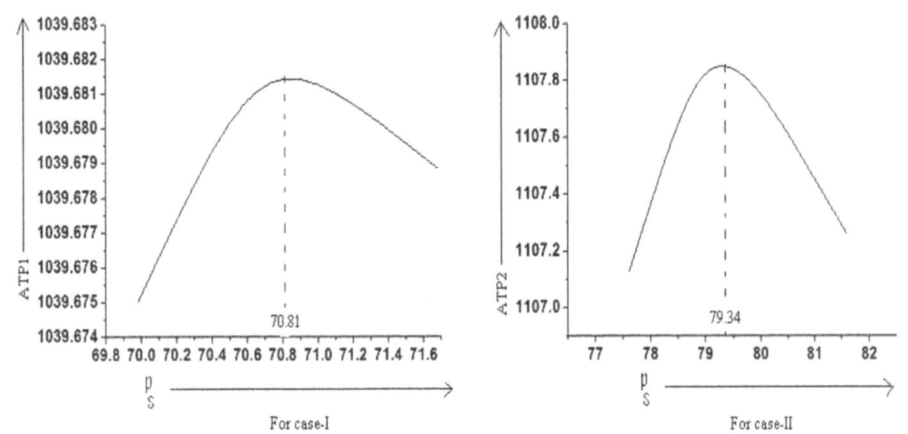

Figure 5 Average profit versus supply rate of suppliers in Case 1 and Case 2.

presented for the supplier, manufacturer, and retailer supply chain under conditionally permissible delay in payments which has developed in both crisp and uncertain environments (Table 3).

Uncertain environment

Input data of different crisp parameters for Case 1 and Case 2 are given in Table 1; the remaining uncertain parameters are depicted in Table 4. Optimal values of objective and decision variables are given in Table 5 and graphically represented in Figure 5.

Achievements and conclusion of the model

In this model, we developed a three-layer production-inventory supply chain model in an uncertain environment. Here, the suppliers are also the manufacturers; they collect the raw material (ore) and produce the raw material of the actual manufacturer. For example, in the petroleum industry, suppliers collect the ore and produce the naphthalene, which is the raw material of the manufacturer. Then manufacturer produces the usable product to sell to the retailer. In this paper, we have developed a production-inventory supply chain model under an uncertain environment. The paper can be extended to imperfect production-inventory system. Deterioration can be allowed for produced items of the retailer and manufacturer. In the case of the retailer, it might be interesting to consider the effect that only a percent of imperfect quality products could be reworked by manufacturers and that other scrap items must be eliminated immediately. In order to show the uncertainties, the present model could be extended, applying stochastic demand and production rate in each member of the supply chain. These are some topics of ongoing and future research, among others.

Author details

[1]Department of Engineering Science, Haldia Institute of Technology, Haldia Purba Midna Pur 721657, West Bengal, India. [2]Department of Mathematics, Mugberia Gangadhar Mahavidyalaya, Mugberia Purba Medinipur 721425, West Bengal, India. [3]Department of Mathematics, Bengal Engineering and Science University, Shibpur Howrah 711103, West Bengal, India.

References

1. Oliver, RK, Webber, MD: Supply-chain management: logistics catches up with strategy (reprint from Outlook) (1982)
2. Cohen, MA, Baghanan, MP: The stabilizing effect of inventory in supply chains. Oper. Res. **46**, 72–83 (1998)
3. Cachon, GP, Zipkin, PH: Competitive and cooperative inventory policies in a two-stage supply chain. Manag. Sci. **45**, 936–953 (1999)
4. Viswanathan, S, Piplani, R: Coordinating supply chain inventories through common replenishment epochs. Eur. J. Oper. Res. **129**, 277–286 (2001)
5. Khouja, M: Optimizing inventory decisions in a multistage multi customer supply chain. Transportation Research. **39**, 193–208 (2003)
6. Agarwal, V, Chao, X, Seshadri, S: Dynamic balancing of inventory in supply chains. Eur. J. Oper. Res. **159**, 296–317 (2004)
7. Rau, H, Wu, MY, Wee, HM: Deteriorating item inventory model with shortage due to supplier in an integrated supply chain. Int. J. Syst. Sci. **35**, 293–303 (2004)
8. Lee, W: A joint economic lot size model for raw material ordering, manufacturing setup,and finished goods delivering. Int. J. Manag. Sci. **33**, 163–174 (2005)
9. Ben-Daya, M, Seliaman, M: An integrated production inventory model with raw material replenishment considerations in a three layer supply chain. Int. J. Prod. Econ. **28**, 11–21 (2010)
10. Sana, S: A production-inventory model of imperfect quality products in a three-layer supply chain. Decision Support Syst. **50**, 539–547 (2011)
11. Pal, B, Sana, SS, Chaudhuri, K: Three-layer supply chain - a production-inventory model for reworkable items. Appl. Math. Comput. **219**, 530–543 (2012)
12. Wang, J, Shu, YF: Fuzzy decision modelling for supply chain management. Fuzzy Sets Syst. **150**, 107–127 (2005)

13. Das, B, Maity, K, Maiti, M: A two warehouse supply-chain model under possibility / necessity / credibility measures. Math. Comput. Model. **46**, 398–409 (2007)
14. Chen, C, Yuan, T, Lee, W: Multi-criteria fuzzy optimization for locating warehouses and distribution centers in a supply chain network. J. Chin. Inst. Chem. Eng. **38**, 393–407 (2007)
15. Peidro, D, Mula, J, Jimenez, M, Botella, MM: A fuzzy linear programming based approach for tactical supply chain planning in an uncertainty environment. Eur. J. Oper. Res. **205**, 65–80 (2010)
16. Chu, SJ: Interactive group decision-making using a fuzzy linguistic approach for evaluating the flexibility in a supply chain. Eur. J. Oper. Res. **213**, 279–289 (2011)
17. Jana, DK, Maity, K, Das, B, Roy, TK: A fuzzy simulation via contractive mapping genetic algorithm approach to an imprecise production inventory model under volume flexibility. J. Simulation. **7**(2), 90–100 (2013)
18. Kristianto, Y, Helo, P, Jiao, J, Sandhu, M: Adaptive fuzzy vendor managed inventory control for mitigating the Bullwhip effect in supply chains. Eur. J. Oper. Res. **216**, 346–355 (2012)
19. Liu, B: Uncertainty Theory. 2nd edn. Springer, Berlin (2007)
20. You, C: Some convergence theorems of uncertain sequences. Math. Comput. Model. **49**, 482–487 (2009)
21. Liu, B: Fuzzy process, hybrid process and uncertain process. J. Uncertain Syst. **2**, 3–16 (2008)
22. Liu, B: Some research problems in uncertainty theory. J. Uncertain Syst. **3**, 3–10 (2009)
23. Goyal, SK: Economic order quantity under conditions of permissible delay in payments. J. Oper. Res. Soc. **36**, 35–38 (1985)
24. Khanra, S, Ghosh, SK, Chaudhuri, KS: An EOQ model for a deteriorating item with time dependent quadratic demand under permissible delay in payment. Appl. Math. Comput. **218**, 1–9 (2011)
25. Maihami, R, Abadi, INK: Joint control of inventory and its pricing for non-instantaneously deteriorating items under permissible delay in payments and partial backlogging. Math. Comput. Model. **55**, 1722–1733 (2012)
26. Liu, B: Uncertainty Theory: A Branch of Mathematics for Modeling Human Uncertainty. Springer, Berlin (2010)

Method of moments for estimating uncertainty distributions

Xiaosheng Wang[1*] and Zixiong Peng[2]

*Correspondence: xswang@hebeu.edu.cn
[1] School of Science, Hebei University of Engineering, Handan, 056038, China
Full list of author information is available at the end of the article

Abstract

Uncertainty theory is a branch of mathematics for modeling human uncertainty, and uncertain statistics is a methodology for collecting and interpreting expert's experimental data by uncertainty theory. In order to estimate uncertainty distributions via experts' experimental data, this paper will present a method of moments and design a numerical method to find moment estimate of unknown parameters.

Keywords: Uncertainty theory; Uncertainty distribution; Uncertain statistics; Moment estimate

Introduction

Probability theory is a branch of mathematics concerned with analysis of random phenomena. The probability distribution describes the range of possible values and the probability of a random variable. In order to determine the probability distribution, random statistics (i.e., the classical mathematical statistics) was proposed as a methodology for collecting and interpreting the test data with correlative information in a system by probability theory. On the other hand, the fuzzy set theory via membership function initiated by Zadeh [1] is a mathematical model of vague qualitative or qualitative data, frequently generated by means of natural language. In order to determine the membership function, fuzzy statistics was presented including fuzzy point estimation by a fuzzy decision-making approach and so on.

However, a lot of surveys showed that some information and knowledge represented usually by human language like 'about 1,000 km,' 'roughly 60 kg,' 'high speed,' and 'small size' are neither randomness nor fuzziness. When the sample size is too small (even no-sample) to estimate a probability distribution, we have to invite some domain experts to evaluate their belief degree that each event will occur. Since human beings usually overweigh unlikely events, the belief degree may have much larger variance than the real frequency. Perhaps some people think that the belief is a subjective probability or fuzzy membership degree. However, the examples [2] tell us that it is inappropriate because probability theory and fuzzy set theory may lead to counterintuitive results in this case.

In order to model this type of imprecise quantities, uncertainty theory was founded by Liu [3] in 2007 and refined by Liu [4] in 2010 and became a branch of mathematics based on the normality, duality, subadditivity, and product axioms. Based on Liu's uncertainty theory, some basic and important theoretical work of uncertainty theory such as uncertain process [5], uncertain calculus [6], uncertain differential equation [6], uncertain logic [7], uncertain inference [8], and uncertain risk analysis [9] have been established. Meanwhile, as an application of uncertainty theory, Liu [10] proposed a spectrum of uncertain programming and applied uncertain programming to system reliability design, facility location problem, vehicle routing problem, project scheduling problem, and so on. Other references related to uncertainty theory are Gao [11], Gao et al. [12], Peng and Iwamura [13], and Liu and Ha [14]. Nowadays, uncertainty theory was well developed on both theory section and practice section. For exploring the recent developments of uncertainty theory, the readers may consult the book [15].

One important issue in uncertainty theory is how to determine uncertainty distribution of an uncertain variable. In order to answer this question, uncertain statistics was firstly presented by Liu [4] in 2010, and is a methodology for collecting and interpreting expert's experimental data by uncertainty theory. In uncertain statistics, Liu [4] first suggested an empirical uncertainty distribution and proposed a principle of least squares as the method for estimating the unknown parameters based on the expert's experimental data. After that, Chen and Ralescu [16] employed uncertain statistics to estimate the travel distance between Beijing and Tianjin. In 2012, Wang et al. recast the Delphi method as a process to determine the uncertainty distributions [17] and presented a method of uncertain hypothesis testing for determining if the views of two domain experts are identical (i.e., wether or not they have the same uncertainty distribution) [18].

Based on experts' experimental data and the empirical uncertainty distribution, the kth sample moment is defined in this paper. Then, a method of moments for estimating the unknown parameters of uncertainty distribution is proposed. The remainder of this paper is organized as follows. The next section is intended to introduce some concepts in uncertainty theory as they are needed. Some basic concepts of uncertain statistics are introduced in Section 'Beginning of uncertain statistics.' The uncertain moment method for estimating parameters is proposed in Section 'Method of moments.' A numerical method to find the moment estimate of unknown parameters is designed in Section 'Numerical method.' Finally, a conclusion is drawn in Section 'Conclusions.'

Preliminaries

In this section, we will introduce some useful definitions about uncertain measure, uncertain variable, uncertain moment, and so on.

Let Γ be a nonempty set, and be a σ-algebra over Γ. Each element $\Lambda \in$ is called an event. A number (Λ) indicates the level that Λ will occur. Uncertain measure was introduced as a set function satisfying the following three axioms (Liu [3]):

Axiom 1. $\{\Gamma\} = 1.$

Axiom 2. $\{\Lambda\} + \{\Lambda^c\} = 1$ for any event Λ.

Axiom 3. For every countable sequence of events $\{\Lambda_i\}$, we have

$$\left\{\bigcup_{i=1}^{\infty} \Lambda_i\right\} \leq \sum_{i=1}^{\infty} \{\Lambda_i\}.$$

Although probability measure satisfies the above three axioms, probability theory is not a special case of uncertainty theory because product probability measure does not satisfy the following product axiom (Liu [3]):

Axiom 4. Let $(\Gamma_k, \quad_k, \quad_k)$ be uncertainty spaces for $k = 1, 2, \ldots$. The product uncertain measure is an uncertain measure satisfying

$$\left\{\prod_{k=1}^{\infty} \Lambda_k\right\} = \bigwedge_{k=1}^{\infty} \quad_k\{\Lambda_k\},$$

where Λ_k are arbitrarily chosen events from $\quad_k$ for $k = 1, 2, \ldots$, respectively.

The concept of uncertain variable was introduced by Liu [3] as a measurable function from an uncertainty space $(\Gamma, \quad, \quad)$ to the set of real numbers. The expected value operator of an uncertain variable was defined by Liu [3] as

$$E[\xi] = \int_0^{+\infty} \quad \{\xi \geq x\}dx - \int_{-\infty}^{0} \quad \{\xi \leq x\}dx,$$

provided that at least one of the two integrals is finite. The kth moment of an uncertain variable ξ is defined by $E\left[\xi^k\right]$, where k is a positive integer.

For any $x \in \Re$, the function $\Phi(x) = \quad \{\xi \leq x\}$ is called the uncertainty distribution of uncertain variable ξ, Peng and Iwamura [13] presented a sufficient and necessary condition of uncertainty distribution that a function $\Phi : \Re \rightarrow [0, 1]$ is an uncertainty distribution if and only if it is an increasing function except $\Phi(x) \equiv 0$ and $\Phi(x) \equiv 1$. Moreover, if the inverse function $\Phi^{-1}(\alpha)$ exists and is unique for each $\alpha \in (0, 1)$, then $\Phi(x)$ is called regular, and the inverse function $\Phi^{-1}(\alpha)$ is called the inverse uncertainty distribution of ξ. If an uncertain variable ξ has regular distribution $\Phi(x)$ and its expected value exists, then the expected value may be presented by the following formula:

$$E[\xi] = \int_0^1 \Phi^{-1}(\alpha)d\alpha.$$

In addition, Liu [4] proved a series of operational laws of independent uncertain variables for calculating the uncertainty distribution of monotone function of uncertain variables. By these operational laws, the uncertainty distributions of sum, difference, product, quotient, maximum, and minimum for n uncertain variables are easy obtained. The detail contents are shown in [4] or [15].

Beginning of uncertain statistics

Uncertain statistics is based on the expert's experimental data rather than historical data. One question is how to obtain expert's experimental data. Liu [4] designed a questionnaire survey for collecting expert's experimental data, that is, we invite one or more domain experts who are asked to complete a questionnaire about the meaning of an uncertain variable ξ like 'about 1,000 km' individually.

We first ask the domain expert to choose a possible value x that the uncertain variable ξ may take. Then, we quiz him 'how likely is ξ less than x?' and denote his belief degree by α. Thus, we obtain an expert's experimental data (x, α) from the domain expert. Repeating the above process, we obtain the expert's experimental data.

Let $(x_1,\alpha_1),(x_2,\alpha_2),\ldots,(x_n,\alpha_n)$ be the expert's experimental data that meet the following condition:

$$x_1 < x_2 < \cdots < x_n,\quad 0 \le \alpha_1 \le \alpha_2 \cdots \le \alpha_n \le 1. \tag{1}$$

Based on the above data, Liu [4] presented the following empirical uncertainty distribution:

$$\Phi(x) = \begin{cases} 0, & \text{if } x < x_1 \\ \alpha_i + \frac{(\alpha_{i+1}-\alpha_i)(x-x_i)}{x_{i+1}-x_i}, & \text{if } x_i \le x \le x_{i+1}, 1 \le i < n \\ 1, & \text{if } x > x_n. \end{cases} \tag{2}$$

Assume that an uncertainty distribution to be determined has a known functional form with one or more unknown parameters like $\Phi(x;\theta_1,\theta_2,\cdots,\theta_p)$ where $\theta_1,\theta_2,\cdots,\theta_p$ are unknown parameters. How do we estimate those unknown parameters? Liu [4] presented the principle of least squares, which says that the unknown parameters $\theta_i, i = 1,2,\ldots,p$ are the solution of the minimization problem,

$$\min_{\theta_1,\ldots,\theta_p} \sum_{i=1}^{n} \left(\Phi\left(x_i;\theta_1,\theta_2,\ldots,\theta_p\right) - \alpha_i\right)^2. \tag{3}$$

Let ξ be an uncertain variable with uncertainty distribution $\Phi(x;\theta_1,\theta_2,\ldots,\theta_p)$ where $\theta_1,\theta_2,\ldots,\theta_p$ are unknown parameters. If the uncertainty distribution $\Phi(x_i;\theta_1,\theta_2,\ldots,\theta_p)$ is regular, then Φ^{-1} is the inverse uncertainty distribution of ξ. Assume that we have some expert's experimental data $(x_1,\alpha_1),(x_2,\alpha_2),\ldots,(x_n,\alpha_n)$ satisfying the condition (1). Then, for any α_i, we have $\left\{\xi \le \Phi^{-1}(\alpha_i)\right\} = \alpha_i, i = 1,2,\ldots,n$. It means that the principle of least squares can be changed into the following form:

$$\min_{\theta_1,\ldots,\theta_p} \sum_{i=1}^{n} \left(\Phi^{-1}\left(\alpha_i;\theta_1,\theta_2,\ldots,\theta_p\right) - x_i\right)^2, \tag{4}$$

where the unknown parameters $\theta_i, i = 1,2,\ldots,p$ are the solution of the minimization problem.

If the inverse uncertainty distribution of uncertain variable ξ is simple and easy to calculate, we should solve the optimization problem (4) rather than the optimization problem (3) to estimate the unknown parameters. For example, let $(x_1,\alpha_1),(x_2,\alpha_2),\cdots,(x_n,\alpha_n)$ be the expert's experimental data, and the uncertain variable ξ to be determined be a normal uncertain variable, which has the following uncertainty distribution:

$$\Phi(x) = \left(1 + \exp\left(\frac{\pi(e-x)}{\sqrt{3}\sigma}\right)\right)^{-1}, \quad x \in \Re,$$

where e and σ are two unknown parameters. The inverse uncertainty distribution of normal uncertain variable ξ is

$$\Phi^{-1}(\alpha) = e + \frac{\sigma\sqrt{3}}{\pi}\ln\frac{\alpha}{1-\alpha}, \quad \alpha \in (0,1).$$

By solving the optimization problem (4), it is easy to obtain the following estimate values:

$$\hat{e} = \bar{x} - \hat{\sigma}\frac{\sqrt{3}}{n\pi}\sum_{i=1}^{n}\ln\frac{\alpha_i}{1-\alpha_i}$$

and

$$\hat{\sigma} = \frac{\sqrt{3}}{3}\pi n \frac{\bar{x}\sum_{i=1}^{n}\ln\frac{\alpha_i}{1-\alpha_i} - \sum_{i=1}^{n} x_i \ln\frac{\alpha_i}{1-\alpha_i}}{\left(\sum_{i=1}^{n}\ln\frac{\alpha_i}{1-\alpha_i}\right)^2 - n\sum_{i=1}^{n}\left(\ln\frac{\alpha_i}{1-\alpha_i}\right)^2}$$

where $\bar{x} = (x_1 + x_2 + \cdots + x_n)/n$.

Method of moments

In this section, a method of moments based on expert's experimental data is proposed to estimate the unknown parameters. Firstly, we present the kth moment of the empirical uncertainty distribution (2), which is the uncertainty distribution of some uncertain variable ξ.

Theorem 1. *Let $(x_i, \alpha_i), i = 1, 2, \ldots, n$ be the expert's experimental data that meet the following condition:*

$$0 \le x_1 < x_2 < \cdots < x_n, \; 0 \le \alpha_1 \le \alpha_2 \le \cdots \le \alpha_n \le 1. \tag{5}$$

Then for any positive integer k, the uncertain variable ξ with the empirical uncertainty distribution (2) has the kth moment

$$E\left[\xi^k\right] = \alpha_1 x_1^k + \frac{1}{k+1}\sum_{i=1}^{n-1}\sum_{j=0}^{k}\left(\alpha_{i+1} - \alpha_i\right) x_i^j x_{i+1}^{k-j} + \left(1 - \alpha_n\right) x_n^k. \tag{6}$$

Proof. Since $0 \le x_1 < x_2 < \cdots < x_n$, by (2), we have

$$\begin{aligned}
E\left[\xi^k\right] &= \int_0^{+\infty} \left\{\xi^k \ge x\right\} dx \\
&= \int_0^{+\infty} kx^{k-1} \{\xi \ge x\}\, dx \\
&= \int_0^{+\infty} kx^{k-1}(1 - \{\xi < x\})dx \\
&= k\int_0^{x_1} x^{k-1}(1 - \{\xi < x\})dx + k\sum_{i=1}^{n-1}\int_{x_i}^{x_{i+1}} x^{k-1}(1 - \{\xi < x\})dx \\
&\quad + k\int_{x_n}^{+\infty} x^{k-1}(1 - \{\xi < x\})dx \\
&= k\int_0^{x_1} x^{k-1}dx + k\sum_{i=1}^{n-1}\int_{x_i}^{x_{i+1}} x^{k-1}(1 - \{\xi < x\})dx \\
&= x_1^k + k\sum_{i=1}^{n-1}\int_{x_i}^{x_{i+1}} x^{k-1}\left(1 - \alpha_i - \frac{(\alpha_{i+1} - \alpha_i)(x - x_i)}{x_{i+1} - x_i}\right) dx \\
&= \frac{k\alpha_1 + \alpha_2}{k+1} x_1^k + \frac{1}{k+1}\sum_{i=1}^{n-1}\sum_{j=1}^{k-1}\left(\alpha_{i+1} - \alpha_i\right) x_i^j x_{i+1}^{k-j} + \frac{1}{k+1}\sum_{i=2}^{n-1}\left(\alpha_{i+1} - \alpha_{i-1}\right) x_i^k \\
&\quad + (1 - \frac{1}{k+1}\alpha_{n-1} - \frac{k}{k+1}\alpha_n)x_n^k \\
&= \alpha_1 x_1^k + \frac{1}{k+1}\sum_{i=1}^{n-1}\sum_{j=0}^{k}\left(\alpha_{i+1} - \alpha_i\right) x_i^j x_{i+1}^{k-j} + \left(1 - \alpha_n\right) x_n^k.
\end{aligned}$$

The theorem is proved. □

Definition 1. *Let ξ be an uncertain variable. For any positive integer k, if the kth moment of ξ exists, then $E\left[\xi^k\right]$ is called the kth theoretical moment of ξ.*

Definition 2. *Let $(x_1,\alpha_1),(x_2,\alpha_2),\cdots,(x_n,\alpha_n)$ be the expert's experimental data that meet the condition (5). For any positive integer k,*

$$\bar{\xi^k}=\alpha_1 x_1^k+\frac{1}{k+1}\sum_{i=1}^{n-1}\sum_{j=0}^{k}\left(\alpha_{i+1}-\alpha_i\right)x_i^j x_{i+1}^{k-j}+\left(1-\alpha_n\right)x_n^k \tag{7}$$

is called the kth sample moment.

Secondly, we introduce a method of moments for estimating the unknown parameter, the description of the method is presented as follows.

Method of moments Let ξ be an uncertain variable with uncertainty distribution $\Phi(x;\theta_1,\theta_2,\ldots,\theta_p)$ where $\theta_1,\theta_2,\ldots,\theta_p$ are unknown parameters. Let (x_1,α_1), $(x_2,\alpha_2),\ldots,(x_n,\alpha_n)$ be the expert's experiment data satisfying the condition (5). Let $E\left[\xi^k\right]$ and $\bar{\xi^k}, k=1,2,\ldots,p$ be the kth theoretical and kth sample moments, respectively. As a general procedure for estimating the unknown parameters $\theta_1,\theta_2,\ldots,\theta_p$, a system of equations is presented:

$$E\left[\xi^k\right]=\bar{\xi^k},\ \ k=1,2,\ldots,p. \tag{8}$$

Since $E\left[\xi^k\right]=k\int_0^{+\infty}x^{k-1}(1-\Phi(x;\theta_1,\theta_2,\ldots,\theta_p))dx$, the above system of equations is equivalent to the following form:

$$k\int_0^{+\infty}x^{k-1}(1-\Phi(x;\theta_1,\theta_2,\ldots,\theta_p))dx=\bar{\xi^k},\ \ k=1,2,\ldots,p. \tag{9}$$

A solution of the equations, $\theta_1,\theta_2,\ldots,\theta_p$, is called a moment estimate. We denote the estimate values by $\hat{\theta_1},\hat{\theta_2},\ldots,\hat{\theta_p}$, respectively.

Example 1. Assume $(x_i,\alpha_i),\ i=1,2,\ldots,n$ are the expert's experimental data satisfying the condition (5) and the uncertainty distribution of uncertain variable ξ has a functional form with one unknown parameter a as follows:

$$\Phi(x;a)=ax^{\frac{1}{2}},\ \ a>0. \tag{10}$$

By Equation 8, the estimate value of the unknown parameter a is the solution of the equation

$$E[\xi]=\bar{\xi},$$

where $\bar{\xi}$ is defined by Equation 7. Since $E[\xi]=\frac{1}{3a^2}$, we have

$$\frac{1}{3a^2}=\frac{\alpha_1+\alpha_2}{2}x_1+\sum_{i=2}^{n-1}\frac{\alpha_{i+1}-\alpha_{i-1}}{2}x_i+(1-\frac{\alpha_{n-1}+\alpha_n}{2})x_n. \tag{11}$$

Hence, we obtain the estimate value of the unknown parameter a,

$$\hat{a}=\left\{3\left[\frac{\alpha_1+\alpha_2}{2}x_1+\sum_{i=2}^{n-1}\frac{\alpha_{i+1}-\alpha_{i-1}}{2}x_i+(1-\frac{\alpha_{n-1}+\alpha_n}{2})x_n\right]\right\}^{-\frac{1}{2}}. \tag{12}$$

Assume that the expert's experiment data are obtained as follows:

$$\left(\frac{1}{16}, \frac{1}{4}\right), \left(\frac{1}{4}, \frac{1}{2}\right), \left(\frac{16}{25}, \frac{4}{5}\right).$$

By Equation 12, we have the moment estimate value

$$\hat{a} = 1.0817,$$

and the moment estimate distribution is

$$\Phi(x) = 1.0817x^{\frac{1}{2}}.$$

Example 2. Let ξ be an uncertain variable with uncertainty distribution

$$\Phi(x; a, b) = ax + b, \quad (a > 0) \tag{13}$$

where a, b are two unknown parameters. Assume $(x_i, \alpha_i), \;\; i = 1, 2, \ldots, n$ are the expert's experimental data satisfying the condition (5). By Equation 8, we will solve the following system of equations:

$$\begin{cases} E[\xi] = \bar{\xi} \\ E\left[\xi^2\right] = \bar{\xi^2} \end{cases},$$

where $\bar{\xi}$ and $\bar{\xi^2}$ are defined by Equation 7. In addition, the inverse uncertainty distribution is $\Phi^{-1}(\alpha) = \frac{\alpha - b}{a}$. We have

$$E[\xi] = \int_0^1 \Phi^{-1}(\alpha) d\alpha = \frac{1 - 2b}{2a},$$

and

$$E\left[\xi^2\right] = \int_0^1 (\Phi^{-1}(\alpha))^2 d\alpha = \frac{1 + 3b^2 - 3b}{3a^2}.$$

Thus, we have the following system of equations

$$\begin{cases} \frac{1-2b}{2a} = \bar{\xi} \\ \frac{1+3b^2-3b}{3a^2} = \bar{\xi^2}. \end{cases}$$

Then the unique solution of the above equations

$$\begin{cases} \hat{a} = \frac{1}{2\sqrt{3}} \left(\bar{\xi^2} - (\bar{\xi})^2\right)^{-\frac{1}{2}} \\ \hat{b} = \frac{1}{2}(1 - 2\hat{a}\bar{\xi}) \end{cases} \tag{14}$$

is the moment estimate value of unknown parameters a and b, and the moment estimate distribution is

$$\Phi(x) = \hat{a}x + \hat{b}.$$

Assume that we have the following expert's experimental data:

$$(0.4, 0.1), \;\; (1.0, 0.2), \;\; (1.5, 0.3), \;\; (2.0, 0.4), \;\; (3.0, 0.7), \;\; (4.0, 0.9).$$

From Equations 7 and 14, we have moment estimate values $\hat{a} = 0.2442$, $\hat{b} = -0.0519$ and the moment estimate distribution is $\Phi(x) = 0.2442x - 0.0519$.

Numerical method

In this section, we introduce Newton's method to estimate the moment estimate value. A general algorithm is proposed to solve the parameter estimate problem as follows.

Assume $(x_i, \alpha_i),\ i = 1, 2, \ldots, n$ are the expert's experimental data satisfying the condition (5) and $\Phi(x; \theta_1, \theta_2, \ldots, \theta_p)$ is the uncertainty distribution of uncertain variable ξ with unknown parameters $\theta_1, \theta_2, \ldots, \theta_p$. The numerical method is to solve the Equations 8,

$$E\left[\xi^k\right] = \bar{\xi^k},\ k = 1, 2, \ldots, p$$

where $\bar{\xi}, \bar{\xi^2}, \ldots, \bar{\xi^p}$ are the sample moments, which are constants calculated by Equation 7.

Denote a function $F : \mathbb{R}^p \to \mathbb{R}^p$,

$$F\left(\theta_1, \theta_2, \cdots, \theta_p\right) = \begin{pmatrix} F_1\left(\theta_1, \theta_2, \cdots, \theta_p\right) \\ F_2\left(\theta_1, \theta_2, \cdots, \theta_p\right) \\ \cdots \\ F_p\left(\theta_1, \theta_2, \cdots, \theta_p\right) \end{pmatrix},$$

where

$$F_k\left(\theta_1, \theta_2, \ldots, \theta_p\right) = k\int_0^{+\infty} x^{k-1}\left(1 - \Phi\left(x; \theta_1, \theta_2, \ldots, \theta_p\right)\right)dx - \bar{\xi^p},\ k = 1, 2, \ldots, n.$$

The Equation 8 is equivalent to $F(\theta_1, \theta_2, \ldots, \theta_p) = 0$.

If the function F is differentiable and the Jacobian matrix of F is easily computed, the procedure can be summarized as follows:

Step 1. Estimate the kth sample moments by Equation (7), $k = 1, 2, \ldots, p$.

Step 2. Give an initial value of parameters $\theta^0 = \left(\theta_1^0, \theta_2^0, \ldots, \theta_p^0\right)$.

Step 3. Compute the converse matrix $J_F\left(\theta^k\right)^{-1}$ of the Jacobian matrix $J_F\left(\theta^k\right),\ k = 0, 1, 2, \ldots$, where

$$J_F\left(\theta^k\right) = \begin{pmatrix} \frac{\partial F_1}{\partial \theta_1} & \frac{\partial F_1}{\partial \theta_2} & \cdots & \frac{\partial F_1}{\partial \theta_p} \\ \frac{\partial F_2}{\partial \theta_1} & \frac{\partial F_2}{\partial \theta_2} & \cdots & \frac{\partial F_2}{\partial \theta_p} \\ \cdots & \cdots & \cdots & \cdots \\ \frac{\partial F_p}{\partial \theta_1} & \frac{\partial F_p}{\partial \theta_2} & \cdots & \frac{\partial F_p}{\partial \theta_p} \end{pmatrix}. \tag{15}$$

The integrals in the equation can be computed by some numerical methods.

Step 4. Calculate θ^{k+1} by the following equation:

$$\theta^{k+1} = \theta^k - J_F\left(\theta^k\right)^{-1} F\left(\theta^k\right). \tag{16}$$

Step 5. Repeat the third to fourth steps when $|\theta^{k+1} - \theta^k| > \alpha$ and $k < M$, where α and M are given positive numbers.

Step 6. Report the last θ^n as the solution, if $|\theta^n - \theta^{n-1}| \leq \alpha$.

Remark 1. If the distribution $\Phi\left(x; \theta_1, \theta_2, \cdots, \theta_p\right)$ is continuous differentiable of the parameters, we have

$$\frac{\partial \int_0^{+\infty}\left(1 - \Phi\left(x; \theta_1, \theta_2, \cdots, \theta_p\right)\right)dx}{\partial \theta_1} = -\int_0^{+\infty} \frac{\partial \Phi\left(x; \theta_1, \theta_2, \cdots, \theta_p\right)}{\partial \theta_1}dx.$$

Remark 2. If the function F is not differentiable or the Jacobian matrix is hard to compute, we can compute

$$\frac{F_i\left(\theta_1,\cdots,\theta_j+h_j,\cdots,\theta_p\right)-F_i\left(\theta_1,\cdots,\theta_j,\cdots,\theta_p\right)}{h_j}$$

instead of

$$\frac{\partial F_i\left(\theta_1,\cdots,\theta_j,\cdots,\theta_p\right)}{\partial\theta_j},$$

where $i,j=1,2,\ldots,p$, and $h_1,h_2,\ldots,h_p$ are given small positive numbers.

Example 3. Assume that an uncertainty distribution has a lognormal form with two unknown parameters e and σ, that is,

$$\Phi(x|e,\sigma)=\left(1+\exp\left(\frac{\pi(e-\ln x)}{\sqrt{3}\sigma}\right)\right)^{-1},$$

and the expert's experimental data are (0.6, 0.1), (1.0, 0.3), (1.5, 0.4), (2.0, 0.6), (2.8, 0.8), (3.6, 0.9). We have

$$F(e,\sigma)=\begin{pmatrix}\displaystyle\int_0^{+\infty}1-\left(1+\exp\left(\frac{\pi(e-\ln x)}{\sqrt{3}\sigma}\right)\right)^{-1}dx-1.86\\ \displaystyle 2\int_0^{+\infty}x\left(1-\left(1+\exp\left(\frac{\pi(e-\ln x)}{\sqrt{3}\sigma}\right)\right)^{-1}\right)dx-3.56\end{pmatrix}$$

and

$$J_F(e,\sigma)=\begin{pmatrix}\displaystyle\int_0^{+\infty}\frac{\pi f(x|e,\sigma)}{\sqrt{3}\sigma\left(1+f(x|e,\sigma)\right)^2}dx & \int_0^{+\infty}\frac{\pi(e-\ln x)f(x|e,\sigma)}{\sqrt{3}\sigma^2(1+f(x|e,\sigma))^2}dx\\ \displaystyle 2\int_0^{+\infty}x\frac{\pi f(x|e,\sigma)}{\sqrt{3}\sigma\left(1+f(x|e,\sigma)\right)^2}dx & 2\int_0^{+\infty}x\frac{\pi(e-\ln x)f(x|e,\sigma)}{\sqrt{3}\sigma^2(1+f(x|e,\sigma))^2}dx\end{pmatrix},$$

where $f(x|e,\sigma)=\exp\left(\frac{\pi(e-\ln x)}{\sqrt{3}\sigma}\right)$. All of these integrals are computed by Simpson formula, and the initial value is $(0.6, 0.5)$.

A run of the algorithm shows that the approximate solution of the equation $F(e,\sigma)=0$ is $\hat{e}=0.50$ and $\hat{\sigma}=0.47$, which leads to the moment estimate distribution

$$\Phi(x|0.50,0.47)=\left(1+\exp\left(\frac{\pi(0.50-\ln x)}{\sqrt{3}\times 0.47}\right)\right)^{-1}.$$

Conclusions

Uncertain statistics is a methodology for collecting and interpreting expert's experimental data by uncertainty theory. The method of moments in this paper is a new way to estimate the unknown parameters in uncertainty distribution. This method depends on the expert's experimental data and is easy to be realistic in the real experiment.

Acknowledgements

This work was supported by National Natural Science Foundation of China (No. 61073121) and Hebei Natural Science Foundation (No. G2013402063 and No. F2012402037).

Author details

[1]School of Science, Hebei University of Engineering, Handan, 056038, China. [2]School of Applied Mathematics, Central University of Finance and Economics, Beijing 100081, China.

References

1. Zadeh, L: Fuzzy sets. Inform. Control **8**, 338–353 (1965)
2. Liu, B: Why is there a need for uncertainty theory? J. Uncertainty Syst. **6**(1), 3–10 (2012)
3. Liu, B: Uncertainty theory. 2nd ed. Springer, Berlin (2007)
4. Liu, B: Uncertainty Theory: a Branch of Mathematics for Modeling Human Uncertainty. Springer, Berlin (2010)
5. Liu, B: Fuzzy process, hybrid process and uncertain process. J. Uncertain Syst. **2**(1), 3–16 (2008)
6. Liu, B: Some research problems in uncertainty theory. J. Uncertain Syst. **3**(1), 3–10 (2009)
7. Li, X, Liu, B: Hybrid logic and uncertain logic. J. Uncertain Syst. **3**(2), 83–94 (2009)
8. Liu, B: Uncertain set theory and uncertain inference rule with application to uncertain control. J. Uncertain Syst. **4**(2), 83–98 (2010)
9. Liu, B: Uncertain risk analysis and uncertain reliability analysis. J. Uncertain Syst. **4**(3), 163–170 (2010)
10. Liu, B: Theory and Practice of Uncertain Programming. 2nd ed. Springer, Berlin (2009)
11. Gao, X: Some properties of continuous uncertain measure. Int. J. Uncertainty, Fuzziness Knowl. Based Syst. **17**(3), 419–426 (2009)
12. Gao, X, Gao, Y, Ralescu, D: On Liu's inference rule for uncertain systems. Int. J. Uncertainty, Fuzziness Knowl. Based Syst. **18**(1), 1–11 (2010)
13. Peng, Z, Iwamura, K: A sufficient and necessary condition of uncertainty distribution. J. Interdiscip. Math. **13**(3), 277–285 (2010)
14. Liu, Y, Ha, M: Expected value of function of uncertain variables. J. Uncertainty Syst. **4**(3), 181–186 (2010)
15. Liu, B: Uncertainty Theory. 4th ed. Springer, Berlin. http://orsc.edu.cn/liu/ut.pdf.
16. Chen, X, Ralescu, D: B-spline method of uncertain statistics with application to estimating distance. J. Uncertain Syst. **6**(4), 256–262 (2012)
17. Wang, X, Gao, Z, Guo, H: Delphi method for estimating uncertainty distributions. Inform.: Int. Interdiscip. J. **15**(2), 449–460 (2012)
18. Wang, X, Gao, Z, Guo, H: Uncertain hypothesis testing for two experts' empirical data. Math. Comput. Modell. **55**, 1478–1482 (2012)

A genetic algorithm-based optimizing approach for project time-cost trade-off with uncertain measure

Hua Ke

Correspondence:
hke@tongji.edu.cn
School of Economics and Management, Tongji University, Shanghai 200092, China

Abstract

Both the trade-off between the project cost and the project completion time and the indeterminacy of the environment are important issues for real-life project managers. In this paper, an uncertain time-cost trade-off problem, where activity cost functions are assumed to be linear and the objective function to be minimized is the project direct cost, is described based on uncertainty theory. Two uncertain time-cost trade-off models are built to satisfy different management requirements. To solve the proposed models, two equivalent crisp mathematical programming models are given, and genetic algorithm is introduced to search for quasi-optimal schedules. For future research, resource constraints or more types of indeterminacy can be included.

Keywords: Time-cost trade-off; Uncertainty theory; Uncertain measure; Genetic algorithm

Introduction

For real-life projects, decision-makers should always consider the trade-offs among the performance goals for project scheduling and control, especially the trade-off between project completion time and project cost. The time-cost trade-off problem (TCTP) takes into account the project time-cost trade-off by crashing or prolonging project activity durations. In 1961, Kelly [1] first did research on the TCTP, which is one branch of the project scheduling problem. In the following years, many researches have been done on the deterministic TCTP [2,3]. For solving the deterministic TCTP, the common analytical methods were linear programming and dynamic programming [4,5]. Besides, some heuristic algorithms, such as genetic algorithm [6] and simulated annealing algorithm [7], were also applied. In recent years, other important factors for project scheduling, such as quality [8], have also been applied in research on TCTP.

As is well known, the real world is indeterminate. In real-life projects, the activity durations may be variational due to many external factors, such as the increase of productivity level and the change of weather. In recent years, many authors have considered the nondeterministic factors for describing the real-life project indeterminacy. In 1985, Wollmer [9] discussed a stochastic version of the deterministic linear TCTP. In 2000, Gutjahr et al. [10]

designed a modified stochastic branch-and-bound approach and applied it to a specific stochastic discrete TCTP. Aghaie and Mokhtari [11] described an approach based on ant colony optimization method and Monte Carlo simulation technique for project crashing problem with exponentially distributed activity durations. Zahraie and Tavakolan [12] embedded two concepts of time-cost trade-off and resource leveling and allocation in a stochastic multiobjective optimization model, where fuzzy set theory was applied to represent different options for each activity. Ke et al. [13] built two models for stochastic TCTP with the philosophies of chance-constrained programming and dependent-chance programming. Mokhtari et al. [14] developed a hybrid optimization approach based on cutting plane method and Monte Carlo simulation for stochastic TCTP in PERT networks. Ke et al. [15] modeled stochastic project time-cost trade-offs with time-dependent activity durations.

For some projects in indeterminate environment, probability theory is no longer valid for describing activity durations for the lack of statistical data, since human beings usually overweigh unlikely events. For this case, the activity durations may be described by fuzzy variables. The first work on the fuzzy TCTP was done by Leu et al. [16]. Jin et al. [17] gave a genetic algorithm (GA)-based fully fuzzy optimal time-cost trade-off model, in which all parameters and variables were characterized by fuzzy numbers and an example in ship building scheduling was demonstrated. Eshtehardian et al. [18] established a multiobjective fuzzy time-cost model. Ghazanfari et al. [19,20] applied possibilistic goal programming to the TCTP to determine optimal duration for each activity in the form of triangular fuzzy numbers. Ke et al. [21] built three fuzzy programming models for TCTP based on credibility theory. Chen and Tsai [22] constructed membership function of fuzzy minimum total crash cost based on Zadeh's extension principle and transformed the time-cost trade-off problem to a pair of parametric mathematical programs.

When the indeterminacy does not behave either randomly or fuzzily, we need a new tool to deal with it. To describe indeterminacy which is neither randomness nor fuzziness, Liu founded uncertainty theory in 2007 [23] and refined it in 2010 [24]. Uncertainty theory is a branch of axiomatic mathematics for modeling human uncertainty. To the knowledge of the authors, no researchers considered time-cost trade-off problem in uncertain environment, which is not either stochastic or fuzzy. In this paper, we introduce uncertainty theory for modeling the TCTP in indeterminate environment. We propose two uncertain time-cost trade-off models according to some different decision-making criteria. As Huang and Ding [25] showed that using standard path algorithms (e.g., the well-known Dijkstra method) was not able to arrive at solutions for searching critical path of this problem, GA is applied in this paper. Some numerical experiments in which the activity durations are assumed to be uncertain variables with known uncertainty distributions are presented.

The remainder of the paper is organized as follows: Section 'Preliminaries' gives some concepts of uncertainty theory as preliminaries for modeling. Section 'Problem description' describes the TCTP with uncertain activity durations. In Section 'Uncertain models of time-cost trade-off problem', based on some decision-making criteria, two types of uncertain models are presented. The following section gives two numerical experiments to illustrate the proposed models. Finally, Section 'Conclusions' draws some concluding statements.

Preliminaries

To better describe and understand uncertain phenomena, Liu [23] proposed uncertainty theory, which has been applied to many fields such as uncertain calculus [26,27], uncertain risk analysis [28], uncertain logic [29], uncertain finance [30], and uncertain differential equation [31]. Based on uncertainty theory, Liu [32] formulated uncertain programming for solving application problems with uncertain factors, which has been applied into some optimization problems, e.g., option pricing [33], facility location [34], inventory problem [35], transportation problem [36], uncertain graph [37], and portfolio selection [38].

In this section, we introduce some basic concepts which will be helpful for establishing some uncertain models for the TCTP. Let Γ be a nonempty set, and $\mathcal{L}$ a σ-algebra over Γ. Each element Λ in $\mathcal{L}$ is called an event.

Definition 1. (Liu [23]) The set function $\mathcal{M}$ is called an uncertain measure if it satisfies:

Axiom 1. $\mathcal{M}\{\Gamma\} = 1$ for the universal set Γ.
Axiom 2. $\mathcal{M}\{\Lambda\} + \mathcal{M}\{\Lambda^c\} = 1$ for any event Λ.
Axiom 3. For every countable sequence of events $\Lambda_1, \Lambda_2, \ldots$, we have

$$\mathcal{M}\left\{\bigcup_{i=1}^{\infty} \Lambda_i\right\} \leq \sum_{i=1}^{\infty} \mathcal{M}\{\Lambda_i\}.$$

Besides, the product uncertain measure on the product σ-algebra Ł was defined by Liu [26] as follows:

Axiom 4. Let $(\Gamma_k, Ł_k, \mathcal{M}_k)$ be uncertainty spaces for $k = 1, 2, \ldots$. The product uncertain measure $\mathcal{M}$ is an uncertain measure satisfying

$$\mathcal{M}\left\{\prod_{k=1}^{\infty} \Lambda_k\right\} = \bigwedge_{k=1}^{\infty} \mathcal{M}_k\{\Lambda_k\},$$

where Λ_k are arbitrarily chosen events from $Ł_k$ for $k = 1, 2, \ldots$, respectively.

Based on the definition of uncertain measure, we can give the concept of an uncertain variable.

Definition 2. (Liu [23]) An *uncertain variable* is a measurable function ξ from an uncertainty space $(\Gamma, \mathcal{L}, \mathcal{M})$ to the set of real numbers, i.e., for any Borel set B of real numbers, the set

$$\{\xi \in B\} = \{\gamma \in \Gamma \mid \xi(\gamma) \in B\}$$

is an event.

With the concept of uncertain variable, we can define the uncertainty distribution of an uncertain variable.

Definition 3. (Liu [23]) The uncertainty distribution Φ of an uncertain variable ξ is defined by

$$\Phi(x) = \mathcal{M}\{\xi \leq x\}$$

for any real number x.

Liu [24] also defined the inverse function Φ^{-1} as the inverse uncertainty distribution of uncertain variable ξ. With inverse uncertainty distribution, Liu [24] gave the operational law of uncertain variables as follows.

Theorem 1. *(Liu [24]) Let $\xi_1, \xi_2, \ldots, \xi_n$ be independent uncertain variables with regular uncertainty distributions $\Phi_1, \Phi_2, \ldots, \Phi_n$, respectively. If the function $f(x_1, x_2, \ldots, x_n)$ is strictly increasing with respect to $x_1, x_2, \ldots, x_m$ and strictly decreasing with respect to $x_{m+1}, x_{m+2}, \ldots, x_n$, then*

$$\xi = f(\xi_1, \xi_2, \ldots, \xi_n)$$

is an uncertain variable with inverse uncertainty distribution

$$\Phi^{-1}(\alpha) = f(\Phi_1^{-1}(\alpha), \ldots, \Phi_m^{-1}(\alpha), \Phi_{m+1}^{-1}(1-\alpha), \ldots, \Phi_n^{-1}(1-\alpha)).$$

For giving out some decision-making criteria for managers, we introduce the following definition:

Definition 4. (Liu [23]) Let ξ be an uncertain variable. The expected value of ξ is defined by

$$E[\xi] = \int_0^{+\infty} \mathcal{M}\{\xi \geq r\}\mathrm{d}r - \int_{-\infty}^{0} \mathcal{M}\{\xi \leq r\}\mathrm{d}r$$

provided that at least one of the above two integrals is finite.

Theorem 2. *(Liu [24]) Let ξ be an uncertain variable with regular uncertainty distribution Φ. If the expected value exists, then*

$$E[\xi] = \int_0^1 \Phi^{-1}(\alpha)\mathrm{d}\alpha.$$

For instance, let $\xi \sim \mathcal{L}(a,b)$ be a linear uncertain variable. Then its inverse uncertainty distribution is $\Phi^{-1}(\alpha) = (1-\alpha)a + \alpha b$, and its expected value is

$$E[\xi] = \int_0^1 ((1-\alpha)a + \alpha b)\mathrm{d}\alpha = \frac{a+b}{2}.$$

Problem description

With the project progress, project managers always need to make a trade-off between the cost and the completion time. Sometimes managers may make a decision in order to finish the project sooner with project cost augmentation by accelerating the project schedule, which is also named as project crashing in project management. In other cases, motivated by reducing project cost, managers may be conscripted to sacrifice by prolonging the project completion time. Therefore, it is naturally desirable for managers to find a schedule to complete a project with the balance of the cost and the completion time.

A project can be represented by an activity-on-arc network $G = (V, A)$, where $V = \{1, 2, \ldots, n\}$ is the set of nodes representing the milestones and A is the set of arcs representing the activities, shown as Figure 1. In the network, nodes 1 and n represent the start and the end of the project, respectively. In this paper, the normal activity durations are assumed to be uncertain variables. Note that the normal duration of activity (i,j) denoted

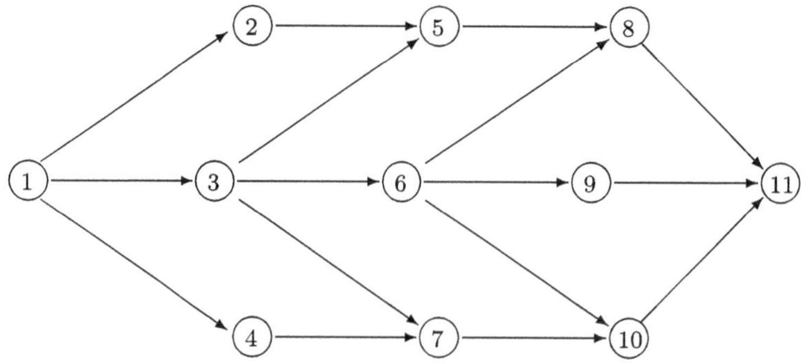

Figure 1 A project. Nodes 1 and n represent the start and the end of the project, respectively.

as ξ_{ij} represents the duration without the influence of the decision made by the manager, and the uncertainty of ξ_{ij} is derived from the uncertain project environment. Correspondingly, the normal cost per time unit of activity (i,j) is denoted by c_{ij}. The decision variable x_{ij} indicates the duration change of activity (i,j) controlled by the manager, such as determining the number change of workers and changing the instruments. The variable x_{ij}, supposed to be an integer for the sake of simplicity, is bounded by some interval $[\,l_{ij}, u_{ij}]$ owing to practical conditions, where l_{ij} and u_{ij} are also assumed to be integers. For each activity (i,j), there also exists another associated cost d_{ij}, regarded as the additional cost of per unit change of x_{ij}. For simplicity, c_{ij} and d_{ij} are both assumed to be constants.

Then, for the trade-off between the completion time and the cost, the goal is to decide the optimal vector $\boldsymbol{x} = \left\{x_{ij} : (i,j) \in A\right\}$ to meet different scheduling requirements.

The uncertain normal activity durations can be concisely written as $\boldsymbol{\xi} = \{\xi_{ij} : (i,j) \in A\}$. The starting time of activity (i,j) is denoted by $T_{ij}(\boldsymbol{x}, \boldsymbol{\xi})$, and the starting time of activity $(1,j) \in A$ is defined as $T_{1j}(\boldsymbol{x}, \boldsymbol{\xi}) = 0$, which means that the starting time of the total project is assumed to be 0. To simplify the problem, we assume that each activity can be processed only if all the foregoing activities are finished and should be processed without interruption, and no lead or lag times are possible. With the assumptions, the starting time of activity (i,j), $i = 2, 3, \ldots, n-1$ can be calculated by

$$T_{ij}(\boldsymbol{x}, \boldsymbol{\xi}) = \max_{(k,i)\in A} \{T_{ki}(\boldsymbol{x}, \boldsymbol{\xi}) + \xi_{ki} + x_{ki}\}.$$

Suppose that $T_{ij}(\boldsymbol{x}, \boldsymbol{\xi})$ has an inverse uncertainty distribution $\Psi_{ij}^{-1}(\boldsymbol{x}, \alpha)$. Then

$$\Psi_{ij}^{-1}(\boldsymbol{x}, \alpha) = \max_{(k,i)\in A} \left\{\Psi_{ki}^{-1}(\boldsymbol{x}, \alpha) + \Phi_{ki}^{-1}(\alpha) + x_{ki}\right\},$$

and the project completion time can be written by

$$T(\boldsymbol{x}, \boldsymbol{\xi}) = \max_{(k,n)\in A} \{T_{kn}(\boldsymbol{x}, \boldsymbol{\xi}) + \xi_{kn} + x_{kn}\} \tag{1}$$

whose inverse uncertainty distribution is

$$\Psi^{-1}(\boldsymbol{x}, \alpha) = \max_{(k,n)\in A} \left\{\Psi_{kn}^{-1}(\boldsymbol{x}, \alpha) + \Phi_{kn}^{-1}(\alpha) + x_{kn}\right\}. \tag{2}$$

The project cost, composed of the normal cost and the additional cost, can be simply computed by

$$C(\boldsymbol{x}, \boldsymbol{\xi}) = \sum_{(i,j)\in A} (c_{ij}\xi_{ij} - d_{ij}x_{ij}) \tag{3}$$

whose inverse uncertainty distribution is

$$\Upsilon^{-1}(\boldsymbol{x}, \alpha) = \sum_{(i,j)\in A} \left(c_{ij}\Phi_{ij}^{-1}(\alpha) - d_{ij}x_{ij} \right). \tag{4}$$

Uncertain models of time-cost trade-off problem

In many researches on indeterminate decision systems, optimizing expected objective value is preferably considered to be the choice for decision-making and expected value model is the most employed model. However, for various practical requirements, other alternative decision-making criteria and optimization models are needed. In this paper, except the expected value model, one more optimization model for the uncertain TCTP is presented with the philosophy of dependent-chance programming.

Expected cost minimization model

As we mentioned above, comparing expected values is the most widely applied decision-making criterion in practice. Risk-averse managers usually want to find the optimal decision with minimum expected project cost subject to some project completion time constraint. With this criterion, we can present the following expected cost minimization model for TCTP:

$$\begin{cases} \min E[C(\boldsymbol{x}, \boldsymbol{\xi})] \\ \text{subject to} \\ \qquad \mathcal{M}\left\{T(\boldsymbol{x}, \boldsymbol{\xi}) \le T^0\right\} \ge \alpha_0 \\ \qquad x_{ij} \in [l_{ij}, u_{ij}], \ \forall (i,j) \in A \\ \qquad x_{ij}, \ \forall (i,j) \in A, \ \text{integers}, \end{cases}$$

where T^0 is the due date of the project, α_0 is a predetermined confidence level, l_{ij} and u_{ij} are integers given in advance, and $T(\boldsymbol{x}, \boldsymbol{\xi})$ and $C(\boldsymbol{x}, \boldsymbol{\xi})$ are defined by (1) and (3), respectively. The above model is equivalent to

$$\begin{cases} \min \int_0^1 \Upsilon^{-1}(\boldsymbol{x}, \alpha) d\alpha \\ \text{subject to} \\ \qquad \Psi^{-1}(\boldsymbol{x}, \alpha_0) \le T^0 \\ \qquad x_{ij} \in [l_{ij}, u_{ij}], \ \forall (i,j) \in A \\ \qquad x_{ij}, \ \forall (i,j) \in A, \ \text{integers}, \end{cases}$$

where $\Psi^{-1}(\boldsymbol{x}, \alpha)$ and $\Upsilon^{-1}(\boldsymbol{x}, \alpha)$ are determined by (2) and (4), respectively.

Chance maximization model

In real-life project, the manager may tend to control the project cost within some budget and meanwhile complete the project in time. However, due to the environmental complexity, these performance goals are not always obtained completely. Then it is natural for the project manager to maximize the chance that the project cost does not exceed the given budget under some project completion time constraint, which follows the philosophy of dependent-chance programming (DCP) introduced by Liu [39]. Based on the DCP philosophy, we can establish the chance maximization model as follows:

$$\begin{cases} \max \mathcal{M}\left\{C(\boldsymbol{x},\boldsymbol{\xi}) \le C^0\right\} \\ \text{subject to} \\ \qquad \mathcal{M}\{T(\boldsymbol{x},\boldsymbol{\xi}) \le T^0\} \ge \alpha_0 \\ \qquad x_{ij} \in [l_{ij}, u_{ij}], \ \forall (i,j) \in A \\ \qquad x_{ij}, \ \forall (i,j) \in A, \ \text{integers,} \end{cases}$$

where α_0 is a predetermined confidence level, T^0 is the due date of the project, C^0 is the budget, l_{ij} and u_{ij} are integers given in advance, and $T(\boldsymbol{x},\boldsymbol{\xi})$ and $C(\boldsymbol{x},\boldsymbol{\xi})$ are defined by (1) and (3), respectively. Note that here so-called chance is measured by uncertain measure. This model is equivalent to

$$\begin{cases} \max \bar{\beta} \\ \text{subject to} \\ \qquad \Upsilon^{-1}(\boldsymbol{x},\bar{\beta}) \le C^0 \\ \qquad \Psi^{-1}(\boldsymbol{x},\alpha_0) \le T^0 \\ \qquad x_{ij} \in [l_{ij}, u_{ij}], \ \forall (i,j) \in A \\ \qquad x_{ij}, \ \forall (i,j) \in A, \ \text{integers,} \end{cases}$$

where $\Psi^{-1}(\boldsymbol{x},\alpha)$ and $\Upsilon^{-1}(\boldsymbol{x},\beta)$ are determined by (2) and (4), respectively.

Numerical experiments

By a simple example, Huang and Ding [25] demonstrated that the standard path algorithms (e.g., the well-known Dijkstra method) were not capable of finding the critical path for the random project scheduling problem, which is applicable to uncertain TCTP. Hence, for solving the above two uncertain time-cost trade-off models, GA is introduced to search for the quasi-optimal solutions.

Consider the project network shown in Figure 1. For every activity (i,j), the duration is assumed to be a linear uncertain variable $\mathcal{L}(2i+3, 2j+2)$. The normal costs and the additional costs of the activities are presented in Table 1, respectively. We also assume that the decision variable x_{ij} is an integer and limited in the interval $[-4, 3]$ for each $(i,j) \in A$.

First, the project manager tends to finish the project in 62 time units with confidence level 0.85 and meanwhile minimizes the project cost in the sense of expected value. With this demand, we can present the following expected cost minimization model:

$$\begin{cases} \min E[C(\boldsymbol{x},\boldsymbol{\xi})] \\ \text{subject to} \\ \qquad \mathcal{M}\{T(\boldsymbol{x},\boldsymbol{\xi}) \le 62\} \ge 0.85 \\ \qquad x_{ij} \in [-4, 3], \ \forall (i,j) \in A \\ \qquad x_{ij}, \ \forall (i,j) \in A, \ \text{integers.} \end{cases}$$

This model is equivalent to

$$\begin{cases} \min \int_0^1 \Upsilon^{-1}(\boldsymbol{x},\alpha)\mathrm{d}\alpha \\ \text{subject to} \\ \qquad \Psi^{-1}(\boldsymbol{x},0.85) \le 62 \\ \qquad x_{ij} \in [-4, 3], \ \forall (i,j) \in A \\ \qquad x_{ij}, \ \forall (i,j) \in A, \ \text{integers.} \end{cases}$$

Table 1 Activity costs of project

Activity (i,j)	Normal cost c_{ij}	Additional cost d_{ij}
(1,2)	170	200
(1,3)	300	280
(1,4)	65	70
(2,5)	270	300
(3,5)	135	150
(3,6)	75	90
(3,7)	150	100
(4,7)	600	400
(5,8)	85	100
(6,8)	300	400
(6,9)	95	90
(6,10)	130	140
(7,10)	200	240
(8,11)	90	120
(9,11)	200	180
(10,11)	320	380

For the above model, we can easily employ GA to search for the quasi-optimal solution. We set the parameters in GA as the population size of one generation pop_size $= 50$, the probability of mutation $P_{\rm m} = 0.5$, and the probability of crossover $P_{\rm c} = 0.7$. After a run of 8,000 generations, we obtain the quasi-optimal solution $\boldsymbol{x}^* = (3, 1, -4, 3, 3, -2, 0, 2, 0, 3, 1, 2, -2, 0, 0, -2)$, $E[C(\boldsymbol{x}^*, \boldsymbol{\xi})] = 43,482.5$ and $\mathcal{M}\{T(\boldsymbol{x}^*, \boldsymbol{\xi}) \leq 62\} = 0.857$.

The other considered model is the chance maximization shown as follows:

$$\begin{cases} \max \mathcal{M}\left\{C(\boldsymbol{x}, \boldsymbol{\xi}) \leq 46,700\right\} \\ \text{subject to} \\ \qquad \mathcal{M}\{T(\boldsymbol{x}, \boldsymbol{\xi}) \leq 65\} \geq 0.85 \\ \qquad x_{ij} \in [-4, 3], \ \forall (i,j) \in A \\ \qquad x_{ij}, \ \forall (i,j) \in A, \ \text{integers.} \end{cases}$$

This model is equivalent to

$$\begin{cases} \max \bar{\beta} \\ \text{subject to} \\ \qquad \Upsilon^{-1}(\boldsymbol{x}, \bar{\beta}) \leq 46,700 \\ \qquad \Psi^{-1}(\boldsymbol{x}, 0.85) \leq 65 \\ \qquad x_{ij} \in [-4, 3], \ \forall (i,j) \in A \\ \qquad x_{ij}, \ \forall (i,j) \in A, \ \text{integers.} \end{cases}$$

After a run of 6,000 generations with pop_size $= 50$, $P_{\rm m} = 0.6$, and $P_{\rm c} = 0.5$, we obtain the quasi-optimal solution $\boldsymbol{x}^* = (2, 0, -4, 3, 2, 0, 1, 2, 3, 3, 0, 3, 0, 1, 3, -1)$ and $\mathcal{M}\left\{C(\boldsymbol{x}^*, \boldsymbol{\xi}) \leq 46,700\right\} = 0.874$.

Conclusions

For real-life project managers, both the trade-off between the cost and the completion time and the indeterminacy environment are considerable issues. In this paper, an uncertain TCTP was formulated with the objective of minimizing the cost with completion time limits based on uncertain measure. With some decision-making criteria, the expected cost minimization model and the chance maximization model were established to satisfy different practical managing requirements. Two equivalent crisp mathematical programming models were also given. To solve the models, GA was introduced.

The TCTP with uncertain activity durations can be regarded as the extension of the fuzzy TCTP. Furthermore, resource constraints or more types of indeterminacies can be included for future research. For real-life application, the models can be applied to many other project optimization problems.

Acknowledgments

The work was partly supported by the National Natural Science Foundation of China (71371141, 71001080).

References

1. Kelley, JE: Critical path planning and scheduling mathematical basis. Oper. Res. **9**, 296–320 (1961)
2. Tareghian, HR, Taheri, SH: On the discrete time, cost and quality trade-off problem. Appl. Math. Comput. **181**, 1305–1312 (2006)
3. Weglarz, J, Jozefowska, J, Mika, M, Waligora, G: Project scheduling with finite or infinite number of activity processing modes - a survey. Eur. J. Oper. Res. **208**, 177–205 (2011)
4. Heberta, JE, Deckro, RF: Combining contemporary and traditional project management tools to resolve a project scheduling problem. Comput. Oper. Res. **38**, 21–32 (2011)
5. Talbot, FB: Resource-constrained project scheduling with time-resource tradeoffs: the non-preemptive case. Manage. Sci. **28**, 1197–1210 (1982)
6. Azaron, A, Perkgoz, C, Sakawa, M: A genetic algorithm approach for the time-cost trade-off in PERT networks. Appl. Math. Comput. **168**, 1317–1339 (2005)
7. Anagnostopoulos, KP, Kotsikas, L: Experimental evaluation of simulated annealing algorithms for the time-cost trade-off problem. Appl. Math. Comput. **217**, 260–270 (2010)
8. Salmasnia, A, Mokhtari, H, Abadi, INK: A robust scheduling of projects with time, cost, and quality considerations. Int. J. Adv. Manuf. Tech. **60**, 631–642 (2012)
9. Wollmer, RD: Critical path planning under uncertainty. Math. Program. Stud. **25**, 164–171 (1985)
10. Gutjahr, WJ, Strauss, C, Wagner, E: A stochastic branch-and-bound approach to activity crashing in project management. INFORMS J. Comput. **12**, 125–135 (2000)
11. Aghaie, A, Mokhtari, H: Ant colony optimization algorithm for stochastic project crashing problem in PERT networks using MC simulation. Int. J. Adv. Manuf. Tech. **45**, 1051–1067 (2009)
12. Zahraie, B, Tavakolan, M: Stochastic time-cost-resource utilization optimization using nondominated sorting genetic algorithm and discrete fuzzy sets. J. Constr. Eng. M. **135**, 1162–1171 (2009)
13. Ke, H, Ma, W, Ni, Y: Optimization models and a GA-based algorithm for stochastic time-cost trade-off problem. Appl. Math. Comput. **215**, 308–313 (2009)
14. Mokhtari, H, Aghaie, A, Rahimi, J, Mozdgir, A: Project time-cost trade-off scheduling: a hybrid optimization approach. Int. J. Adv. Manuf. Tech. **50**, 811–822 (2010)
15. Ke, H, Ma, W, Chen, X: Modeling stochastic project time-cost trade-offs with time-dependent activity durations. Appl. Math. Comput. **218**, 9462–9469 (2012)
16. Leu, SS, Chen, AT, Yang, CH: A GA-based fuzzy optimal model for construction time-cost trade-off. Int. J. Proj. Manage. **19**, 47–58 (2001)
17. Jin, C, Ji, Z, Lin, Y, Zhao, Y, Huang, Z: Research on the fully fuzzy time-cost trade-off based on genetic algorithms. J. Mar. Sci. Appl. **4**, 18–23 (2005)
18. Eshtehardian, E, Afshar, A, Abbasnia, R: Time-cost optimization: using GA and fuzzy sets theory for uncertainties in cost. Constr. Manage. Econ. **26**, 679–691 (2008)
19. Ghazanfari, M, Shahanaghi, K, Yousefli, A: An application of possibility goal programming to the time-cost trade off problem. J. Uncertain Syst. **2**, 22–28 (2008)
20. Ghazanfari, M, Yousefli, A, Ameli, MSJ, Bozorgi-Amiri, A: A new approach to solve time-cost trade-off problem with fuzzy decision variables. Int. J. Adv. Manuf. Tech. **42**, 408–414 (2009)
21. Ke, H, Ma, W, Gao, X, Xu, W: New fuzzy models for time-cost trade-off problem. Fuzzy Optim. Decis. Ma. **9**, 219–231 (2010)
22. Chen, S, Tsai, M: Time-cost trade-off analysis of project networks in fuzzy environments. Eur. J. Oper. Res. **212**, 386–397 (2011)
23. Liu, B: Uncertainty Theory. 2nd ed. Springer, Berlin (2007)
24. Liu, B: Uncertainty Theory: a Branch of Mathematics for Modeling Human Uncertainty. Springer, Berlin (2010)
25. Huang, W, Ding, L: Project-scheduling problem with random time-dependent activity duration times. IEEE T. Eng. Manage. **58**, 377–387 (2011)
26. Liu, B: Some research problems in uncertainty theory. J. Uncertain Syst. **3**, 3–10 (2009)
27. Chen, XW, Ralescu, DA: Liu process and uncertain calculus. J. Uncertain. Anal. Appl. **1**, Article 3 (2013)
28. Liu, B: Uncertain risk analysis and uncertain reliability analysis. J. Uncertain Syst. **4**, 163–170 (2010)
29. Liu, B: Uncertain logic for modeling human language. J. Uncertain Syst. **5**, 3–20 (2011)

30. Liu, B: Toward uncertain finance theory. J. Uncertain. Anal. Appl. **1**, Article 1 (2013)
31. Yao, K: Extreme values and integral of solution of uncertain differential equation. J. Uncertain. Anal. Appl. **1**, 2 (2013)
32. Liu, B: Theory and Practice of Uncertain Programming. 2nd ed. Springer, Berlin (2009)
33. Chen, XW: American option pricing formula for uncertain financial market. Int. J. Oper. Res. **8**, 32–37 (2011)
34. Gao, Y: Uncertain models for single facility location problems on networks. Appl. Math. Model. **36**, 2592–2599 (2012)
35. Rong, LX: Two new uncertainty programming models of inventory with uncertain costs. J. Inf. Comput. Sci. **8**, 280–288 (2011)
36. Sheng, Y, Yao, K: Fixed charge transportation problem in uncertain environment. Ind. Eng. Manage. Sys. **11**, 183–187 (2012)
37. Zhang, B, Peng, J: Euler index in uncertain graph. Appl. Math. Comput. **218**, 10279–10288 (2012)
38. Zhu, Y: Uncertain optimal control with application to a portfolio selection model. Cy. Sys. **41**, 535–547 (2010)
39. Liu, B: Dependent-chance programming: a class of stochastic programming. Comput. Math. Appl. **34**, 89–104 (1997)

4

A new approach for tuning interval type-2 fuzzy knowledge bases using genetic algorithms

Praveen Kumar Shukla[1*] and Surya Prakash Tripathi[2]

* Correspondence: praveenshuklaniec@yahoo.co.in
[1]Department of Information Technology, Babu Banarasi Das Northern India Institute of Technology, Lucknow 226028, India
Full list of author information is available at the end of the article

Abstract

Fuzzy knowledge-based systems (FKBS) are significantly applicable in the area of control, classification, and modeling, having knowledge in the form of fuzzy if-then rules. Type-2 fuzzy theory is used to make these systems more capable of dealing with inherent uncertainties in real-world problems. In this paper, the authors have proposed a genetic tuning approach named lateral displacement and expansion/compression (LDEC) in which α and β parameters are calculated to adjust the parameters of interval type-2 membership functions. α tuning deals with lateral displacement, whereas β tuning carries out compression/expansion operation. The interpretability and accuracy features are considered during the development of this approach. The experimental results show the performance of the proposed approach.

Introduction

Fuzzy systems, more specifically fuzzy knowledge-based systems (FKBS) or fuzzy rule-based systems (FRBS), are significantly applicable in areas like control [1], classification [2], and modeling [3]. The essential feature of FKBS is the incorporation of human expert knowledge which is in the form of fuzzy [4] extended if-then rules. The major components of FKBS are fuzzification interface, inference engine, knowledge base, and defuzzification interface [5]. Knowledge base (KB) is composed of two components: data base (DB) and rule base (RB). DB is the repository of membership functions (MFs) and scaling functions (SFs) representing linguistic values, whereas RB is the collection of knowledge related to problems in terms of fuzzy if-then rules.

The design and implementation of KB can be assumed as an optimization task. Hence, genetic algorithms (GAs) are used for learning and tuning of various parameters of KB due to their strong capacity of searching in a complicated and poorly defined search space. Such an application of GAs in developing FKBS is specifically named as genetic fuzzy systems (GFS) [5-8]. GFS have been used for handling various types of applications like predicting surface finish in ultraprecision diamond [9], bioaerosol detector [10], classification of intrusion attacks from a network traffic data [11], tool wear monitoring [12], smart base isolation system [13], etc.

Fuzzy systems for applications like in economics, medicine, etc. are to be developed such that the users may understand how they work by inspecting their KB and functioning. Technically, this feature is called 'interpretability' [14] which is the subjective feature of a fuzzy system showing how much the system is readable/understandable to the users by observing its functionality. Accuracy [15] is another feature showing the closeness between the real model and the developed model. Interpretability and accuracy are contradictory with each other, i.e., one can be improved at the cost of the other, denoted by 'interpretability-accuracy trade-off' (I-A Trade-Off) [16-19]. For the above applications, interpretability as well as accuracy is required to be maintained at the higher level by maintaining a good I-A Trade-Off.

Interpretability and accuracy features are directly related to the approaches of developing FKBS which are domain expert method and experimental data method. In the first method, domain experts of the problem are contributing their knowledge to develop the RB of the FKBS. Such FKBS are much more interpretable. In the second method, RB is generated by using some machine learning method applied on the data set of the particular problem. The FKBS developed by the second method are less interpretable but are more generic. An idea of generating FKBS with the experimental data method guided by the domain expert method is good enough toward achieving an I-A Trade-Off with higher levels of interpretability as well as accuracy.

The special interest of this paper is the use of interval type-2 fuzzy systems (IT2FS) [20]. The membership functions are tuned using GAs, which leads toward a new system, the 'type-2 genetic fuzzy system' (T2GFS).

The paper continues with the 'Interpretability issues in FKBS' section in which the interpretability issues of FKBS are discussed. The 'Tuning and learning operations in FKBS' section introduces the basics of tuning and learning approaches. The fundamentals of type-2 fuzzy systems are discussed in the 'Type-2 fuzzy systems' section. A new lateral displacement and expansion/compression (LDEC) tuning approach is discussed in the 'Proposed LDEC tuning approach' section. The genetic representation of KB and the proposed tuning approach is discussed in the 'Genetic representation of knowledge base' section. Experimental results are discussed in the 'Experiments and results' section.

Interpretability issues in FKBS

Interpretability [14,21-23] and accuracy [15] are the two important features considered during the design of fuzzy systems. Basically, interpretability is identified as a feature to understand the significance of something [21], and it is also known with other names like comprehensibility, intelligibility, transparency, readability, understandability, etc. Also, the quantification of interpretability is a highly subjective task depending on various parameters like experience, preference, and the knowledge of the person who interprets the system functionality.

Linguistic fuzzy modeling (LFM) and precise fuzzy modeling (PFM) [24] are two modeling approaches of fuzzy systems. In LFM, fuzzy models are developed by means of linguistic FRBS which are called Mamdani-type FKBS [25] mainly focusing on interpretability. On the other hand, PFM is developed considering the accuracy parameter and called Takagi-Sugeno FKBS [26]. Accuracy improvement in LFM [15] and interpretability improvement in PFM [14] are carried out to achieve the desired I-A Trade-Off.

Various approaches have been developed to deal with different issues of the interpretability of fuzzy systems. These are discussed in Table 1.

Many other indexes and methodologies have been developed for assessing the interpretability, which are considered in this paper. These are (1) number of rules (NOR),

Table 1 Interpretability in type-1 FKBS

Year	Authors	Description of work	Reference
2000	Y. Jin	Interpretability improvement in high-dimensional fuzzy systems	[23]
2001	S. Guillaume	Automatic rule generation and structure optimization for maintaining interpretability	[27]
2005	R. Mikut et al.	Maintaining interpretability in data-based fuzzy system development along with user-controllable trade-off in between interpretability and accuracy	[28]
2006	R. Alcala et al.	Seven hybrid techniques for developing accurate and interpretable FKBS	[16]
2008, 2012	J. M. Alonso et al.	Highly interpretable linguistic knowledge (HILK) utilizing the features of LFM and PFM	[29,30]
2008	S. M. Zhou and J. A. Gan	Identification of two interpretability levels: low level on the fuzzy set and high level on the fuzzy rule	[31]
2008	C. Mencar and A. M. Fanelli	Introduction of semantic constraints, distinguishability, coverage, convexity, and normality	[32]
2009	J. M. Alonso et al.	Conceptual framework for assessing the interpretability based on two issues: 'description' and 'explanation'	[22]
2011	M. J. Gacto et al.	A proposal of double-axis taxonomy: 'complexity and semantic interpretability' and 'rule base and fuzzy partition'	[33]
2013	M. Fazzolari et al.	I-A Trade-Off handling with instance selection techniques	[34]

(2) total rule length (TRL) - the sum of the number of premises in all the rules, and (3) average rule length (ARL) - calculated by TRL/NOR.

Nauck's index (NI) [35] has been proposed to assess the interpretability of fuzzy rule-based classifiers. It is given by

$$I_{\text{Nauck}} = \text{comp} \times \text{part} \times \text{cov}$$

where $\text{comp} = \frac{\text{number of classes}}{\text{total number of premises}}$ (it measures the complexity), $\text{part} = \frac{1}{\text{number of labels}-1}$ (it is the average normalized partition index), and cov is the average normalized coverage degree of the fuzzy partition. For strong fuzzy partition (SFP), it is equal to 1.

Similarly, a new global fuzzy index has been proposed in [36]. In this approach, the index has been computed as the outcomes of the inference of hierarchical fuzzy system.

Tuning and learning operations in FKBS

During the design of genetic FKBS, tuning and learning operations (Figure 1) are carried out to improve the performance of FKBS [5,6]. In the tuning operation, the parameters of DB constituents, MFs and SFs, are adjusted, maintaining no change in the

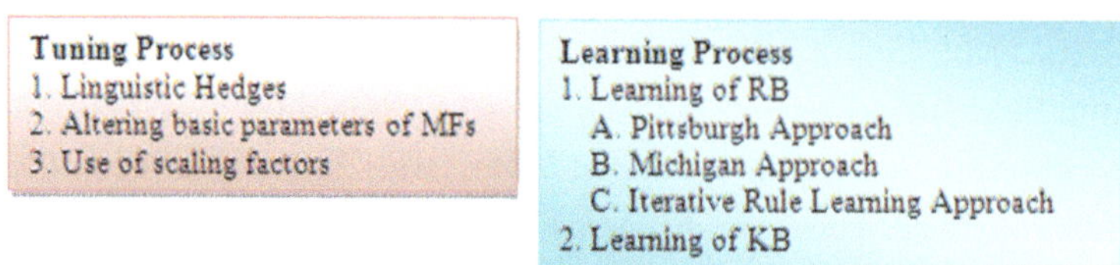

Figure 1 Tuning and learning approaches for the FKBS.

previously defined RB, whereas in the learning operation, the parameters of RB are changed simultaneously with the DB. There are three main approaches for carrying out learning operations: the Pittsburgh approach [37], Michigan approach [38], and iterative rule learning approach [39].

In the literature, two types of approaches are found for tuning operations: one is related to applying SFs for handling linguistic hedges and the other is the tuning of the MF parameters. In this paper, the second approach of MF tuning is considered.

The scaling functions are responsible for adjusting the universe of discourse of input and output variables to the domain. The parameters used for tuning the scaling functions are scaling factor, upper and lower bounds (linear scaling functions), and contraction/dilation parameters (non-linear scaling function). The linguistic hedges are used and applied on the tuned MFs as discussed in [40-42]. The main linguistic hedges are as follows: very, more-or-less, extremely, very-very, positively, and negatively. Linguistic hedges are playing the role of adjectives and adverbs in the languages responsible for changing the qualitative linguistic statements.

Apart from tuning, learning, and interpretability issues in the design of FKBS, several other burning issues are like dealing with the high dimensionality of the data along with handling imbalanced data sets (Figure 2).

Type-2 fuzzy systems

To implement FKBS, type-2 fuzzy sets (T2FS) [43,44] are used having more capacity to deal with inherent uncertainties in the system to be developed. General type-2 fuzzy sets require high computational cost and type reduction complexity; hence, interval type 2 fuzzy sets [45-48] are preferred to model and implement various problems.

T2FS which is denoted by A^* is characterized by MF $\mu_{A^*}(x,u)$, where $x \in X$ and $u \in J_x \subseteq [0,1]$:

$$A^* = \{(x,u), \mu_{A^*}(x,u) | \forall x \in X, \forall u \in J_x \subseteq [0,1]\}$$

Here, $0 \leq \mu_{A^*}(x,u) \leq 1$; when all $\mu_{A^*}(x,u) = 1$, then A^* is an interval type-2 fuzzy set.

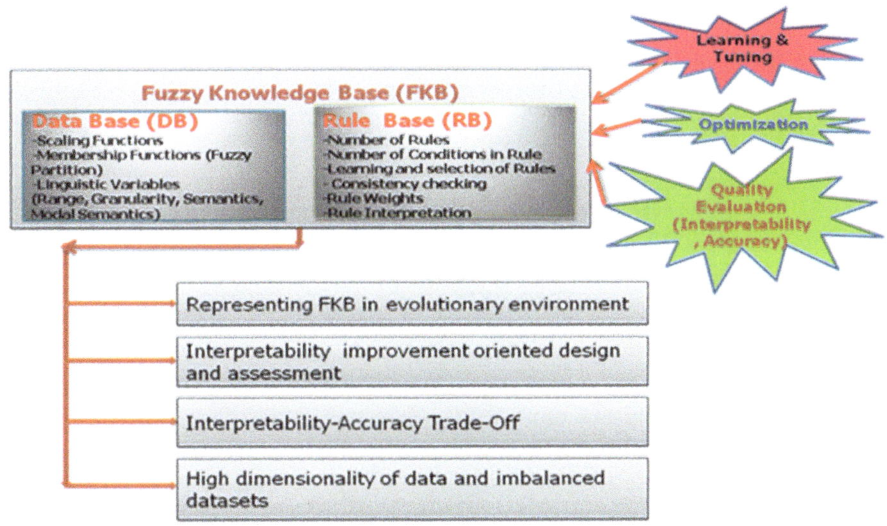

Figure 2 Issues and challenges in fuzzy knowledge-based system design.

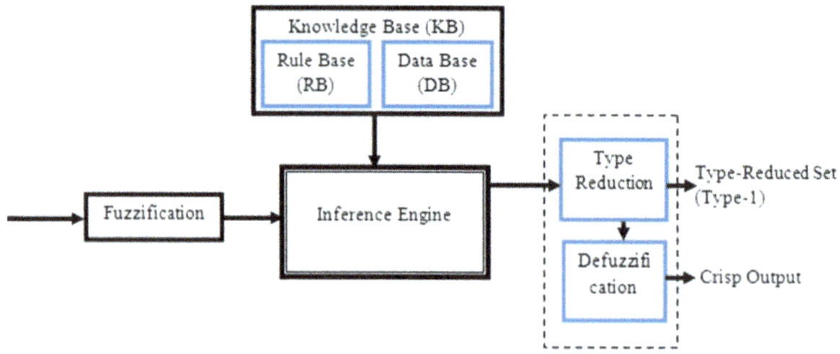

Figure 3 Block diagram of type 2 fuzzy systems [49].

A type-2 fuzzy system [49] is identified as a FLS with if-then rules in which at least one linguistic term is a T2FS. Normally, a type-2 fuzzy system differs from a type-1 fuzzy system by having one extra component at the output processing, which is called type reducer (Figure 3). Also, in the type 2 fuzzy system, the antecedent and consequent parts of the rule must have at least one T2FS.

Proposed LDEC tuning approach

In this section, the authors have proposed a LDEC tuning approach for adjusting the parameters of interval type-2 fuzzy MFs. The two-phase procedure of the tuning approach is given in Figure 4. The first phase includes α tuning operation, and in the second phase, β tuning operation is performed.

α tuning operation

In the α tuning operation, all the coordinates of IT2MF are shifted by parameter α and the new coordinates would be as follows: $a' = a \pm \alpha$, $b' = b \pm \alpha$, $c' = c \pm \alpha$, $d' = d \pm \alpha$, $e' = e \pm \alpha$, depending on the positive and negative values of α. When the value of parameter α is positive, it leads to a tuned MF with forward lateral displacement (Figure 5a), and the negative value of α leads to backward lateral displacement (Figure 5b). The value of α is calculated as given below.

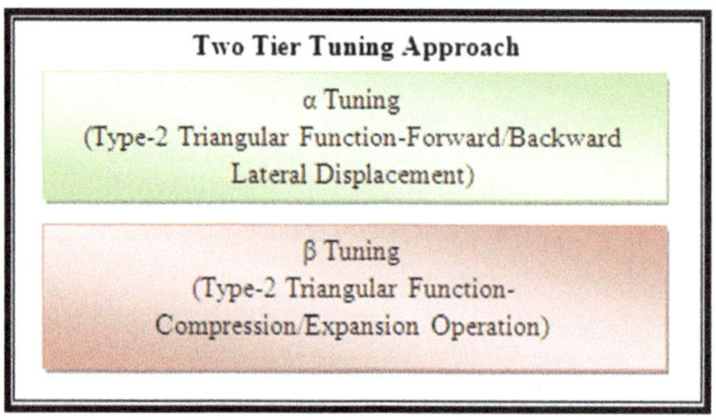

Figure 4 Two-tier tuning approach.

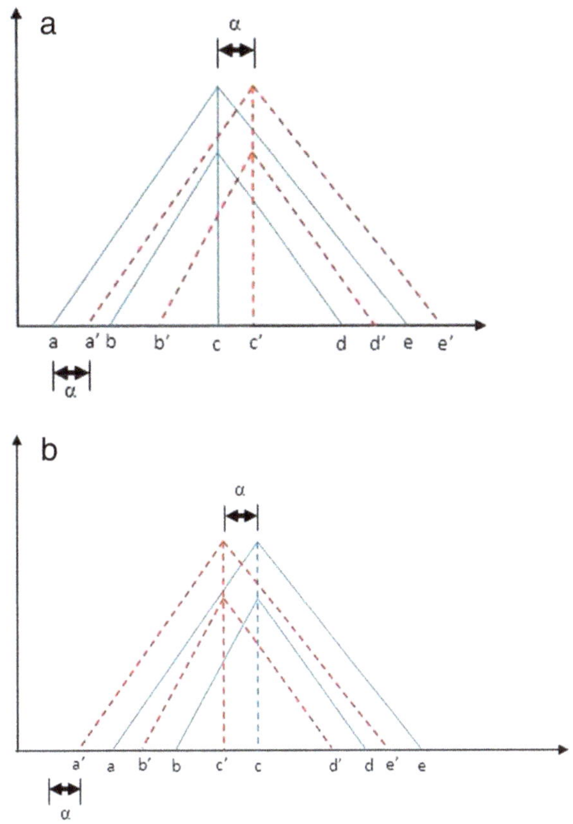

Figure 5 ***α*** **tuning approach. (a)** Forward lateral displacement. **(b)** Backward lateral displacement.

$$\mathrm{val}_{\alpha} = \frac{1}{2}\left[\frac{\mathrm{val}_e - \mathrm{val}_a}{\mathrm{val}_e} \times (\mathrm{val}_c - \mathrm{val}_a)\right]$$

β tuning operation

In the β tuning approach, parameter β is applicable on parameters a, b, d, and e. After the tuning operation, the coordinates would be as follows:

$$a' = a+\beta, b' = b+\beta, d' = d-\beta, e' = e-\beta \quad \text{if } \beta > 0 \text{ then compression or if } \beta < 0 \text{ then expansion}$$

The position of c is assumed to be fixed. The value of β is calculated as follows:

$$\mathrm{val}_{\beta} = \frac{1}{4}\left(\frac{\mathrm{val}_c - \mathrm{val}_a}{\mathrm{val}_e}\right) \times \mathrm{val}_c$$

A positive value of β leads to compression (Figure 6a), whereas a negative value performs the expansion operation (Figure 6b).

Genetic representation of knowledge base

GAs [50,51] are popular search techniques for ill-defined and complex search spaces. They are based on natural evolution. The initial population $G(0)$ is generated with chromosomes representing DB and RB information and subsequently goes under evolution. During evolution, the next generation $G(n + 1)$ is generated by applying crossover and mutation operators

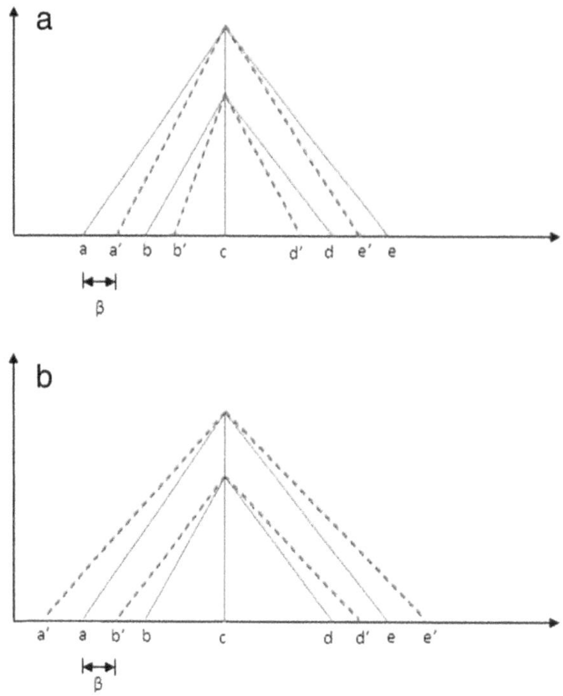

Figure 6 ***β*** **tuning approach. (a)** Compression. **(b)** Expansion.

on the generation $G(n)$. On each generation, each individual is evaluated by a fitness function. A termination condition is set to stop the evolution process.

In [52], inter-valued fuzzy sets (IVFS) have been used to implement a linguistic fuzzy rule-based classification system based on a new interval fuzzy reasoning method along with a new fuzzy rule learning process, called IVTURS-FARC.

In [53], the performance of a fuzzy rule-based classification system is improved using an interval-valued fuzzy set and a tuning approach using genetic algorithm. The uncertainty is modeled by the function 'weak ignorance.'

Various parameters of type-2 fuzzy systems are optimized using GAs and other bio-inspired optimization algorithms. Few of these works are summarized in Table 2.

New proposed KB representation using GA

Encoding scheme

A two-folded encoding scheme has been presented here to represent the DB information:

$$\mathrm{CR_{GA}} = \mathrm{CR_M} + \mathrm{CR_T}$$

where $\mathrm{CR_M}$ encodes the membership function and $\mathrm{CR_T}$ encodes the tuning information for the membership function.

Each MF would be represented by a five-tuple representation scheme (Figure 7). The ith MF of the jth input will be represented by $\mathrm{MF}_i(x_j)$ and mathematically would be expressed as shown in Figure 7.

The following rule is encoded as shown in Figure 8: IF x_1 is $\mathrm{MF}_{i1}(x_1)$ ... and x_n is $\mathrm{MF}_{in}(x_n)$, THEN y is $\mathrm{MF}_{in+1}(y)$. It is represented by $\mathrm{CR_R}$.

Table 2 Type-2 fuzzy system

Year	Authors	Description	Reference
2006	D. Wu and W. W. Tan	Less computational expensive type-2 FLC is developed for real-time applications	[54]
2006	D. Wu and W. W. Tan	GAs are used to evolve type-2 FLC	[55]
2007	R. Sepulveda et al.	Feedback control systems for a non-linear plant using type-1 and type-2 fuzzy logic controllers	[56]
2009	R. Martinez et al.	Type-2 fuzzy systems and GAs are used to implement track controller for unicycle mobile robot	[57]
2009	M. H. F. Zarandi et al.	An interval type-2 fuzzy system has been developed for stock price analysis	[58]
2011	O. Castillo et al.	An interval type-2 fuzzy logic controller has been developed using evolutionary algorithms	[59]
2012	O. Castillo et al.	Ant colony optimization (ACO), particle swarm optimization (PSO), and GAs are used to optimize the MF parameters of a fuzzy logic controller	[60]
2012	D. Hidalgo et al.	A footprint of uncertainty (FoU)-based type-2 fuzzy system optimization has been developed	[61]
2012	O. Castillo and P. Mellin	A review on the optimization methods of type-2 fuzzy systems using bio-inspired computing	[62]
2012	R. Hosseini et al.	Automatic tuning and learning approach for type-2 fuzzy systems has been proposed applied to lung CAD classification system	[63]

The real coded chromosomes are used to encode the DB tuning information (CR_T) (Figure 9). For the ith input variable, the chromosome will be represented as shown in Figure 9 if there are n MFs for one variable.

Figure 10 gives the description of the tuning operation on MFs using α and β parameters.

Fitness function

The chromosomes are evaluated with the fitness function that considers the minimization of mean squared error (MSE):

$$\mathrm{MSE} = \frac{1}{2.M}\sum_{i=1}^{M}\left(\left(F\left(a^i\right)-b^i\right)^2\right.$$

where the size of the data set is M. $F(a^i)$ is the output obtained from FRBS for the ith example. The desired output is b^i.

GA operators

To perform GA operations, the following GA operators are used:

- Selection: Tournament selection has been used for the selection operation.
- Crossover: Crossover is the operator that generates new offspring by integrating multiple parents. A simple two-point crossover has been applied to all the chromosomes.
- Mutation: This operator is used to maintain the diversity in the solutions from one generation to another generation. This operator changes the values of one or more bits in the chromosomes. In this proposed approach, a uniform mutation operator has been used in which the bits of chromosomes are altered within uniform random values at user-specified ranges.

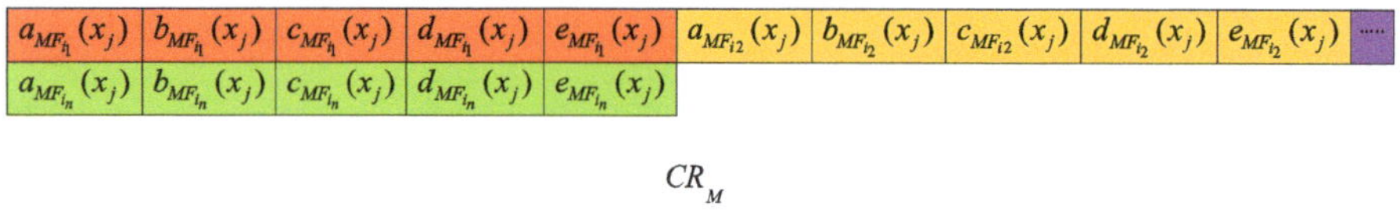

Figure 7 Chromosome encoding for the MF representation.

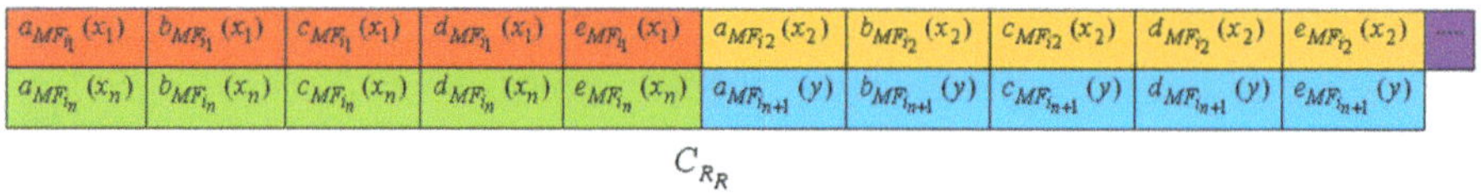

Figure 8 Chromosome encoding in the RB.

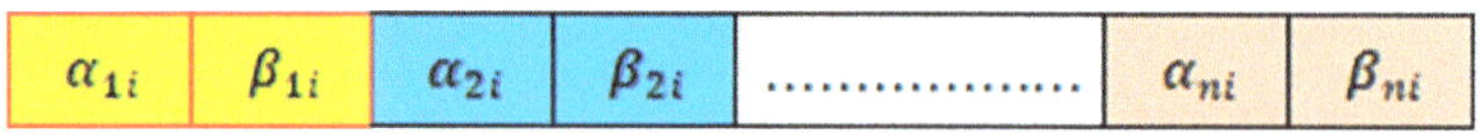

Figure 9 Chromosome representation for DB tuning information.

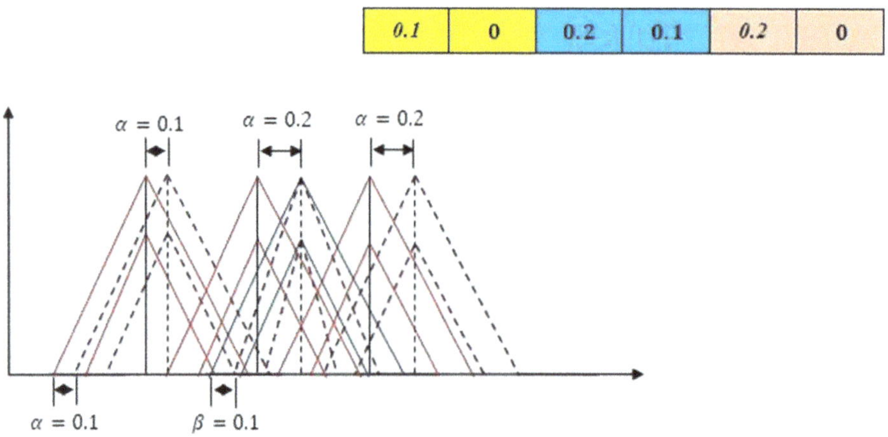

Figure 10 α-β (LDEC) tuning shown with MFs.

Table 3 Description of data set

Serial number	Characteristics	Value
1	Type	Classification
2	Number of attributes	3
3	Number of instances	306
4	Attribute characteristics	Integer

Experiments and results

The RB generation methods used in the experiments are the decision tree (DT) method, Wang-Mendel method [64], and fast prototyping algorithms. The experiments are supported by the open-access free software tool 'Guaje' [29,65] for type-1 fuzzy system implementation.

The proposed approach has been tested on Haberman's Survival Data Set. This data set is available at the UCI Machine Learning Repository [66]. The data set is prepared on behalf of the test cases of survivals of patients who have undergone breast cancer surgery. The major characteristics of the data set are tabulated in Table 3.

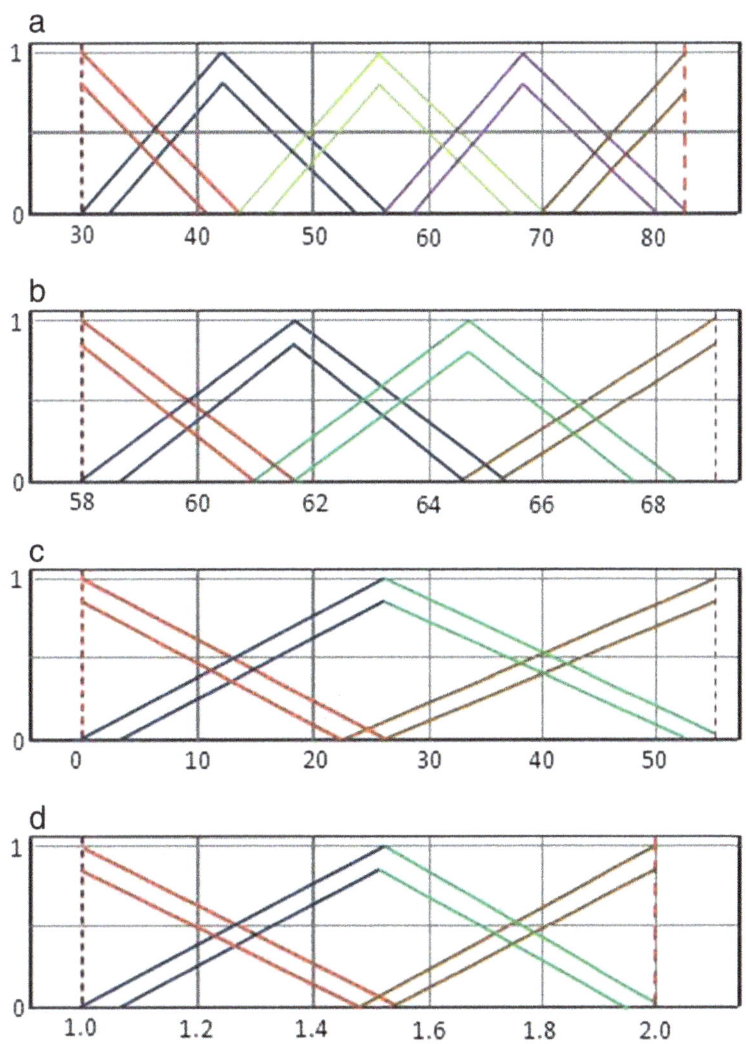

Figure 11 Input (a, b, c) and output (d) variables. (a) Age. **(b)** Year of operation. **(c)** Number of positive auxiliary nodes detected. **(d)** Survival.

Table 4 Accuracy and interpretability measures

Parameter	E1	E2	E3	E4
Accuracy				
MSE	0.121	0.112	0.117	0.092
RMSE	0.491	0.474	0.483	0.428
Interpretability				
NI	0.016	0.016	0.009	0.016
ARL	2.839	2.773	2.69	2.652
NOR	31	22	42	23
TRL	88	61	113	61
AIFR	3.902	4.964	6.199	4.075

MSE, mean square error; RMSE, root-mean-square error; NI, Nauck's index; ARL, average rule length; NOR, number of rules; TRL, total rule length; AIFR, average inferential fired rule.

The IT2MF for the data set input and output are given in Figure 11a,b,c,d.

Type-1 fuzzy system implementation

The values of accuracy and interpretability measures calculated in the following experiments are given in Table 4 and Figure 12:

- Experiment 1 (E1)
  Fuzzy partition method: hierarchical fuzzy partition (HFP) and rule generation method: Wang-Mendel method
- Experiment 2 (E2)
  Fuzzy partition method: strong fuzzy partition (SFP) and rule generation method: Wang-Mendel method
- Experiment 3 (E3)
  Fuzzy partition method: HFP and rule generation method: fuzzy decision trees
- Experiment 4 (E4)
  Fuzzy partition method: SFP and rule generation method: Wang-Mendel method

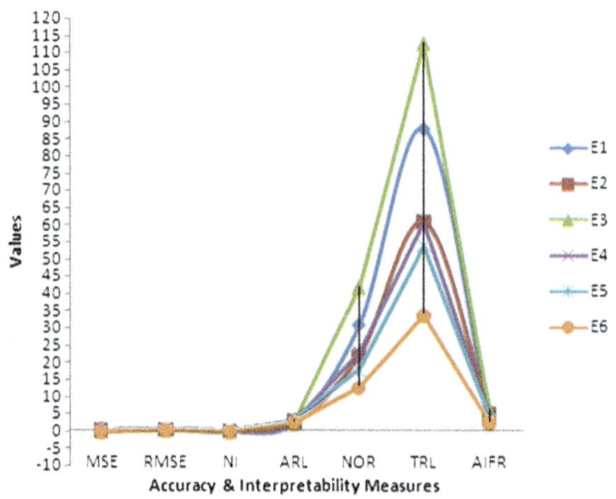

Figure 12 Interpretability and accuracy parameters.

Table 5 α and β parameters

Serial number	Variable name	α value	β value
1	Age	2.785	1.12
2	Year of operation	0.184	0.918
3	Number of auxiliary nodes	13.75	3.42

Table 6 Results of experiment 5

Parameter	Value
Accuracy	
MSE	0.094
RMSE	0.432
Interpretability	
NI	0.019
ARL	18
NOR	54
TRL	2.98
AIFR	3.876

Table 7 Results of experiment 6

Parameter	Value
Accuracy	
MSE	0.081
RMSE	0.398
Interpretability	
NI	0.029
ARL	2.615
NOR	13
TRL	34
AIFR	2.02

Table 8 Comparative results

Method	Number of rules	Total number of instances	Number of wrong classification	MSE
WM	18	100	22	0.11
FDT	14	100	20	0.10
WM + LDEC	24	100	16	0.08
FDT + LDEC	22	100	14	0.07

WM, Wang-Mendel; FDT, fuzzy decision trees.

Type 2 fuzzy system implementation

The values of tuning parameters α and β calculated in the experiment are given in Table 5.

- Experiment 5 (E5)
  In this experiment (Table 6), the parameters of the genetic algorithm are as follows:
  - Number of generations = 2,000
  - Size of population = 70
  - Tournament size = 2
  - Size of population = 70
  - Mutation probability = 0.1
  - Crossover probability =0.5

  Initial rules are generated by using the Wang-Mendel method.
- Experiment 6 (E6)
  In this experiment (Table 7), the initial rules are generated by a fuzzy decision tree with the following parameter settings:
  - Minimum cardinality of leaf = 1
  - Coverage threshold = 0.9
  - Minimum deviance gain = 0.001
  - Minimum significant level = 0.2
  - Pruning condition = yes

The genetic algorithm parameters are the same as those in experiment 5.

The result comparisons of the proposed approach are outlined in Table 8.

Conclusions

Type-2 fuzzy systems are strongly capable of modeling uncertainties in FKBS than type1 fuzzy systems using three-dimensional membership function representation. General type-2 fuzzy systems are deteriorating the interpretability of the systems, so IT2FS have been preferred to implement the proposed model with good interpretability.

The tuning and learning operations in the development of fuzzy systems playa vital role in improving their performance. This is considered as an optimization task and dealt properly with the application of evolutionary approaches, like GAs. The proposed tuning approach LDEC adjusts the parameters of interval type-2 fuzzy membership functions. This approach is based on the lateral displacement, expansion, and compression operations on the MFs. The proposed tuning approach is interpretable and the experimental results are found satisfactory.

Abbreviations

DB: data base; FKBS: fuzzy knowledge-based system; GAs: genetic algorithms; GFS: genetic fuzzy systems; KB: knowledge base; MFs: membership functions; RB: rule base; SFs: scaling functions.

Author details

[1]Department of Information Technology, Babu Banarasi Das Northern India Institute of Technology, Lucknow 226028, India. [2]Department of Computer Science & Engineering, Institute of Engineering & Technology, Lucknow 226021, India.

References

1. Palm, R, Drainkov, D, Hellendorn, H: Model Based Fuzzy Control. Springer, Berlin (1997)
2. Kuncheva, LI: Fuzzy Classifier Design. Studies in Fuzziness and Soft Computing. Springer, Berlin (2000)

3. Pedrycz, W: Fuzzy Modelling: Paradigms and Practices. Kluwer, Boston (1996)
4. Ross, TJ: Fuzzy Logic with Engineering Applications. Wiley, Chichester (2009)
5. Cordon, O, Herrera, F, Hoffmann, F, Magdalena, L: Genetic Fuzzy Systems: Evolutionary Tuning and Learning of Fuzzy Knowledge Bases. World Scientific, Singapore (2001)
6. Herrera, F: Genetic fuzzy systems: taxonomy, current research trends and prospects. Evol. Intel. **1**, 27–46 (2008)
7. Herrera, F: Genetic fuzzy systems: status, critical considerations and future directions. Int. J. Comput. Intell. Res. **1**(1), 59–67 (2005)
8. Cordon, O, Gomide, F, Herrera, F, Hoffmann, F, Magdalena, L: Ten years of genetic fuzzy systems: current framework and new trends. Fuzzy Set. Syst. **141**, 5–31 (2005)
9. Roy, SS: Design of genetic fuzzy expert system for predicting surface finish in ultra-precision diamond tuning of metal matrix composite. J. Mater. Process. Technol. **173**, 337–344 (2006)
10. Pulkkinen, P, Hytonen, J, Koivisto, H: Developing a bioaerosol detector using hybrid genetic fuzzy systems. Eng. Appl. Artif. Intel. **21**, 1330–1346 (2008)
11. Tseng, CH, Kwong, S, Wang, H: Genetic fuzzy rule mining approach and evaluation of feature selection techniques for anomaly intrusion detection. Pattern Recogn. **40**, 2373–2391 (2007)
12. Achiche, S, Balazinski, M, Baron, L, Jemielniak, K: Tool wear monitoring using genetically-generated fuzzy knowledge bases. Eng. Appl. Artif. Intel. **15**, 303–314 (2002)
13. Kim, HS, Roschke, PN: Design of fuzzy logic controller for smart base isolation system using genetic algorithms. Eng. Struct. **28**, 84–96 (2006)
14. Cassilas, J, Cordon, O, Herrera, F, Magdalena, L: Interpretability Issues in Fuzzy Modeling. Studies in Fuzziness and Soft Computing. Springer, Berlin (2003)
15. Cassilas, J, Cordon, O, Herrera, F, Magdalena, L: Accuracy Improvements in Linguistic Fuzzy Modeling. Studies in Fuzziness and Soft Computing. Springer, Berlin (2003)
16. Alcala, R, A-Fdez, J, Cassilas, J, Cordon, O, Herrera, F: Hybrid learning models to get the interpretability-accuracy trade-off in fuzzy modeling. Soft. Comput. **10**, 717–734 (2006)
17. Shukla, PK, Tripathi, SP: A survey on interpretability-accuracy (I-A) trade-off in evolutionary fuzzy systems. In: Proceedings of 5th International Conference on Genetic and Evolutionary Computing (ICGEC 2011), Kitakyushu, 29 Aug–1 Sept 2011
18. Shukla, PK, Tripathi, SP: A review on the interpretability-accuracy trade-off in evolutionary multi-objective fuzzy systems (EMOFS). Information **3**(3), 256–277 (2012)
19. Shukla, PK, Tripathi, SP: Interpretability issues in evolutionary multi-objective fuzzy knowledge base systems. In: Bansal, JC (ed.) Proceedings of 7th International Conference on Bio-inspired Computing: Theories and Applications (BIC-TA 2012) Advances in Intelligent Systems and Computing, vol. 201, pp. 473–484. Springer, New Delhi (2012)
20. Liang, Q, Mendel, JM: Interval type-2 fuzzy logic systems: theory and design. IEEE Trans. Fuzzy Syst. **8**(5), 535–550 (2000)
21. Alonso, JM, Magdalena, L: Special issue on interpretable fuzzy systems. Inform. Sci. **181**, 4331–4339 (2011)
22. Alonso, JM, Magdalena, L, Gonzalez-Rodriguez, G: Looking for a good fuzzy system interpretability index: an experimental approach. Int. J. Approx. Reason. **51**, 115–134 (2009)
23. Jin, Y: Fuzzy modeling of high dimensional systems: complexity reduction and interpretability improvement. IEEE Trans. Fuzzy Syst. **8**(2), 212–221 (2000)
24. Cassilas, J, Cordon, O, Herrera, F, Magdalena, L: Interpretability improvements to find the balance interpretability-accuracy in fuzzy modeling: an overview. In: Cassilas, J, Cordon, O, Herrera, F, Magdalena, L (eds.) Interpretability Issues in Fuzzy Modeling, Studies in Fuzziness and Soft Computing, pp. 3–22. Springer, Heidelberg (2003)
25. Mamdani, EH: Applications of fuzzy algorithms for controlling a simple dynamic plant. Proceedings of Institution of Electrical Engineers **121**(12), 1585–1588 (1974)
26. Takagi, T, Sugeno, M: Fuzzy identification of systems and its application to modeling and control. IEEE Trans. Syst. Man Cybern. **15**, 116–132 (1985)
27. Guillaume, S: Designing fuzzy inference system from data: an interpretability oriented review. IEEE Trans. Fuzzy Syst. **9**(3), 426–443 (2001)
28. Mikut, R, Jakel, J, Groll, L: Interpretability issues in data based learning of fuzzy systems. Fuzzy Set. Syst. **150**, 179–197 (2005)
29. Alonso, JM, Magdalena, L: HILK++: an interpretability guided fuzzy modeling methodology for learning readable and comprehensible fuzzy rule based classifiers. Soft. Comput. **15**(10), 1959–1980 (2011)
30. Alonso, JM, Magdalena, L, Guillaume, S: HILK: a new methodology for designing highly interpretable linguistic knowledge bases using fuzzy logic formalism. Int. J. Intell. Syst. **23**(7), 761–794 (2008)
31. Zhou, SM, Gan, JQ: Low level interpretability and high level interpretability: a unified view of data-driven interpretable fuzzy system modeling. Fuzzy Set. Syst. **159**, 3091–3131 (2008)
32. Mencar, C, Fanelli, AM: Interpretability constraints for fuzzy information granulation. Inform. Sci. **178**, 4585–4618 (2008)
33. Gacto, MJ, Alcala, R, Herrera, F: Interpretability of linguistic fuzzy rule based systems: an overview of interpretability measures. Inform. Sci. **181**, 4340–4360 (2011)
34. Fazzolari, M, Giglio, B, Alcala, R, Marcelloni, F, Herrera, F: A study on the application of instance selection techniques in genetic fuzzy rule based classification systems: accuracy-complexity trade-off. Knowledge Based Syst **54**, 32–41 (2013)
35. Nauck, DD: Measuring interpretability in rule based classification systems. In: Proceedings of FUZZ-IEEE. Missouri. 25–28 May 2003
36. Alonso, JM, Guillaume, S, Magdalena, L: A hierarchical fuzzy system for assessing interpretability of linguistic knowledge bases in classification problems. In: Proceedings of IPMU 2006, Information Processing and Management of Uncertainty in Knowledge Based Systems, Paris, 2–7 July 2006, pp. 348–355
37. Smith, SF: A learning system based on genetic adaptive algorithms. Dissertation. Department of Computer Science, University of Pittsburgh (1980)
38. Booker, LB: Intelligent behavior as an adaptation to the task environment. Dissertation. Department of Computer and Communication Sciences, University of Michigan (1982)

39. Venturini, G: SIA: A supervised inductive algorithm with genetic search for learning attribute based concepts. In: Proceedings of European Conference on Machine Learning, Vienna. 5–7 Apr 1993
40. Shi, H, Ward, R, Kharma, N: Expanding the definitions of linguistic hedges. In: Proceedings of Joint 9th IFSA World Congress & 20th NAFIPS, Vancouver, 25–28 July 2001
41. Zadeh, LA: A fuzzy set theoretic interpretation of linguistic hedges. J Cybernetics **2**(3), 4–34 (1972)
42. Cox, E: The Fuzzy Systems Handbook. AP Professional, New York (1998)
43. Mizumoto, M, Tanaka, K: Some properties of fuzzy sets of type 2. Inf. Control. **31**, 312–340 (1976)
44. Mendel, JM, John, RIB: Type-2 fuzzy sets made simple. IEEE Trans. Fuzzy Syst. **10**(2), 117–127 (2002)
45. Wu, H, Mendel, JM: Uncertainty bounds and their use in the design of interval type-2 fuzzy logic systems. IEEE Trans. Fuzzy Syst. **10**(5), 622–639 (2002)
46. Wu, D: On the fundamental differences between interval type-2 and type-1 fuzzy logic controllers. IEEE Trans. Fuzzy Syst. **20**(5), 832–848 (2012)
47. Wu, D: Approaches for reducing the computational cost of interval type-2 fuzzy logic systems: overview and comparison. IEEE Trans. Fuzzy Syst. **21**(1), 80–99 (2013)
48. Chen, SM, Chang, YC, Pan, JS: Fuzzy rules interpolation for sparse fuzzy rule based systems based on interval type-2 Gaussian fuzzy sets and genetic algorithms. IEEE Trans. Fuzzy Syst. **21**(3), 412–425 (2013)
49. Karnik, NN, Mendel, JM: Type-2 fuzzy logic systems. IEEE Trans. Fuzzy Syst. **7**(6), 643–658 (1999)
50. Goldberg, DE: Genetic Algorithms in Search, Optimization and Machine Learning. Addison-Wesley, Reading (1989)
51. Michalewicz, Z: Genetic Algorithms + Data Structures = Evolution Programs. Springer Verlag, Berlin (1996)
52. Sanz, JA, Fernandez, A, Bustince, H: IVTURS: A linguistic fuzzy rule based classification system based on a new interval valued fuzzy reasoning method with tuning and rule selection. IEEE Trans. Fuzzy Syst. **21**(3), 399–411 (2013)
53. Sanz, J, Fernandez, A, Bustince, H, Herrera, F: A genetic tuning to improve the performance of fuzzy rule based classification systems with interval valued fuzzy sets: degree of ignorance and lateral position. Int. J. Approx. Reason. **52**(6), 751–766 (2011)
54. Wu, D, Tan, WW: A simplified type-2 fuzzy logic controller for real time control. ISA Trans. **45**(4), 503–516 (2006)
55. Wu, D, Tan, WW: Genetic learning and performance evaluation of interval type-2 fuzzy logic controllers. Eng. Appl. Artif. Intel. **19**(8), 829–841 (2006)
56. Sepulveda, R, Castillo, O, Melin, P, R–Diaz, A, Montiel, O: Exponential study of intelligent controllers under uncertainty using type-1 and type-2 fuzzy logic. Inform. Sci. **177**(10), 2023–2048 (2007)
57. Martinez, R, Castillo, O, Aguilar, LT: Optimization of interval type-2 fuzzy logic controllers for a perturbed autonomous wheeled mobile robot using genetic algorithms. Inform. Sci. **179**(13), 2158–2174 (2009)
58. Zarandi, MHF, Rezaee, B, Turksen, IB, Neshat, E: A type-2 fuzzy rule-based expert system model for stock price analysis. Expert Syst. Appl. **36**(1), 139–154 (2009)
59. Castillo, O, Melin, P, Alanis, A, Montiel, O, Sepulveda, R: Optimization of interval type-2 fuzzy logic controllers using evolutionary algorithms. Soft. Comput. **15**(6), 1145–1160 (2011)
60. Castillo, O, M-Marroquin, R, Melin, P, Valdez, F, Soria, J: Comparative study of bio-inspired algorithms applied to optimization of type-1 and type-2 fuzzy controllers for an autonomous mobile robot. Inform. Sci. **192**(1), 19–38 (2012)
61. Hidalgo, D, Melin, P, Castillo, O: An optimization method for designing type-2 fuzzy inference systems based on the footprint of uncertainty using genetic algorithms. Expert Syst. Appl. **39**(4), 4590–4598 (2012)
62. Castillo, O, Melin, P: Optimization of type-2 fuzzy systems based on bio-inspired methods: a concise review. Inform. Sci. **205**(1), 1–19 (2012)
63. Hosseini, R, Qanadli, SD, Barman, S, Mazinani, M, Ellis, T, Dehmeshki, J: An automatic approach for learning and tuning Gaussian interval type-2 fuzzy membership functions applied to lung CAD classification system. IEEE Trans. Fuzzy Syst. **20**(2), 224–234 (2012)
64. Wang, LX, Mendel, JM: Generating fuzzy rule by learning from examples. IEEE Trans. Syst. Man Cybern. **22**(6), 1414–1427 (1992)
65. Alonso, JM, Magdalena, L: Generating understandable and accurate fuzzy rule based systems in a Java environment. In: Fanelli, AM, Pedrycz, W, Petrosino, A (eds.) Fuzzy Logic and Applications, 9th International Workshop, WILF 2011, Trani, Italy, August 29–31, 2011. Lecture Notes in Artificial Intelligence, vol. 6857, pp. 212–219. Springer, Berlin (2011)
66. Bache, K, Lichman, M: UCI Machine Learning Repository. School of Information and Computer Science, University of California, Irvine, CA. http://archive.ics.uci.edu/ml (2013). Accessed 15 June 2013

Uncertainty analysis of the coefficients of friction during the tightening process of bolted joints

Arthur Seibel*, Andreas Japing and Josef Schlattmann

*Correspondence:
arthur.seibel@tuhh.de
Workgroup on System Technologies and Engineering Design Methodology, Hamburg University of Technology, 21073 Hamburg, Germany

Abstract

In this paper, we introduce closed-form symbolic expressions for the possibility distributions of the coefficients of friction during the tightening process of bolted joints. The parameters in the distribution functions are then identified by a standardized fastener testing system according to ISO 16047. An uncertainty analysis finally shows that the total amount of uncertainty in the coefficient of bearing friction is almost 40% larger than in the coefficient of thread friction. Furthermore, the real value of the coefficent of bearing friction is likely to be about 5% higher than the expected value.

Keywords: Bolted joints tribology; Coefficient of bearing friction; Coefficient of thread friction; Uncertainty analysis; Possibility theory

Introduction

Bolted joints are among the most used joints in mechanical engineering. But failures still occur during the tightening process and in operation. In order to improve the quality and reliability of bolted joints, a main goal is to achieve a proper clamping force. With a certain effort [1], it is possible to measure the clamping force directly. But most of the bolts and screws still are assembled by an indirect measurement of the clamping force [2].

The required torque T for tightening a bolted joint is determined by [3]

$$T = T_\mathrm{b} + T_\mathrm{t}, \tag{1}$$

where T_b denotes the bearing friction torque component and T_t the thread friction torque component. These components are in a direct relationship with the clamping force and the coefficients of friction of the particular material pairing. Hence, the information about the coefficients of friction has a great influence on the quality and reliability of bolted joint connections.

Suppliers of bolts and screws often give only parameter windows for the desired coefficients of friction. These intervals are rather general and may not be suitable for individual applications. Furthermore, no information about the distribution functions of the coefficients of friction is provided.

Basically, two main approaches for modeling distribution functions are available: probability [4] and possibility theory [5]. In probability theory, the integral over the distribution function $p(x)$ is equal to one:

$$\int_{-\infty}^{+\infty} p(x)\,\mathrm{d}x = 1, \tag{2}$$

whereas in possibility theory, the distribution functions $\pi(x)$ are always normalized:

$$\max\{\pi(x)\} = 1. \tag{3}$$

For example, the Gaussian distribution function can be generally expressed by

$$f(x) = C \exp\left[-\frac{1}{2}\left(\frac{x-\bar{x}}{\sigma}\right)^2\right],$$

where $\bar{x}$ denotes the mean value and σ the standard deviation of $f(x)$. In probability theory, the constant scaling factor is set to

$$C = \frac{1}{\sigma\sqrt{2\pi}}$$

to satisfy the property (2). However, in possibility theory, we set

$$C = 1$$

to meet the requirement (3).

The general disadvantage in working with probability distribution functions is that in order to construct a probabilistic model, a large amount of statistical data is needed, which, in reality, is not always the case. In contrast to that, when only little information about the uncertain parameters is available, possibility theory is most appropriate. For this reason, we use possibility theory in this paper to model parametric uncertainty.

The main contributions of this paper are the following. First, closed-form symbolic expressions for the possibility distributions of the coefficients of friction during the tightening process of bolted joints are introduced for the first time. The parameters in the distribution functions are then identified by a standardized fastener testing system according to ISO 16047. An uncertainty analysis finally compares the total amount of uncertainty in the coefficient of bearing friction with that in the coefficient of thread friction. Furthermore, the deviations of the real values of the coefficients of friction from the expected values are given.

Possibility theory

Historically, possibility theory emerged from the theory of fuzzy sets [6] because of 'the need for a systematic way of dealing with [...] forms of uncertainty which are not probabilistic in nature' [5]. In fact, the membership function $\mu_{\tilde{x}}(x)$ of a fuzzy number $\tilde{x}$ can be viewed as a possibility distribution $\pi(x)$ of the parameter x [7]. More specifically, the possibility distribution $\pi(x)$ is a function that maps a value x to the possibility of the singleton event $\{x\}$ [8]:

$$\pi(x) = \Pi(\{x\}).$$

However, a high possibility of some event does not automatically mean that this event has also a high probability. In fact, the probability $P(A)$ of an event A can be viewed as a lower bound for the corresponding possibility $\Pi(A)$ [9]:

$$\Pi(A) \geq P(A).$$

This is in accordance with intuition since 'before an event becomes probable, it must be possible' [8]. For a detailed overview of possibility theory, the reader is referred to [5].

Sensitivity analysis

For the analysis of models with uncertain parameters, it is important to know to which extent the uncertainties of the individual model parameters contribute to the overall uncertainty of the model answer. Such a sensitivity analysis can be defined by the total differential [10]

$$\mathrm{d}f = \sum_{i=1}^{n} (\mathrm{d}f)_i = \sum_{i=1}^{n} \frac{\partial f(\bar{\boldsymbol{x}})}{\partial x_i}\, \mathrm{d}x_i \tag{4}$$

of the model function f at the point $\bar{\boldsymbol{x}} = (\bar{x}_1, \ldots, \bar{x}_n)$, where $\mathrm{d}f$ represents the total change of f in a neighborhood of $\bar{\boldsymbol{x}}$ if all parameters are changed simultaneously. The change $\mathrm{d}x_i$ of a single parameter x_i contributes to the amount $(\mathrm{d}f)_i$ to the total change $\mathrm{d}f$.

If the change rates $\mathrm{d}x_i$ of the parameters x_i are assumed to be a constant fraction c of the corresponding modal values $\bar{x}_i$, the total differential from Equation (4) can be written as [10]

$$\mathrm{d}f = c \sum_{i=1}^{n} \frac{\partial f(\bar{\boldsymbol{x}})}{\partial x_i}\, \bar{x}_i. \tag{5}$$

Using Equation (5), we can define the following relative sensitivity measures [10]:

$$\rho_i = \frac{\left| \frac{\partial f(\bar{\boldsymbol{x}})}{\partial x_i} \right| \bar{x}_i}{\sum_{j=1}^{n} \left| \frac{\partial f(\bar{\boldsymbol{x}})}{\partial x_j} \right| \bar{x}_j}$$

satisfying the consistency condition

$$\sum_{i=1}^{n} \rho_i = 1.$$

Uncertainty propagation

In order to propagate the uncertainties through the computations, we introduce the following transformation:

Let

$$\pi(x) = \begin{cases} \pi^{\mathrm{L}}(x), & x \leq \bar{x}, \\ \pi^{\mathrm{R}}(x), & x > \bar{x}, \end{cases}$$

be the possibility distribution of the parameter x in the x domain, where $\pi^{\mathrm{L}}(x)$ denotes the left branch, $\pi^{\mathrm{R}}(x)$ the right branch, and $\bar{x}$ the modal value of $\pi(x)$. The transformation of $\pi(x)$ into the π domain leads to the (generalized) interval $x(\pi) = \left[x^{\mathrm{L}}(\pi), x^{\mathrm{R}}(\pi)\right]$ with

$$\begin{aligned} x^{\mathrm{L}}(\pi) &= \pi^{\mathrm{L}}(x)^{-1}, \quad & 0 < \pi \leq 1, \\ x^{\mathrm{R}}(\pi) &= \pi^{\mathrm{R}}(x)^{-1}, \quad & 0 < \pi \leq 1. \end{aligned}$$

For example, the transformation of the Gaussian possibility distribution leads to

$$\begin{aligned} x^{\mathrm{L}}(\pi) &= \bar{x} - \sigma\sqrt{-2\ln(\pi)}, \quad & 0 < \pi \leq 1, \\ x^{\mathrm{R}}(\pi) &= \bar{x} + \sigma\sqrt{-2\ln(\pi)}, \quad & 0 < \pi \leq 1. \end{aligned}$$

The Gaussian possibility distribution in the x and the π domain is illustrated in Figure 1.

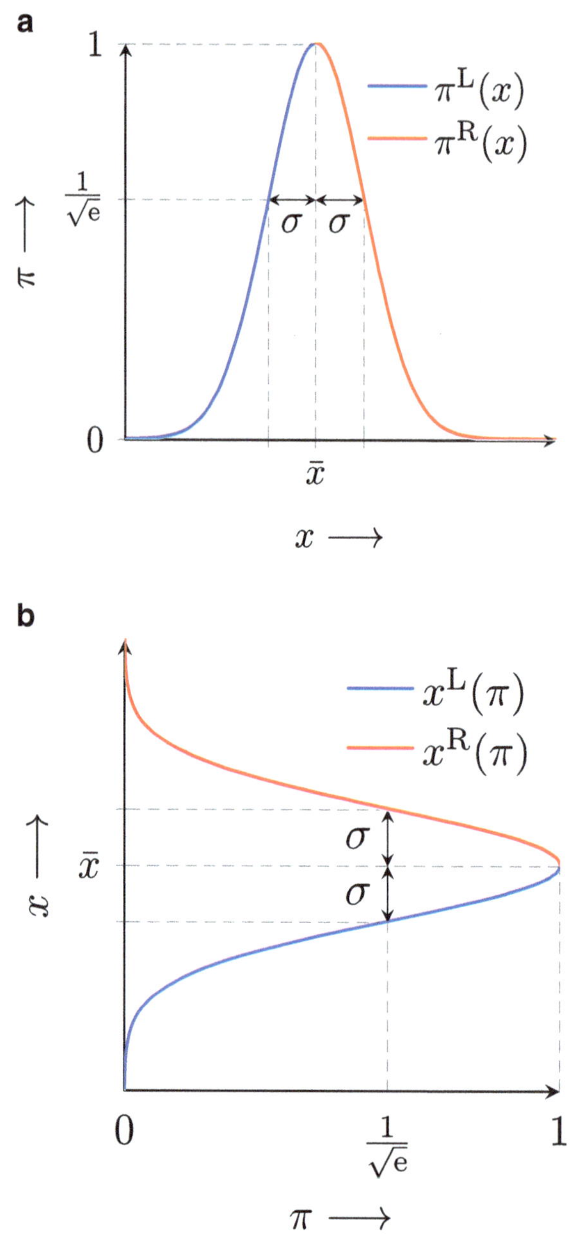

Figure 1 Gaussian possibility distribution. (a) x domain. **(b)** π domain.

Using the above transformation, we can formulate the following analytical approach [11]:

Let $\pi(x_1), \ldots, \pi(x_n)$ be the possibility distributions of the n independent parameters $x_1, \ldots, x_n$, and let $f : \mathbb{R}^n \to \mathbb{R}$ be a continuous function with $y = f(x_1, \ldots, x_n)$. Furthermore, let f be (strictly) monotonic increasing in x_i, $i = 1, \ldots, k$, and (strictly) monotonic decreasing in x_j, $j = k+1, \ldots, n$, in the domain of interest. Then, the possibility distribution $y(\pi) = \left[y^L(\pi), y^R(\pi)\right]$ of y in the π domain is determined by

$$\begin{aligned} y^L(\pi) &= f\left(x_1^L(\pi), \ldots, x_k^L(\pi), x_{k+1}^R(\pi), \ldots, x_n^R(\pi)\right), \qquad 0 < \pi \leq 1, \\ y^R(\pi) &= f\left(x_1^R(\pi), \ldots, x_k^R(\pi), x_{k+1}^L(\pi), \ldots, x_n^L(\pi)\right), \qquad 0 < \pi \leq 1. \end{aligned} \tag{6}$$

If Equations (6) are invertible with respect to π, then the possibility distribution of y in the y domain yields

$$\pi(y) = \begin{cases} y^{\mathrm{L}}(\pi)^{-1}, & y^{\mathrm{L}}(0) < y \le y^{\mathrm{L}}(1), \\ y^{\mathrm{R}}(\pi)^{-1}, & y^{\mathrm{R}}(1) < y < y^{\mathrm{R}}(0). \end{cases}$$

Distribution functions

In this section, we use the above analytical approach to derive closed-form symbolic expressions for the distribution functions of the coefficients of friction during the tightening process of bolted joints.

Coefficient of bearing friction

The coefficient of friction between the turning head or nut and the bearing surface μ_{b} can be computed from [3]

$$\mu_{\mathrm{b}} = \frac{T_{\mathrm{b}}}{Fr_{\mathrm{b}}},$$

where T_{b} denotes the bearing friction torque component, F the clamping force, and r_{b} the effective bearing radius.

During the tightening process, the clamping force F should be achieved exactly. Hence, it exhibits no uncertainty. The parameters T_{b} and r_{b}, on the other hand, are assumed to be normally distributed.

With the partial derivatives

$$\frac{\partial \mu_{\mathrm{b}}}{\partial T_{\mathrm{b}}} = +\frac{1}{Fr_{\mathrm{b}}},$$
$$\frac{\partial \mu_{\mathrm{b}}}{\partial r_{\mathrm{b}}} = -\frac{T_{\mathrm{b}}}{Fr_{\mathrm{b}}^2},$$

the sensitivity measures for μ_{b} are

$$\rho_{T_{\mathrm{b}}} = \rho_{r_{\mathrm{b}}} = 0.5.$$

Hence, the uncertainties of T_{b} and r_{b} contribute to the same amount to the overall uncertainty of μ_{b}.

We can see from the signs of the partial derivatives that μ_{b} is (strictly) monotonic increasing in T_{b} and (strictly) monotonic decreasing in r_{b} for positive values. Hence, according to Equations (6), the possibility distribution of μ_{b} in the π domain is $\mu_{\mathrm{b}}(\pi) = \left[\mu_{\mathrm{b}}^{\mathrm{L}}(\pi), \mu_{\mathrm{b}}^{\mathrm{R}}(\pi)\right]$ with

$$\mu_{\mathrm{b}}^{\mathrm{L}}(\pi) = \frac{T_{\mathrm{b}}^{\mathrm{L}}(\pi)}{Fr_{\mathrm{b}}^{\mathrm{R}}(\pi)} = \frac{\bar{T}_{\mathrm{b}} - \sigma_{T_{\mathrm{b}}}\sqrt{-2\ln(\pi)}}{F\left(\bar{r}_{\mathrm{b}} + \sigma_{r_{\mathrm{b}}}\sqrt{-2\ln(\pi)}\right)}, \qquad 0 < \pi \le 1,$$
$$\mu_{\mathrm{b}}^{\mathrm{R}}(\pi) = \frac{T_{\mathrm{b}}^{\mathrm{R}}(\pi)}{Fr_{\mathrm{b}}^{\mathrm{L}}(\pi)} = \frac{\bar{T}_{\mathrm{b}} + \sigma_{T_{\mathrm{b}}}\sqrt{-2\ln(\pi)}}{F\left(\bar{r}_{\mathrm{b}} - \sigma_{r_{\mathrm{b}}}\sqrt{-2\ln(\pi)}\right)}, \qquad 0 < \pi \le 1,$$

and in the μ_{b} domain,

$$\pi(\mu_{\mathrm{b}}) = \exp\left[-\frac{1}{2}\left(\frac{F\bar{r}_{\mathrm{b}}\mu_{\mathrm{b}} - \bar{T}_{\mathrm{b}}}{F\sigma_{r_{\mathrm{b}}}\mu_{\mathrm{b}} + \sigma_{T_{\mathrm{b}}}}\right)^2\right]. \tag{7}$$

Introducing the abbreviations

$$\bar{\mu}_{\mathrm{b}} = \frac{\bar{T}_{\mathrm{b}}}{F\bar{r}_{\mathrm{b}}},$$

$$\sigma_{\mu_{\mathrm{b}}} = \frac{\sigma_{T_{\mathrm{b}}}}{F\sigma_{r_{\mathrm{b}}}},$$

Equation (7) can be simplified to

$$\pi(\mu_{\mathrm{b}}) = \exp\left[-\frac{1}{2}\frac{\bar{r}_{\mathrm{b}}^2}{\sigma_{r_{\mathrm{b}}}^2}\left(\frac{\mu_{\mathrm{b}} - \bar{\mu}_{\mathrm{b}}}{\mu_{\mathrm{b}} + \sigma_{\mu_{\mathrm{b}}}}\right)^2\right].$$

Coefficient of thread friction

For metric ISO threads with a thread profile angle of 60°, the coefficient of friction between male and female threads μ_{t} can be approximated by [3]

$$\mu_{\mathrm{t}} = \frac{\sqrt{3}}{d_{\mathrm{t}}}\left(\frac{T_{\mathrm{t}}}{F} - \frac{P}{2\pi}\right),$$

where d_{t} denotes the effective thread diameter, T_{t} the thread friction torque component, F the clamping force, and P the thread pitch. (Note that the possibility degree π, which is used throughout the paper, should not be confused with the mathematical constant $\pi = 3.14\ldots$).

Again, the clamping force F is assumed to exhibit no uncertainty, and the other parameters are assumed to be normally distributed.

With the partial derivatives

$$\frac{\partial \mu_{\mathrm{t}}}{\partial d_{\mathrm{t}}} = -\frac{\sqrt{3}}{d_{\mathrm{t}}^2}\left(\frac{T_{\mathrm{t}}}{F} - \frac{P}{2\pi}\right),$$

$$\frac{\partial \mu_{\mathrm{t}}}{\partial T_{\mathrm{t}}} = +\frac{\sqrt{3}}{d_{\mathrm{t}}}\frac{1}{F},$$

$$\frac{\partial \mu_{\mathrm{t}}}{\partial P} = -\frac{\sqrt{3}}{d_{\mathrm{t}}}\frac{1}{2\pi},$$

the sensitivity measures for μ_{t} are

$$\rho_{d_{\mathrm{t}}} = \frac{2\pi\bar{T}_{\mathrm{t}} - F\bar{P}}{4\pi\bar{T}_{\mathrm{t}}},$$

$$\rho_{T_{\mathrm{t}}} = 0.5,$$

$$\rho_P = \frac{F\bar{P}}{4\pi\bar{T}_{\mathrm{t}}},$$

with

$$\rho_{d_{\mathrm{t}}} + \rho_P = 0.5.$$

However, since $2\pi\bar{T}_{\mathrm{t}} \gg F\bar{P}$, we can deduce:

$$\rho_{d_{\mathrm{t}}} \approx 0.5,$$

$$\rho_{T_{\mathrm{t}}} = 0.5,$$

$$\rho_P \approx 0.$$

Hence, the uncertainties of d_{t} and T_{t} contribute to the nearly same amount to the overall uncertainty of μ_{t}, whereas the uncertainty of P has a negligible influence.

We can see from the signs of the partial derivatives that μ_{t} is (strictly) monotonic increasing in T_{t} and (strictly) monotonic decreasing in d_{t} for positive values. Hence, according to Equations (6), the possibility distribution of μ_t in the π domain is $\mu_t(\pi) = \left[\mu_t^L(\pi), \mu_t^R(\pi)\right]$ with

$$\begin{aligned}
\mu_t^L(\pi) &= \frac{\sqrt{3}}{d_t^R(\pi)}\left(\frac{T_t^L(\pi)}{F} - \frac{P}{2\pi}\right)\\
&= \frac{\sqrt{3}}{2\pi F}\frac{(2\pi\bar{T}_t - FP) - 2\pi\sigma_{T_t}\sqrt{-2\ln(\pi)}}{\bar{d}_t + \sigma_{d_t}\sqrt{-2\ln(\pi)}}, \qquad 0 < \pi \leq 1,\\
\mu_t^R(\pi) &= \frac{\sqrt{3}}{d_t^L(\pi)}\left(\frac{T_t^R(\pi)}{F} - \frac{P}{2\pi}\right)\\
&= \frac{\sqrt{3}}{2\pi F}\frac{(2\pi\bar{T}_t - FP) + 2\pi\sigma_{T_t}\sqrt{-2\ln(\pi)}}{\bar{d}_t - \sigma_{d_t}\sqrt{-2\ln(\pi)}}, \qquad 0 < \pi \leq 1,
\end{aligned}$$

and in the μ_t domain,

$$\pi(\mu_t) = \exp\left[-\frac{1}{2}\left(\frac{a\mu_t^2 + b\mu_t + c}{d\mu_t^2 + e\mu_t + f}\right)\right]$$

with

$$\begin{aligned}
a &= 4\pi^2\bar{d}_t^2F^2,\\
b &= -4\sqrt{3}\pi\bar{d}_tF\left(2\pi\bar{T}_t - FP\right),\\
c &= 12\pi^2\bar{T}_t^2 - 12\pi\bar{T}_tFP + 3F^2P^2,\\
d &= 4\pi^2\sigma_{d_t}^2F^2,\\
e &= 8\sqrt{3}\pi^2\sigma_{d_t}\sigma_{T_t}F,\\
f &= 12\pi^2\sigma_{T_t}^2.
\end{aligned}$$

Parameter identification

In order to identify the parameters in the distribution functions of μ_b and μ_t, practical experiments had been carried out on a fastener testing system according to ISO 16047 [12] as illustrated in Figure 2. This system contains a multi-parameter sensor that enables a measurement of the fastener preload, the tightening torque, and one of the friction torque components at the same time.

Figure 2 Fastener testing system used in our experiments.

Materials and methods

The descriptions and materials of the test parts are summarized in Table 1. The dimensions of the flat steel bar are illustrated in Figure 3.

For carrying out the experiments, we followed the instructions from ISO 16047 [12]. A tightening torque was steadily applied until a clamping force of 16 kN was reached. Here, F, T, and T_b were measured, and T_t was computed according to Equation (1). The sample size was 32 bolts.

Experimental results

Since the effective bearing radius is usually hard to determine in practice [13], r_b is assumed to be normally distributed with the mean value

$$\bar{r}_b = \frac{d_o + d_i}{4} = 6.625\,\text{mm}$$

and the standard deviation

$$\sigma_{r_b} = \frac{d_o - d_i}{12} = 0.625\,\text{mm},$$

where d_o denotes the outer diameter and d_i the inner diameter of the bearing surface.

Furthermore, the effective thread diameter is also usually hard to determine in practice [13]. For this reason, d_t is also assumed to be normally distributed with the mean value

$$\bar{d}_t = d_2 = 7.188\,\text{mm}$$

and the standard deviation

$$\sigma_{d_t} = \frac{d - D_1}{6} = 0.226\,\text{mm},$$

where d_2 denotes the thread pitch diameter, d the nominal thread diameter, and D_1 the minor nut thread diameter. The standard deviations are chosen such that the intervals $[d_i, d_o]$ and $[D_1, d]$ correspond to $6 \cdot 2\sigma_{r_b}$ and $6\sigma_{d_t}$, respectively. (The factor two at σ_{r_b} results from the fact that the standard deviation of a diameter is twice as large as the standard deviation of the corresponding radius). Beyond the interval boundaries, the possibility values are smaller than 1% [14] and can be thus neglected. In fact, they are physically impossible.

Using the above assumptions, the distribution functions of μ_b and μ_t for $F \in [2, 16]$ kN are illustrated in Figure 4. We can see that after an initial shakedown, a steady-state distribution is always reached.

According to ISO 16047 [12], the determination of the coefficients of friction shall be carried out at 75% of the maximum clamping force. In our experiments, this is

Table 1 Descriptions and materials of the test parts

Test part	Description/material	Remark
Bolt	DIN 6921 M8 × 50 10.9	Black chromated
Nut	DIN 934 M8 8.8	Black chromated
Flat bar	S235JRC+C (EN 1.0122)	Surface polished

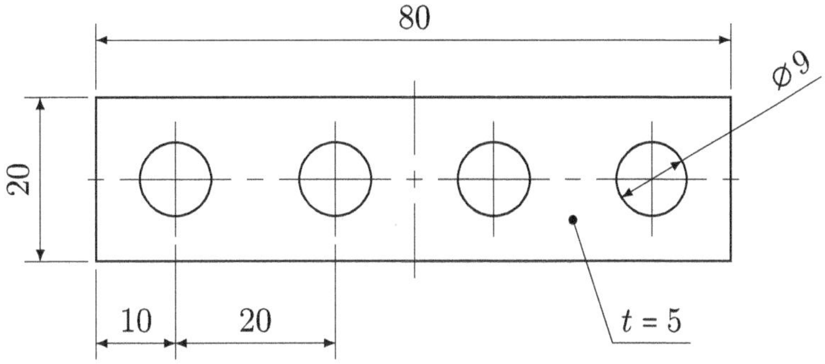

Figure 3 Dimensions of the flat steel bar. Here, t denotes the thickness.

the case at $F = 12$ kN. The distribution functions of μ_b and μ_t at this point are

$$\pi(\mu_b) = \exp\left[-56\left(\frac{\mu_b - 0.114}{\mu_b + 0.091}\right)^2\right],$$
$$\pi(\mu_t) = \exp\left[-\frac{1}{2}\left(\frac{2.94\cdot 10^5\mu_t^2 - 1.13\cdot 10^5\mu_t + 0.109\cdot 10^5}{7.19\cdot 10^{-3}\mu_t^2 + 152\mu_t + 19.9}\right)\right]. \quad (8)$$

The plots of Equations (8) are illustrated in Figure 5.

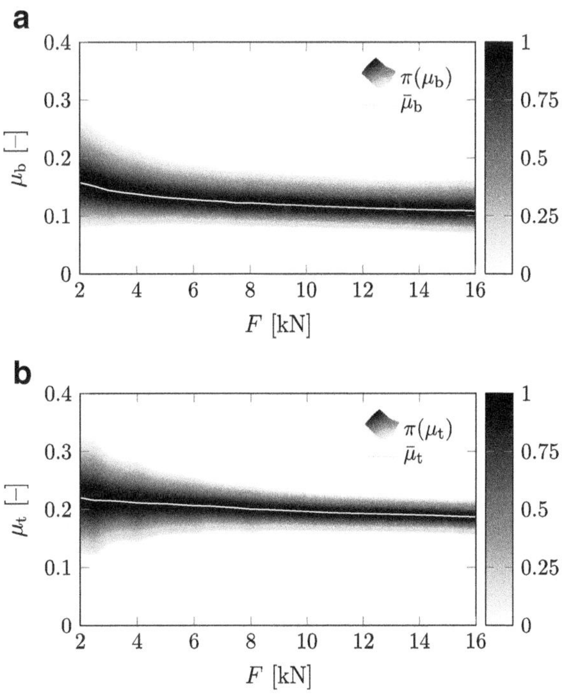

Figure 4 Distribution functions of μ_b and μ_t for $F \in [2, 16]$ kN. (a) Distribution function of μ_b. **(b)** Distribution function of μ_t.

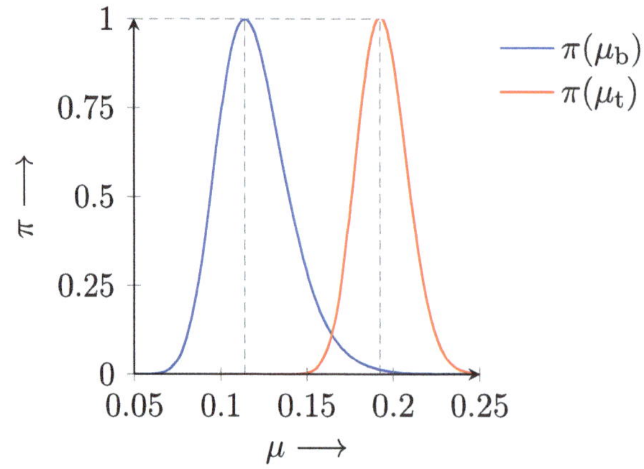

Figure 5 Distribution functions of μ_b and μ_t at $F = 12$ kN.

Uncertainty analysis

In order to quantify the total amount of uncertainty of a possibility distribution $\pi(x)$, we use the (absolute) cardinality [14]:

$$\text{card}\,(\pi(x)) = \int_{-\infty}^{+\infty} \pi(x)\,\mathrm{d}x.$$

Since the coefficients of friction are all elements of $[0, 1]$, the total amount of uncertainty can be defined as

$$\text{uncer}\,(\pi(x)) = \int_{0}^{1} \pi(x)\,\mathrm{d}x.$$

Numerical integration of Equations (8) yields

$$\text{uncer}\,(\pi(\mu_b)) = 0.050,$$
$$\text{uncer}\,(\pi(\mu_t)) = 0.036,$$

meaning that in our application, the total amount of uncertainty in μ_b is almost 40% larger than in μ_t.

In the next step, the (relative) eccentricities of the possibility distributions of the coefficients of friction are analyzed [14]:

$$\text{ecc}\,(\pi(x)) = \frac{\text{defuzz}\,(\pi(x)) - \bar{x}}{\bar{x}}.$$

Here,

$$\text{defuzz}\,(\pi(x)) = \frac{\int_0^1 x\,\pi(x)\,\mathrm{d}x}{\int_0^1 \pi(x)\,\mathrm{d}x}$$

denotes the defuzzified value of $\pi(x)$.

Further numerical integration yields

$$\text{defuzz}\,(\pi(\mu_b)) = 0.120,$$
$$\text{defuzz}\,(\pi(\mu_t)) = 0.194.$$

Hence, the desired eccentricities are

$$\mathrm{ecc}\,(\pi(\mu_b)) = \frac{0.120 - 0.114}{0.114} = 0.05,$$
$$\mathrm{ecc}\,(\pi(\mu_t)) = \frac{0.194 - 0.193}{0.193} = 0.01.$$

Both eccentricities are positive, that is, the real values of the coefficients of friction will be potentially higher than the expected values. More specifically, the real value of μ_b is likely to be about 5% and the real value of μ_t to be about 1% higher than the corresponding expected value.

Conclusions

We introduced closed-form symbolic expressions for the possibility distributions of the coefficients of friction during the tightening process of bolted joints. This relieves the engineer from the burden of propagating the uncertainties through the computations to obtain the uncertain output. The parameters in the distribution functions can be identified by a standardized fastener testing system according to ISO 16047 as has been demonstrated in this paper. An uncertainty analysis also revealed that the total amount of uncertainty in the coefficient of bearing friction is almost 40% larger than in the coefficient of thread friction. This finding suggests that in practice, it is more important to control the coefficient of bearing friction than the coefficient of thread friction. The uncertainty analysis also showed that the real value of the coefficient of bearing friction is likely to be about 5% and the real value of the coefficient of thread friction to be about 1% higher than the corresponding expected value.

Competing interests

The authors declare that they have no competing interests.

References

1. Baker, R: Ultrasonic tightening. Sealing Technol. **2011**(8), 10–12 (2011)
2. Kloos, K-H, Thomala, W: Schraubenverbindungen: Grundlagen, Berechnung, Eigenschaften, Handhabung. Springer, Berlin (2007)
3. Kellermann, R, Klein, H-C: Untersuchungen über den Einfluß der Reibung auf Vorspannung und Anzugsmoment von Schraubenverbindungen. Konstruktion. **7**(2), 54–68 (1955)
4. Marczyk, J: Principles of Simulation-Based Computer-Aided Engineering. FIM Publications, Barcelona (1999)
5. Dubois, D, Prade, H: Possibility Theory: An Approach to Computerized Processing of Uncertainty. Plenum, New York (1988)
6. Zadeh, LA: Fuzzy sets. Inform Contr. **8**, 338–353 (1965)
7. Zadeh, LA: Fuzzy sets as a basis for a theory of possibility. Fuzzy Set Syst. **1**, 3–28 (1978)
8. Degrauwe, D: Uncertainty propagation in structural analysis by fuzzy numbers. PhD Thesis, Katholieke Universiteit Leuven, Leuven, Belgium (2007)
9. Bothe, H-H: Fuzzy Logic: Einführung in Theorie und Anwendungen. Springer, Berlin (1993)
10. Hanss, M, Klimke, A: On the reliability of the influence measure in the transformation method of fuzzy arithmetic. Fuzzy Set Syst. **143**(3), 371–390 (2004)
11. Seibel, A, Schlattmann, J: An analytical approach to evaluating monotonic functions of fuzzy numbers. In: Proceedings of the 8th Conference of the European Society for Fuzzy Logic and Technology, Milano, Italy, 11–13 Sept 2013, pp. 289–293 (2013)
12. International Organization for Standardization: ISO 16047: Fasteners—Torque/Clamp Force Testing. International Organization for Standardization, Geneva (2005)
13. Zou, Q, Zhu, D, Sun, TS, Nassar, S, Barber, GC, El-Khiamy, H: Contact mechanics approach to determine effective radius in bolted joints. J Tribol. **127**(1), 30–36 (2005)
14. Hanss, M: Applied Fuzzy Arithmetic: An Introduction with Engineering Applications. Springer, Berlin (2005)

6

Uncertain systems are universal approximators

Zixiong Peng[1] and Xiaowei Chen[2*]

*Correspondence:
chenx@nankai.edu.cn
[2]Department of Risk Management and Insurance, Nankai University, Tianjin 300071, China
Full list of author information is available at the end of the article

Abstract

Uncertain inference is a process of deriving consequences from uncertain knowledge or evidences via the tool of conditional uncertain set. Based on uncertain inference, uncertain system is a function from its inputs to outputs. This paper proves that uncertain systems are universal approximators, which means that uncertain systems are capable of approximating any continuous function on a compact set to arbitrary accuracy. This result can be viewed as an existence theorem of an optimal uncertain system for a wide variety of problems.

Keywords: Uncertainty theory; Uncertain inference; Uncertain system; Universal approximation

Introduction

Fuzzy systems were developed from fuzzy set theory initialized by Zadeh [1] in 1965. In recent decades, fuzzy systems have been successfully applied to a wide variety of practical problems, such as fuzzy control, fuzzy identification, fuzzy expert system. Generally speaking, there are two types of widely used fuzzy systems: Mamdani fuzzy systems [2] and Takagi-Sugeno fuzzy systems [3]. The main difference between these fuzzy systems lies on inference rules, such as Zadeh's compositional rule of inference, Lukasiewicz's inference rule [4], Mamdani's inference rules [2,5] for Mamdani fuzzy systems. On the other hand, Wang [6] proved that fuzzy systems are universal approximators which is an important theoretical basis for the application of it.

Although fuzzy systems are successfully used in engineering problems, it isn't perfect in its theoretical framework. For example, membership functions are adopted to describe the fuzzy sets used in fuzzy inference rules, however, the overuse maximum operations in fuzzy set theory is impeached. Surveys have shown that the inputs and outputs of practical systems should not be fuzzy sets. This promotes Liu [7] to define uncertain set and introduced uncertain systems based Liu's inference rules. Then, Gao Gao and Ralescu [8] extended the Liu's inference rule to multi-antecedent. As an application of uncertain system, Gao [9] used the uncertain controller to successfully control the inverted pendulum. Besides, uncertain logic was introduced by Liu [10] via uncertain set. Universal approximation capability of uncertain systems is the basis of almost all the theoretical research and practical applications of uncertain systems. In this paper, we will prove that uncertain systems are universal approximators, and this result can be viewed as an existence theorem of an optimal uncertain system for a wide variety of problems.

The rest of the paper is organized as follows. Section 'Preliminaries' introduces some basic concepts and results on uncertain sets as preliminaries. The framework of Liu's inference rule and uncertain systems are introduced in Section 'Inference rule and uncertain systems'. It is shown that uncertain systems are universal approximators in Section 'Uncertain systems are universal approximators'. Section 'Conclusions' concludes this paper with a brief summary.

Preliminaries

Liu [11] founded an uncertainty theory that is based on normality, duality, subadditivity and product axioms. Gao [12] studied some properties of uncertain measure. Since then, considerable work has been done based on uncertainty theory. Liu [13] founded uncertain programming to model optimization problems in uncertain environments. Liu [14] proposed the concept of uncertain process and uncertain calculus. Li and Liu [15] proposed uncertain logic. And Liu [16] studied entailment based on uncertain proposition logic. Recently, uncertain systems are suggested by Liu [7] in 2010 via uncertain inference and uncertain set theory. Uncertainty theory has become a new tool to describe belief degree and has applications in practical use. For a detailed exposition, the interested reader may consult the book [17] to know more about uncertainty theory. In this section, we will introduce some definitions related on uncertain systems. Let Γ be a nonempty set, and let L be a σ-algebra over Γ. Each element $\Lambda \in \Gamma$ is called an event. In order to measure uncertain events, Liu [11] introduced an uncertain measure $\mathcal{M}$ as a set function satisfying normality, self-duality, and countable subadditivity axioms. Then the triplet $(\Gamma, \mathrm{L}, \mathcal{M})$ is called an uncertainty space.

Let Γ be a nonempty set, and $\mathcal{L}$ a σ-algebra over Γ. An uncertain measure $\mathcal{M}$ (Liu [11]) is a set function defined on $\mathcal{L}$ satisfying the following four axioms:

Axiom 1. (Normality Axiom) $\mathcal{M}\{\Gamma\} = 1$;

Axiom 2. (Duality Axiom) $\mathcal{M}\{\Lambda\} + \mathcal{M}\{\Lambda^c\} = 1$ for any event Λ;

Axiom 3. (Subadditivity Axiom) For every countable sequence of events $\{\Lambda_i\}$, we have

$$\mathcal{M}\left\{\bigcup_{i=1}^{\infty} \Lambda_i\right\} \leq \sum_{i=1}^{\infty} \mathcal{M}\{\Lambda_i\}.$$

Axiom 4. (Product Axiom [18]) Let $(\Gamma_k, \mathrm{L}_k, \mathcal{M}_k)$ be uncertainty spaces for $k = 1, 2, \ldots$. The product uncertain measure $\mathcal{M}$ is an uncertain measure satisfying

$$\mathcal{M}\left\{\prod_{i=1}^{\infty} \Lambda_i\right\} = \bigwedge_{i=1}^{\infty} \mathcal{M}\{\Lambda_i\}$$

where Λ_k are arbitrarily chosen events from L_k for $k = 1, 2, \ldots$, respectively.

Definition 1. (Liu [7]) An *uncertain set* is a measurable function ξ from an uncertainty space $(\Gamma, \mathrm{L}, \mathcal{M})$ to a collection of sets, i.e., both $\{B \subset \xi\}$ and $\{\xi \subset B\}$ are events for any Borel set B.

Let ξ and η be two nonempty uncertain sets. What is the appropriate event that ξ is included in η? We may suggest that $\{\xi \subset \eta\} = \{\gamma | \xi(\gamma) \subset \eta\}$ intuitively. The set $\{B \not\subset \xi\} = \{\gamma | B \not\subset \xi(\gamma)\}$ and $\{\xi \not\subset B\} = \{\gamma | \xi(\gamma) \not\subset B\}$ are also events.

Definition 2. (Liu [19]) An uncertain set ξ is said to have a membership function μ if for any Borel set B of real numbers, we have

$$\mathcal{M}\{B \subset \xi\} = \inf_{x \in B} \mu(x), \qquad \mathcal{M}\{\xi \subset B\} = 1 - \sup_{x \in B^c} \mu(x).$$

The above equations will be called measure inversion formulas.

Example 1. By a *rectangular uncertain set* we mean the uncertain set fully determined by the pair (a, b) of crisp numbers with $a < b$, whose membership function is

$$\mu(x) = 1, \quad a \leq x \leq b.$$

Example 2. By a *triangular uncertain set* we mean the uncertain set fully determined by the triplet (a, b, c) of crisp numbers with $a < b < c$, whose membership function is

$$\mu(x) = \begin{cases} \dfrac{x-a}{b-a}, & \text{if } a \leq x \leq b \\ \dfrac{x-c}{b-c}, & \text{if } b \leq x \leq c. \end{cases}$$

Example 3. By a *trapezoidal uncertain set* we mean the uncertain set fully determined by the quadruplet (a, b, c, d) of crisp numbers with $a < b < c < d$, whose membership function is

$$\mu(x) = \begin{cases} \dfrac{x-a}{b-a}, & \text{if } a \leq x \leq b \\ 1, & \text{if } b \leq x \leq c \\ \dfrac{x-d}{c-d}, & \text{if } c \leq x \leq d. \end{cases}$$

Liu [17] proved that a real-valued function μ is a membership function if and only if $0 \leq \mu(x) \leq 1$.

Definition 3. (Liu [19]) Let ξ be an uncertain set with membership function μ. Then the set-valued function

$$\mu^{-1}(\alpha) = \left\{x \in \Re \;\middle|\; \mu(x) \geq \alpha\right\}$$

is called the *inverse membership function* of ξ. Sometimes, the set $\mu^{-1}(\alpha)$ is called the *α-cut of μ*.

The rectangular uncertain set $\xi = (a, b)$ has an inverse membership function

$$\mu^{-1}(\alpha) \equiv [a, b].$$

The triangular uncertain set $\xi = (a, b, c)$ has an inverse membership function

$$\mu^{-1}(\alpha) = [(1-\alpha)a + \alpha b, \alpha b + (1-\alpha)c].$$

The trapezoidal uncertain set $\xi = (a,b,c,d)$ has an inverse membership function

$$\mu^{-1}(\alpha) = [\,(1-\alpha)a + \alpha b, \alpha c + (1-\alpha)d\,].$$

Liu [17] proposed the conditional membership function of an uncertain set ξ after some event A has occurred:

$$\mathcal{M}\{B \subset \xi | A\} = \begin{cases} \dfrac{\mathcal{M}\{(B \subset \xi) \bigcap (\xi \subset A)\}}{\mathcal{M}\{\xi \subset A\}}, & \text{if } \dfrac{\mathcal{M}\{(B \subset \xi) \bigcap (\xi \subset A)\}}{\mathcal{M}\{\xi \subset A\}} < 0.5 \\ 1 - \dfrac{\mathcal{M}\{(B \not\subset \xi) \cap (\xi \subset A)\}}{\mathcal{M}\{\xi \subset A\}}, & \text{if } \dfrac{\mathcal{M}\{(B \not\subset \xi) \cap (\xi \subset A)\}}{\mathcal{M}\{\xi \subset A\}} < 0.5 \\ 0.5, & \text{otherwise} \end{cases}$$

and

$$\mathcal{M}\{\xi \subset B | A\} = \begin{cases} \dfrac{\mathcal{M}\{(\xi \subset B) \bigcap (\xi \subset A)\}}{\mathcal{M}\{\xi \subset A\}}, & \text{if } \dfrac{\mathcal{M}\{(\xi \subset B) \bigcap (\xi \subset A)\}}{\mathcal{M}\{\xi \subset A\}} < 0.5 \\ 1 - \dfrac{\mathcal{M}\{(\xi \not\subset B) \bigcap (\xi \subset A)\}}{\mathcal{M}\{\xi \subset A\}}, & \text{if } \dfrac{\mathcal{M}\{(\xi \not\subset B) \bigcap (\xi \subset A)\}}{\mathcal{M}\{\xi \subset A\}} < 0.5 \\ 0.5, & \text{otherwise.} \end{cases}$$

Definition 4. (Liu [20]) The uncertain sets $\xi_1, \xi_2, \ldots, \xi_n$ are said to be independent if for any Borel sets $B_1, B_2, \ldots, B_n$, we have

$$\mathcal{M}\left\{\bigcap_{i=1}^{n} (\xi_i^* \subset B_i)\right\} = \bigwedge_{i=1}^{n} \mathcal{M}\left\{\xi_i^* \subset B_i\right\}$$

and

$$\mathcal{M}\left\{\bigcup_{i=1}^{n} (\xi_i^* \subset B_i)\right\} = \bigvee_{i=1}^{n} \mathcal{M}\left\{\xi_i^* \subset B_i\right\}$$

where ξ_i^* are arbitrarily chosen from $\{\xi_i, \xi_i^c\}$, $i = 1, 2, \ldots, n$, respectively.

Definition 5. (Liu [20]) Let ξ and η be independent uncertain sets with membership functions μ and ν, respectively. Then their union $\xi \cap \eta$ has a membership function

$$\lambda(x) = \mu(x) \vee \nu(x).$$

Definition 6. (Liu [20]) Let ξ and η be independent uncertain sets with membership functions μ and ν, respectively. Then their union $\xi \cup \eta$ has a membership function

$$\lambda(x) = \mu(x) \wedge \nu(x).$$

Definition 7. (Liu [20]) Let ξ be an uncertain sets with membership functions μ. Then their union ξ^c has a membership function

$$\lambda(x) = 1 - \nu(x).$$

Definition 8. (Liu [20]) Let $\xi_1, \xi_2, \ldots, \xi_n$ be independent uncertain sets with inverse membership functions $\mu_1(x), \mu_2(x), \ldots, \mu_n(x)$, respectively. If f is a measurable function, then the uncertain set $\xi = f(\xi_1, \xi_2, \ldots, \xi_n)$ has an inverse membership function,

$$\mu^{-1}(\alpha) = f(\mu_1^{-1}(\alpha), \mu_2^{-1}(\alpha), \ldots, \mu_n^{-1}(\alpha)).$$

Inference rule and uncertain systems

Uncertain inference was proposed by Liu [7] as a process of deriving consequences from uncertain knowledge or evidence via the tool of conditional uncertain set. Then Gao, Gao and Ralescu [8] extended Liu's inference rule to the one with multiple antecedents and with multiple if-then rules. Besides, Gao [9] applied uncertain inference to invert pendulum. Furthermore, uncertain set was extended to the field of logic (Liu [10]) where the uncertain quantifier, uncertain subject and uncertain predicate are described by uncertain sets. In this section, we will introduce the concept and some properties of uncertain inference.

Liu's Inference Rule [7]. Let $\mathbb{X}$ and $\mathbb{Y}$ be two concepts. Assume rules 'if $\mathbb{X}$ is an uncertain set ξ then $\mathbb{Y}$ is an uncertain set η'. From $\mathbb{X}$ is a constant a, we infer that $\mathbb{Y}$ is an uncertain set

$$\eta* = \eta|_{a\in\eta}$$

which is the conditional uncertain set of η given $a \in \eta$. The inference rule is represented by

Rule: if $\mathbb{X}$ is ξ then $\mathbb{Y}$ is η
From: $\mathbb{X}$ is a constant a

Infer: $\mathbb{Y}$ is $\eta^* = \eta|_{a\in\xi}$

It has been proved by Liu [7] that the membership function of η^* satisfying

$$\nu^*(y) = \begin{cases} \dfrac{\nu(y)}{\mu(a)}, & \text{if } \nu(y) < \mu(a)/2 \\ \dfrac{\nu(y) + \mu(a) - 1}{\mu(a)}, & \text{if } \nu(y) > 1 - \mu(a)/2 \\ 0.5, & \text{otherwise.} \end{cases}$$

Liu's Inference Rule (multiple antecedent [8]). Let $\mathbb{X}$, $\mathbb{Y}$, and $\mathbb{Z}$ be three concepts. Assume rules 'if $\mathbb{X}$ is an uncertain set ξ and $\mathbb{Y}$ is η then $\mathbb{Z}$ is an uncertain set τ.' From $\mathbb{X}_1$ is a constant a and $\mathbb{Y}$ is a constant b, respectively, we infer that $\mathbb{Z}$ is an uncertain set.

$$\eta* = \eta|_{(a\in\eta)\bigcap(b\in\eta)}$$

which is conditional uncertain set of τ of given $a \in \xi$ and $b \in \eta$. The inference rule is represented by

Rule: if $\mathbb{X}$ is ξ and $\mathbb{Y}$ is η then $\mathbb{Z}$ is τ
From: $\mathbb{X}$ is a constant a and $\mathbb{Y}$ is a constant b
---
Infer: $\mathbb{Z}$ is $\eta^* = \eta|_{(a\in\eta)\bigcap(b\in\eta)}$

It has been proved by Gao, Gao, and Ralescu [8] that

$$\lambda^*(z) = \begin{cases} \dfrac{\lambda(z)}{\mu(a) \wedge \nu(b)}, & \text{if } \lambda(z) < \dfrac{\mu(a) \wedge \nu(b)}{2} \\ \dfrac{\lambda(z) + \mu(a) \wedge \nu(b) - 1}{\mu(a) \wedge \nu(b)}, & \text{if } \lambda(z) > 1 - \dfrac{\mu(a) \wedge \nu(b)}{2} \\ 0.5, & \text{otherwise.} \end{cases}$$

Liu's Inference Rule (Multiple If-Then Rules). Let $\mathbb{X}$ and $\mathbb{Y}$ be two concepts. Assume two rules 'if $\mathbb{X}$ is an uncertain set ξ_1 then $\mathbb{Y}$ is an uncertain set η_1' and 'if $\mathbb{X}$ is an uncertain set ξ_2 then $\mathbb{Y}$ is an uncertain set η_2'. From $\mathbb{X}$ is a constant a, we infer that $\mathbb{Y}$ is an uncertain set

$$\eta^* = \frac{\mathcal{M}\{a \in \xi_1\} \cdot \eta_1|_{a\in\xi_1}}{\mathcal{M}\{a \in \xi_1\} + \mathcal{M}\{a \in \xi_2\}} + \frac{\mathcal{M}\{a \in \xi_2\} \cdot \eta_2|_{a\in\xi_2}}{\mathcal{M}\{a \in \xi_1\} + \mathcal{M}\{a \in \xi_2\}}.$$

The inference rule is represented by

Rule 1: if $\mathbb{X}$ is ξ_1 then $\mathbb{Y}$ is η_1
Rule 2: if $\mathbb{X}$ is ξ_2 then $\mathbb{Y}$ is η_2
From: $\mathbb{X}$ is a constant a

Infer: $\mathbb{Y}$ is η^*

If $\xi_1, \xi_2, \eta_1, \eta_2$ are independent uncertain sets with continuous membership functions $\mu_1, \mu_2, \nu_1, \nu_2$, respectively, then the inference rule yields

$$\eta^* = \frac{\mu_1(a)}{\mu_1(a) + \mu_2(a)}\eta_1^* + \frac{\mu_2(a)}{\mu_1(a) + \mu_2(a)}\eta_2^*$$

where η_i^* are uncertain sets whose membership functions are

$$\nu_i^*(y) = \begin{cases} \dfrac{\nu_i(y)}{\mu_i(a)}, & \text{if } \nu_i(y) < \mu_i(a)/2 \\ \dfrac{\nu_i(y) + \mu_i(a) - 1}{\mu_i(a)}, & \text{if } \nu_i(y) > 1 - \mu_i(a)/2 \\ 0.5, & \text{otherwise.} \end{cases}$$

For a general system, Liu [17] proposed the following inference rule.

Liu's Inference Rule (general form) [17]. Let $\mathbb{X}_1, \mathbb{X}_2, \ldots, \mathbb{X}_m$ be m concepts. Assume the rules 'if $\mathbb{X}_1$ is ξ_{i1} and $\cdots$ and $\mathbb{X}_m$ is ξ_{im}, then $\mathbb{Y}$ is η_i' for $i = 1, 2, \ldots, k$. From $\mathbb{X}_1$ is a_1 and $\cdots$ and $\mathbb{X}_m$ is a_m, we infer that $\mathbb{Y}$ is an uncertain set

$$\eta^* = \sum_{i=1}^{k} \frac{c_i \cdot \eta_i|_{(a_1\in\xi_{i1})\cap(a_2\in\xi_{i2})\cap\cdots\cap(a_m\in\xi_{im})}}{c_1 + c_2 + \cdots + c_k}$$

where the coefficients are determined by

$$c_i = \mathcal{M}\left\{(a_1 \in \xi_{i1}) \cap (a_2 \in \xi_{i2}) \cap \cdots \cap (a_m \in \xi_{im})\right\}, \ i = 1, 2, \ldots, k.$$

The inference rule is represented by

Rule 1: If $\mathbb{X}_1$ is ξ_{11} and $\cdots$ and $\mathbb{X}_m$ is ξ_{1m} then $\mathbb{Y}$ is η_1
Rule 2: If $\mathbb{X}_1$ is ξ_{21} and $\cdots$ and $\mathbb{X}_m$ is ξ_{2m} then $\mathbb{Y}$ is η_2
$\cdots$
Rule k: If $\mathbb{X}_1$ is ξ_{k1} and $\cdots$ and $\mathbb{X}_m$ is ξ_{km} then $\mathbb{Y}$ is η_k
From: $\mathbb{X}_1$ is a_1 and $\cdots$ and $\mathbb{X}_m$ is a_m

---

Infer: $\mathbb{Y}$ is η^*

Assume $\xi_{i1}, \xi_{i2}, \ldots, \xi_{im}, \eta_i$ are independent uncertain sets with membership functions $\mu_{i1}, \mu_{i2}, \ldots, \mu_{im}, \nu_i$, $i = 1, 2, \ldots, k$, respectively. If $\xi_1^*, \xi_2^*, \ldots, \xi_m^*$ are constants $a_1, a_2, \ldots, a_m$, respectively. It has been proved by Liu [17] that

$$\eta^* = \sum_{i=1}^{k} \frac{c_i \cdot \eta_i^*}{c_1 + c_2 + \cdots + c_k}$$

where η_i^* are uncertain sets whose membership functions are given by

$$\nu_i^*(y) = \begin{cases} \dfrac{\nu_i(y)}{c_i}, & \text{if } \nu_i(y) < c_i/2 \\ \dfrac{\nu_i(y) + c_i - 1}{c_i}, & \text{if } \nu_i(y) > 1 - c_i/2 \\ 0.5, & \text{otherwise.} \end{cases} \qquad c_i = \min_{1 \le l \le m} \mu_{il}(a_l)$$

Uncertain systems are universal approximators

An uncertain system, introduced by Liu [7], is a function from its inputs to outputs via Liu's inference rule, in which five parts are contained: input, a rule-base, Liu's inference rule, an expected value operator and outputs. An uncertain system is a function $f : (\alpha_1, \alpha_2, \cdots, \alpha_m) \to (\beta_1, \beta_2, \cdots, \beta_n)$, namely,

$$(\beta_1, \beta_2, \cdots, \beta_n) = f(\alpha_1, \alpha_2, \cdots, \alpha_m).$$

Then we get an uncertain system f. Gao [9] has succeeded in controlling inverted pendulum. In this section, we will prove that uncertain systems are universal approximators. That is, uncertain systems are capable of approximating any continuous function on a compact set to arbitrary accuracy. Since there are great flexibilities in constructing rules, we have different methods to construct uncertain systems. The quantity of Liu's inference rules in an uncertain system is under consideration.

Theorem 1. *For any given continuous function g on a compact set $D \subset \Re^m$, and any given $\epsilon > 0$, there exists an uncertain system f such that*

$$\sup_{(\alpha_1,\alpha_2,\ldots,\alpha_m)\in D} ||f(\alpha_1, \alpha_2, \ldots, \alpha_m) - g(\alpha_1, \alpha_2, \ldots, \alpha_m)|| < \epsilon.$$

Proof. Without loss of generality, assume g is a real-valued function with two variables α_1 and α_2 and $D = [0, 1]^2$. Since D is a compact set and g is a continuous function, we know that g is uniformly continuous on D. Thus, for any given number $\epsilon > 0$, there exists a number $\delta > 0$ such that

$$|g(\alpha_1, \alpha_2) - g(\alpha_1', \alpha_2') < \epsilon, (\alpha_1, \alpha_2), (\alpha_1', \alpha_2') \in D$$

where $||(\alpha_1, \alpha_2) - (\alpha_1', \alpha_2')|| < \delta$. Let k be the minimum integer larger than $1/(\sqrt{2}\delta)$. Let

$$D_{i,j} = \left\{(\alpha_1, \alpha_2)|\frac{i-1}{k} < \alpha_1 \le \frac{i}{k}, \frac{j-1}{k} < \alpha_2 \le \frac{j}{k}\right\}, \ i, j = 1, 2, \ldots, k.$$

Note that $D_{i,j}$ are disjoint sets and $||(\alpha_1,\alpha_2)-(i/k,j/k)|| \le \delta$, wherever $(\alpha_1,\alpha_2)\in D_{i,j}$. Let ξ_i and η_j be independent rectangular uncertain sets with membership functions μ_i and ν_i

$$\mu_i(x)=\begin{cases}1, & \text{if } \frac{i-1}{k}<x\le\frac{i}{k}\\ 0, & \text{otherwise}\end{cases} \text{ and } \nu_i(x)=\begin{cases}1, & \text{if } \frac{j-1}{k}<x\le\frac{j}{k}\\ 0, & \text{otherwise}\end{cases}$$

where $i,j=1,2,\ldots,k$, respectively. Next, we construct a rule-base like,

If $\mathbb{X}_1$ is ξ_i and $\mathbb{X}_2$ is η_j then $\mathbb{Y}$ is $g(i/k,j/k)$

where $i,j=1,2,\ldots,k$, respectively. Hence, we get an uncertain system f. From a detailed investigation using definition of uncertain system and Liu's inference rule, we know that the uncertain system is

$$f(\alpha_1,\alpha_2)=g(i/k,j/k), \text{if}(\alpha_1,\alpha_2)\in D_{i,j}.$$

Then we have

$$\sup_{(\alpha_1,\alpha_2)\in D}|f(\alpha_1,\alpha_2)-g(\alpha_1,\alpha_2)| = \max_{i,j\in\{1,2,\ldots,k\}}\left\{\sup_{(\alpha_1,\alpha_2)\in D_{i,j}}|f(\alpha_1,\alpha_2)-g(\alpha_1,\alpha_2)|\right\} \quad (1)$$

$$= \max_{i,j\in\{1,2,\ldots,k\}}\left\{\sup_{(\alpha_1,\alpha_2)\in D_{i,j}}|g(i/k,j/k)-g(\alpha_1,\alpha_2)|\right\} \quad (2)$$

$$\le \max_{i,j\in\{1,2,\ldots,k\}}\epsilon=\epsilon. \quad (3)$$

The theorem is proved. □

Theorem 1 shows that the uncertain systems can approximate continuous functions. Next, we will extend the result in the sense of mean square convergence.

Theorem 2. *For any given function g on a compact set $D\subset\Re^m$ and any given $\epsilon>0$, there exists an uncertain system f such that*

$$\left(\int_D|f(\alpha_1,\alpha_2,\ldots,\alpha_m)-g(\alpha_1,\alpha_2,\ldots,\alpha_m)|^2\mathrm{d}\alpha_1\mathrm{d}\alpha_2\ldots\mathrm{d}\alpha_m\right)^{1/2}<\epsilon.$$

Proof. Without loss of generality, let g be a real-valued function with two variables. Note that D is a compact set, and we have $\int_D|g(\alpha,\alpha_2)|^2\mathrm{d}\alpha_1\mathrm{d}\alpha_2<\infty$. Then there exists a continuous function h on D such that

$$\left(\int_D|f(\alpha_1,\alpha_2)-g(\alpha_1,\alpha_2)|^2\mathrm{d}\alpha_1\mathrm{d}\alpha_2\right)^{1/2}<\epsilon/2.$$

It follows from Theorem 1 that there exists an uncertain system f satisfying

$$\sup_{(\alpha_1,\alpha_2)\in D}|f(\alpha_1,\alpha_2)-h(\alpha_1,\alpha_2)|<\epsilon/(2V^{1/2})$$

where $V = \int_D d\alpha_1 d\alpha_2$. Hence,

$$\begin{aligned}\left(\int_D |f(\alpha_1,\alpha_2) - g(\alpha_1,\alpha_2)|^2 d\alpha_1 d\alpha_2\right)^{1/2} &\le \left(\int_D |f(\alpha_1,\alpha_2) - h(\alpha_1,\alpha_2)|^2 d\alpha_1 d\alpha_2\right)^{1/2} \\ &\quad + \left(\int_D |h(\alpha_1,\alpha_2) - g(\alpha_1,\alpha_2)|^2 d\alpha_1 d\alpha_2\right)^{1/2} \\ &< \left(\int_D \frac{\epsilon^2}{4V} d\alpha_1 d\alpha_2\right)^{1/2} + \epsilon/2 \\ &= \epsilon.\end{aligned}$$

Theorem 2 shows that uncertain systems are capable of L^2 approximating functions which are not required continuously, including simple functions and piecewise continuous function. □

Conclusions

In this paper, we prove that uncertain systems are universal approximators, and a method to construct an uncertain system to approximate any continuous function on a compact set to arbitrary accuracy is given. Hence, uncertain systems are capable of approximating any continuous function on a compact set. Furthermore, uncertain systems are capable of approximating functions, which are not continuous, in the sense of mean square convergence.

Acknowledgements

This work was supported by National Natural Science Foundation of China Grant No. 61304182.

Author details

[1]School of Applied Mathematics, Central University of Finance and Economics, Beijing 100081, China. [2]Department of Risk Management and Insurance, Nankai University, Tianjin 300071, China.

References

1. Zadeh, L: Fuzzy sets. Inf. Control. **8**, 338–353 (1965)
2. Mamdani, EH: Applications of fuzzy algorithms for control of a simple dynamic plant. Proc. IEEE. **121**(12), 1585–1588 (1974)
3. Takagi, T, Sugeno, M: Fuzzy identification of system and its applications to modeling and control. IEEE Trans. Syst. Man Cybernatics. **15**(1), 116–132 (1985)
4. Lukasiewicz, J: Selected Works - Studies in Logic and the Foundations of Mathematics. North-Holland, Amsterdam, Warsaw (1970)
5. Mamdani, EH: Application of fuzzy logic to approximate reasoning using linguistic synthesis. IEEE Trans. Comput. **C-26**(12), 1182–1191 (1977)
6. Wang, L: Fuzzy systems are universal approximators. In: IEEE International Conference on Fuzzy Systems, pp. 1163–1170, San Diego, 8–12 March 1992
7. Liu, B: Uncertain set theory and uncertain inference rule with application to uncertain control. J. Uncertain Syst. **4**(2), 83–98 (2010)
8. Gao, X, Gao, Y, Ralescu, DA: On Liu's inference rule for uncertain systems. Int. J. Uncertainty Fuzziness Knowledge-Based Syst. **18**(1), 1–11 (2010)
9. Gao, Y: Uncertain inference control for balancing an inverted pendulum. Fuzzy Optimization Decis. Making. **11**(4), 481–492 (2012)
10. Liu, B: Uncertain logic for modeling human language. J. Uncertain Syst. **5**(1), 3–20 (2011)
11. Liu, B: Uncertainty Theory. 2nd ed. Springer, Berlin (2007)
12. Gao, X: Some properties of continuous uncertain measure. Int. J. Uncertainty Fuzziness Knowledge-Based Syst. **17**(3), 419–426 (2009)
13. Liu, B: Theory and Practice of Uncertain Programming. 2nd ed. Springer, Berlin (2009)
14. Liu, B: Fuzzy process, hybrid process and uncertain process. J. Uncertain Syst. **2**(1), 3–16 (2008)
15. Li, X, Liu, B: Hybrid logic and uncertain logic. J. Uncertain Syst. **3**(2), 83–94 (2009)

16. Liu, B: Uncertain entailment and modus ponens in the framework of uncertain logic. J. Uncertain Syst. **3**(4), 243–251 (2009)
17. Liu, B: Uncertainty Theory: a Branch of Mathematics for Modeling Human Uncertainty. Springer, Berlin (2010)
18. Liu, B: Some research problems in uncertainty theory. J. Uncertain Syst. **3**(1), 3–10 (2009)
19. Liu, B: Membership functions and operational law of uncertain sets. Fuzzy Optimization Decis. Making. **11**(4), 387–410 (2012)
20. Liu, B: A new definition of independence of uncertain sets. Fuzzy Optimization Decis. Making. **12**(4), 451–461 (2013)

7

Some formulas of variance of uncertain random variable

Yuhong Sheng[1*] and Kai Yao[2]

*Correspondence: sheng-yh12@mails.tsinghua.edu.cn
[1] Department of Mathematical Sciences, Tsinghua University, Beijing 100084, China
Full list of author information is available at the end of the article

Abstract

Uncertainty and randomness are two basic types of indeterminacy. Chance theory was founded for modeling complex systems with not only uncertainty but also randomness. As a mixture of randomness and uncertainty, an uncertain random variable is a measurable function on the chance space. It is usually used to deal with measurable functions of uncertain variables and random variables. There are some important characteristics about uncertain random variables. The expected value is the average value of uncertain random variable in the sense of chance measure and represents the size of uncertain random variable. The variance of uncertain random variable provides a degree of the spread of the distribution around its expected value. In order to describe the variance of uncertain random variable, this paper provides some formulas to calculate the variance of uncertain random variables through chance distribution and inverse chance distribution. Several practical examples are also provided to calculate the variance for through chance distribution inverse chance distribution.

Keywords: Chance theory; Uncertain random variable; Inverse chance distribution; Variance

Introduction

Probability theory is a branch of mathematics for studying the behavior of random phenomena. In order to deal with randomness, Kolmogorov [1] defined a probability measure as a set function satisfying the following three axioms: (i) normality, (ii) nonnegativity, and (iii) additivity. Before applying it in practice, we should first obtain the probability distribution *via* statistics or test the probability distribution to make sure it is close enough to the real frequency, either of which is based on a lot of observed data. However, due to the technological or economical difficulties, we sometimes have no observed data. In this case, we have to rely on domain experts evaluating their belief degree about the chances that the possible events happen. According to Kahneman and Tversky [2], human tends to overweight unlikely events, so the belief degree generally has a much larger range than the real frequency. As a result, the probability theory is not applicable in this case; otherwise, it may lead to counterintuitive results [3]. Liu [4] gave some examples of it.

The uncertainty theory is a branch of mathematics for modeling belief degrees. In order to deal with human uncertainty, Liu [5] first presented uncertain measure as a set function satisfying four measure axioms: (i) normality, (ii) duality, (iii) subadditivity, and (iv) product axiom [6]. As a fundamental concept in uncertainty theory, the uncertain variable was presented by Liu [5] in 2007. In order to describe the uncertain variable, Liu [5] introduced the concept of uncertainty distribution. Liu [3] proposed the concept of inverse uncertainty distribution, and Liu [4] verified a sufficient and necessary condition for it. After that, many researchers widely studied the uncertainty theory and made significative progress. Gao [7] studied the properties of continuous uncertain measure. Peng and Iwamura [8] proved a sufficient and necessary condition for uncertainty distribution. Furthermore, a measure inversion theorem was given by Liu [3] that may yield uncertain measures of some special events from the uncertainty distribution of the corresponding uncertain variable. In addition, the concept of independence was proposed by Liu [6]. After, Liu [3] presented the operational law of uncertain variables. In order to rank uncertain variables, Liu [5] proposed the concept of expected value of uncertain variables. The linearity of expected value operator was verified by Liu [3]. As an important contribution, Liu and Ha [9] derived a useful formula for calculating the expected values of strictly monotone functions of independent uncertain variables. Meanwhile, Liu [5] presented the concept of variance of uncertain variables and also proposed some formulas to calculate the variance through uncertainty distribution. Recently, Yao [10] proposed a formula to calculate the variance using inverse uncertainty distribution. Sheng and Kar [11] verified some results of moment of uncertain variable through inverse uncertainty distribution.

However, in many cases, randomness and uncertainty exist simultaneously in a complex system. In order to describe such a system, Liu [12] pioneered the concepts of uncertain random variable and chance measure. Meanwhile, Liu [12] defined the concepts of chance distribution, expected value and variance for uncertain random variable. Following that, Liu [13] presented the operational law of uncertain random variable, and proposed uncertain random programming as a branch of mathematical programming involving uncertain random variables. In addition, Guo and Wang [14] proved a formula for calculating the variance of uncertain random variables. Yao and Gao [15] verified a law of large numbers for uncertain random variables of independent and identical distribution. Gao and Yao [16] proposed uncertain random process.

This paper further studies the variance of uncertain random variables and provides some formulas to calculate the variance through inverse chance distribution. The rest of this paper is organized as follows: Section 'Preliminary' presents some basic concepts and theorems about uncertain random variables and recalled some concepts of expected value, variance, and so on. Some formulas about the variance are derived through chance distribution and inverse chance distribution in Section 'Variance of uncertain random variable'. Finally, some conclusions are given in Section 'Conclusions'.

Preliminary

In this section, we will introduce some useful definitions and theorems about random variable, uncertain variable and uncertain random variable.

Random variable

Let $\mathcal{A}$ be a σ-algebra on a nonempty set Ω. A set function $\Pr : \mathcal{A} \rightarrow [0, 1]$ is called a probability measure if it satisfies the following axioms:

Axiom 1: (Normality axiom) $\Pr\{\Omega\} = 1$ for the universal set Ω.

Axiom 2: (Nonnegativity axiom) $\Pr\{A\} \geq 0$ for any event A.

Axiom 3: (Additivity axiom) For every countable sequence of events $\mathcal{A}_1, \mathcal{A}_2, \cdots$, we have

$$\Pr\left\{\bigcup_{i=1}^{\infty} \mathcal{A}_i\right\} = \sum_{i=1}^{\infty} \Pr\{\mathcal{A}_i\}.$$

Definition 1. *(Kolmogorov [1]) The set function* Pr *is called a probability measure if it satisfies the normality, nonnegativity, and additivity axioms.*

Besides, the product probability on the product σ-algebra $\mathcal{A}$ is defined as follows:
Let $(\Omega_k, \mathcal{A}_k, \Pr_k)$ be probability spaces for $k = 1, 2, \cdots$. It has been proven that there is a unique probability measure Pr on the product σ-algebra $\mathcal{A}$ such that

$$\Pr\left\{\prod_{i=1}^{\infty} \mathcal{A}_k\right\} = \prod_{k=1}^{\infty} \Pr_k\{\mathcal{A}_k\}$$

where $\mathcal{A}_k$ are arbitrarily chosen events from $\mathcal{A}_k$ for $k = 1, 2, \cdots$, respectively. This conclusion is called *product probability theorem.* Such a probability measure is called *product probability measure,* denoted by

$$\Pr = \Pr_1 \times \Pr_2 \times \cdots$$

Uncertain variable

Definition 2. *(Liu [5]) Let $\mathcal{L}$ be a σ-algebra on a nonempty set Γ. A set function $\mathcal{M} : \mathcal{L} \to [0, 1]$ is called an uncertain measure if it satisfies the following axioms:*

Axiom 1: (Normality axiom) $\mathcal{M}\{\Gamma\} = 1$ for the universal set Γ.
Axiom 2: (Duality axiom) $\mathcal{M}\{\Lambda\} + \mathcal{M}\{\Lambda^c\} = 1$ for any event Λ.
Axiom 3: (Subadditivity axiom) For every countable sequence of events $\Lambda_1, \Lambda_2, \cdots$, we have

$$\mathcal{M}\left\{\bigcup_{i=1}^{\infty} \Lambda_i\right\} \leq \sum_{i=1}^{\infty} \mathcal{M}\{\Lambda_i\}.$$

Besides, the product uncertain measure on the product σ-algebra $\mathcal{L}$ is defined by Liu [6] as follows:

Axiom 4: (Product axiom) Let $(\Gamma_k, \mathcal{L}_k, \mathcal{M}_k)$ be uncertainty spaces for $k = 1, 2, \cdots$. Then the product uncertain measure $\mathcal{M}$ is an uncertain measure satisfying

$$\mathcal{M}\left\{\prod_{i=1}^{\infty} \Lambda_k\right\} = \bigwedge_{k=1}^{\infty} \mathcal{M}_k\{\Lambda_k\}$$

where Λ_k are arbitrarily chosen events from $\mathcal{L}_k$ for $k = 1, 2, \cdots$, respectively.

An uncertain variable is essentially a measurable function from an uncertainty space to the set of real numbers. In order to describe an uncertain variable, a concept of uncertainty distribution is defined as follows.

Definition 3. *(Liu [5]) Let ξ be an uncertain variable. Then its uncertainty distribution is defined by*

$$\Phi(x) = \mathcal{M}\{\xi \leq x\}$$

for any real number x.

An uncertainty distribution Φ is said to be regular if its inverse function $\Phi^{-1}(\alpha)$ exists and is unique for each $\alpha \in (0,1)$. Inverse uncertainty distribution plays an important role in the operation of independent uncertain variables.

The expected value of uncertain variable is the average value of uncertain variable in the sense of uncertain measure and it represents the size of uncertain variable. The variance of uncertain variable provides a degree of the spread of the distribution around its expected value.

Definition 4. *(Liu [5]) The expected value of an uncertain variable ξ is defined by*

$$E[\xi] = \int_0^{+\infty} \mathcal{M}\{\xi \geq x\}\mathrm{d}x - \int_{-\infty}^{0} \mathcal{M}\{\xi \leq x\}\mathrm{d}x$$

provided that at least one of the two integrals is finite.

Theorem 1. *(Liu [5]) Let ξ be an uncertain variable with uncertainty distribution Φ. If the expected value exists, then*

$$E[\xi] = \int_0^{+\infty} (1 - \Phi(x))\mathrm{d}x - \int_{-\infty}^{0} \Phi(x)\mathrm{d}x.$$

The expected value of the linear uncertain variable $\xi \sim \mathcal{L}(a,b)$ is $E[\xi] = (a+b)/2$; the expected value of the normal uncertain variable $\xi \sim \mathcal{N}(e,\sigma)$ is $E[\xi] = e$; and the expected value of the lognormal uncertain variable $\xi \sim \mathcal{LOGN}(e,\sigma)$ is

$$E[\exp(\xi)] = \begin{cases} \sqrt{3}\sigma \exp(e) \csc\left(\sqrt{3}\sigma\right), & \text{if } \sigma \leq \pi/\sqrt{3} \\ +\infty, & \text{otherwise.} \end{cases}$$

In 2010, Liu [3] first introduced a formula for expected value by inverse uncertainty distribution, that is

$$E[\xi] = \int_0^1 \Phi^{-1}(\alpha)\mathrm{d}\alpha.$$

Liu and Ha [9] proposed a generalized formula for expected value by inverse uncertainty distribution.

Theorem 2. *(Liu and Ha [9]) Let $\xi_1, \xi_2, \cdots, \xi_n$ be independent uncertain variables with uncertainty distributions $\Phi_1, \Phi_2, \cdots, \Phi_n$, respectively. If $f(\xi_1, \xi_2, \cdots, \xi_n)$ is strictly increasing with respect to $\xi_1, \xi_2, \cdots, \xi_m$ and strictly decreasing with respect to $\xi_{m+1}, \xi_{m+2}, \cdots, \xi_n$, then the uncertain variable $\xi = f(\xi_1, \xi_2, \cdots, \xi_n)$ has an expected value*

$$E[\xi] = \int_0^1 f(\Phi_1^{-1}(\alpha), \cdots, \Phi_m^{-1}(\alpha), \Phi_{m+1}^{-1}(1-\alpha), \cdots, \Phi_n^{-1}(1-\alpha))\mathrm{d}\alpha.$$

Based on this result, Liu [6] proved the linearity property of the expected value operator. For two independent uncertain variables ξ and η, we have $E[a\xi + b\eta] = aE[\xi] + bE[\eta]$ where a and b are real numbers.

Definition 5. *(Liu [5]) Let ξ be an uncertain variable with finite expected value $E[\xi]$. Then the variance of ξ is*

$$V[\xi] = E[(\xi - E[\xi])^2].$$

Since the uncertain measure is a subadditivity measure, the variance of uncertain variable cannot be derived simply by the uncertainty distribution. In this case, Liu [5] suggested to accept the following formulas as a stipulation

$$V[\xi] = \int_0^{+\infty} [1 - \Phi(E[\xi] + \sqrt{x}) + \Phi(E[\xi] - \sqrt{x})]\, dx.$$

Based on this stipulation, the variance of the linear uncertain variable $\xi \sim \mathcal{L}(a, b)$ is $V[\xi] = (b - a)^2/12$ and the variance of the normal uncertain variable $\xi \sim \mathcal{N}(e, \sigma)$ is $V[\xi] = \sigma^2$.

In 2013, Yao [10] proved a formula to calculate the variance of an uncertain variable *via* the inverse uncertainty distribution.

Theorem 3. *(Yao [10]) Let ξ be an uncertain variable with a regular uncertainty distribution Φ. If the expected value $E[\xi]$ exists, then the variance is*

$$V[\xi] = \int_0^1 (\Phi^{-1}(\alpha) - E[\xi])^2 d\alpha.$$

Uncertain random variable

In 2013, Liu [12] first proposed chance theory, which is a mathematical methodology for modeling complex systems with both uncertainty and randomness, including chance measure, uncertain random variable, chance distribution, operational law, expected value, and so on. The chance space refers to the product $(\Gamma, \mathcal{L}, \mathcal{M}) \times (\Omega, \mathcal{A}, \Pr)$, in which $(\Gamma, \mathcal{L}, \mathcal{M})$ is an uncertainty space and $(\Omega, \mathcal{A}, \Pr)$ is a probability space.

Definition 6. (Liu [12]) *Let $(\Gamma, \mathcal{L}, \mathcal{M}) \times (\Omega, \mathcal{A}, \Pr)$ be a chance space, and let $\Theta \in \mathcal{L} \times \mathcal{A}$ be an uncertain random event. Then the chance measure of Θ is defined as*

$$\mathrm{Ch}\{\Theta\} = \int_0^1 \Pr\{\omega \in \Omega \mid \mathcal{M}\{\gamma \in \Gamma | (\gamma, \omega) \in \Theta\} \geq r\} dr.$$

Liu [12] proved that a chance measure satisfies normality, duality, and monotonicity properties, that is

(i) $\mathrm{Ch}\{\Gamma \times \Omega\} = 1$;
(ii) $\mathrm{Ch}\{\Theta\} + \mathrm{Ch}\{\Theta^c\} = 1$ for and event Θ;
(iii) $\mathrm{Ch}\{\Theta_1\} \leq \mathrm{Ch}\{\Theta_2\}$ for any real number set $\Theta_1 \subset \Theta_2$.

Besides, Hou [17] proved the subadditivity of chance measure, that is,

$$\mathrm{Ch}\left\{\bigcup_{i=1}^{\infty} \Theta_i\right\} \leq \sum_{i=1}^{\infty} \mathrm{Ch}\{\Theta_i\}$$

for a sequence of events $\Theta_1, \Theta_2, \cdots$.

Definition 7. *(Liu [12]) An uncertain random variable is a measurable function ξ from a chance space $(\Gamma, \mathcal{L}, \mathcal{M}) \times (\Omega, \mathcal{A}, \Pr)$ to the set of real numbers, i.e., $\xi \in B$ is an event for any Borel set B.*

Random variables and uncertain variables can be regarded as special cases of uncertain random variables. Let η be a random variable, τ be an uncertain variable. Then $\eta + \tau$ and $\eta \times \tau$ are both uncertain random variables.

Theorem 4. *(Liu [12]) Let $f : R^n \to R$ be a measurable function, and $\xi_1, \xi_2, \cdots, \xi_n$ uncertain random variables on the chance space $(\Gamma, \mathcal{L}, \mathcal{M}) \times (\Omega, \mathcal{A}, \Pr)$. Then $\xi = f(\xi_1, \xi_2, \cdots, \xi_n)$ is an uncertain random variable determined by*

$$\xi(\gamma, \omega) = f(\xi_1(\gamma, \omega), \xi_2(\gamma, \omega), \cdots, \xi_n(\gamma, \omega))$$

for all $(\gamma, \omega) \in \Gamma \times \Omega$.

To calculate the chance measure, Liu [13] presented a definition of chance distribution.

Definition 8. *(Liu [13]) Let ξ be an uncertain random variable. Then its chance distribution is defined by*

$$\Phi(x) = \mathrm{Ch}\{\xi \leq x\}$$

for any $x \in \mathcal{R}$.

The chance distribution of a random variable is just its probability distribution, and the chance distribution of an uncertain variable is just its uncertainty distribution.

Theorem 5. *(Liu [13]) Let $\eta_1, \eta_2, \cdots, \eta_m$ be independent random variables with probability distributions $\Psi_1, \Psi_2, \cdots, \Psi_m$, respectively, and let $\tau_1, \tau_2, \cdots, \tau_n$ be uncertain variables. Then the uncertain random variable $\xi = f(\eta_1, \eta_2, \cdots, \eta_m, \tau_1, \tau_2, \cdots, \tau_n)$ has a chance distribution*

$$\Phi(x) = \int_{\Re^m} F(x, y_1, \cdots, y_m) \mathrm{d}\Psi_1(y_1) \cdots \mathrm{d}\Psi_m(y_m)$$

where $F(x, y_1, \cdots, y_m)$ is the uncertainty distribution of uncertain variable

$$f(\eta_1, \eta_2, \cdots, \eta_m, \tau_1, \tau_2, \cdots, \tau_n)$$

for any real numbers $y_1, y_2, \cdots, y_m$.

Let η be a random variable with probability distribution Ψ, and let τ be an uncertain variable with uncertainty distribution Υ, by Theorem 5 we proved that the sum $\xi = \eta + \tau$ has a chance distribution

$$\Phi(x) = \int_{-\infty}^{+\infty} \Upsilon(x - y) \mathrm{d}\Psi(y).$$

Let η be a positive random variable with probability distribution Ψ, and let τ be a positive uncertain variable with uncertainty distribution Υ; by Theorem 5, we proved that the product $\xi = \eta\tau$ has a chance distribution

$$\Phi(x) = \int_{0}^{+\infty} \Upsilon(x/y) \mathrm{d}\Psi(y).$$

Let η be a random variable with probability distribution Ψ, and let τ be an uncertain variable with uncertainty distribution Υ; by Theorem 5, we proved that the minimum $\xi = \eta \wedge \tau$ has a chance distribution

$$\Phi(x) = \Psi(x) + \Upsilon(x) - \Psi(x)\Upsilon(x).$$

Let η be a random variable with probability distribution Ψ, and let τ be an uncertain variable with uncertainty distribution Υ; by Theorem 5, we proved that the maximum $\xi = \eta \vee \tau$ has a chance distribution

$$\Phi(x) = \Psi(x)\Upsilon(x).$$

Definition 9. *(Liu [13]) Let ξ be an uncertain random variable. Then its expected value is defined by*

$$E[\xi] = \int_0^{+\infty} Ch\{\xi \geq r\}\mathrm{d}r - \int_{-\infty}^{0} Ch\{\xi \leq r\}\mathrm{d}r$$

provided that at least one of the two integrals is finite.

Let Φ denote the chance distribution of ξ. Liu [13] proved a formula to calculate the expected value of uncertain random variable with chance distribution, that is,

$$E[\xi] = \int_0^{+\infty} (1 - \Phi(x))\mathrm{d}x - \int_{-\infty}^{0} \Phi(x)\mathrm{d}x.$$

For a random variable η and an uncertain variable τ, Liu [13] proved that $E[\eta + \tau] = E[\eta] + E[\tau]$ and $E[\eta \times \tau] = E[\eta] \times E[\tau]$. In fact, we have the following theorem about the expected value of uncertain random variables.

Theorem 6. *(Liu [13]) Let $\eta_1, \eta_2, \cdots, \eta_m$ be independent random variables with probability distributions $\Psi_1, \Psi_2, \cdots, \Psi_m$, respectively, and let $\tau_1, \tau_2, \cdots, \tau_n$ be uncertain variables (not necessarily independent), then the uncertain random variable $\xi = f(\eta_1, \eta_2, \cdots, \eta_m, \tau_1, \tau_2, \cdots, \tau_n)$ has an expected value*

$$E[\xi] = \int_{\Re^m} E[f(y_1, y_2, \cdots, y_m, \tau_1, \tau_2, \cdots, \tau_n)]\, \mathrm{d}\Psi_1(y_1) \cdots \mathrm{d}\Psi_m(y_m)$$

where $E[f(y_1, y_2, \cdots, y_m, \tau_1, \tau_2, \cdots, \tau_n)]$ is the expected value of the uncertain variable $f(y_1, y_2, \cdots, y_m, \tau_1, \tau_2, \cdots, \tau_n)$ for any real numbers $y_1, y_2, \cdots, y_m$.

Theorem 7. *(Liu [13]) Let $\eta_1, \eta_2, \cdots, \eta_m$ be independent uncertain random variables with probability distributions $\Psi_1, \Psi_2, \cdots, \Psi_m$, and let $\tau_1, \tau_2, \cdots, \tau_n$ be independent uncertain random variables with uncertainty distributions $\Upsilon_1, \Upsilon_2, \cdots, \Upsilon_n$, respectively. If $f(\eta_1, \eta_2, \cdots, \eta_m, \tau_1, \tau_2, \cdots, \tau_n)$ is a strictly increasing or strictly decreasing function with respect to $\tau_1, \tau_2, \cdots, \tau_n$, then the uncertain random variable*

$$\xi = f(\eta_1, \eta_2, \cdots, \eta_m, \tau_1, \tau_2, \cdots, \tau_n)$$

has an expected value

$$E[\xi] = \int_{\Re^m} \int_0^1 f(y_1, \cdots, y_m, \Upsilon_1^{-1}(\alpha), \cdots, \Upsilon_n^{-1}(\alpha))\mathrm{d}\alpha \mathrm{d}\Psi_1(y_1) \cdots \mathrm{d}\Psi_m(y_m).$$

Let η be a random variable with probability distribution Ψ, and let τ be an uncertain variable with uncertainty distribution Υ. With these, we proved that

$$E[\eta \vee \tau] = \int_{\Re} \int_0^1 (y \vee \Upsilon^{-1}(\alpha))\mathrm{d}\alpha \mathrm{d}\Psi(y)$$

and

$$E[\eta \wedge \tau] = \int_{\Re} \int_0^1 (y \wedge \Upsilon^{-1}(\alpha))\mathrm{d}\alpha \mathrm{d}\Psi(y).$$

Variance of uncertain random variable

This section will introduce the concept of variance for uncertain random variable as well as a stipulation on the formula to calculate the variance.

Definition 10. *(Liu [13]) Let ξ be an uncertain random variable with a finite expected value $E[\xi]$. Then the variance of ξ is*

$$V[\xi] = E[(\xi - E[\xi])^2].$$

Since $(\xi - E[\xi])^2$ is a nonnegative uncertain random variable, we also have

$$V[\xi] = \int_0^{+\infty} \mathrm{Ch}\{(\xi - E[\xi])^2 \geq x\} \mathrm{d}x.$$

How to obtain variance from chance distribution? Since the chance measure is a subadditivity measure, the variance of uncertain random variable cannot be derived simply by the chance distribution. In this case, Guo and Wang [14] suggest to a stipulation as follows:

Stipulation 1. (Guo and Wang [14]) *Let ξ be an uncertain random variable with a chance distribution Φ. If ξ has a finite expected value $E[\xi]$, then*

$$V[\xi] = \int_0^{+\infty} \left(1 - \Phi\left(E[\xi] + \sqrt{x}\right) + \Phi\left(E[\xi] - \sqrt{x}\right)\right) \mathrm{d}x.$$

Based on this stipulation, we will give some formulas to calculate the variance of an uncertain random variable with the chance distribution and the inverse chance distribution.

Theorem 8. *Let ξ be an uncertain random variable with a chance distribution Φ. If ξ has a finite expected value $E[\xi]$, then*

$$V[\xi] = \int_{-\infty}^{+\infty} (x - E[\xi])^2 \,\mathrm{d}\Phi(x).$$

Proof: This theorem is based on Stipulation 1 that says the variance of ξ is

$$V[\xi] = \int_0^{+\infty} (1 - \Phi(E[\xi] + \sqrt{y}))\mathrm{d}y + \int_0^{+\infty} \Phi(E[\xi] - \sqrt{y})\mathrm{d}y.$$

Substituting $E[\xi] + \sqrt{y}$ with x and y with $(x - E[\xi])^2$, the change of variables and integration by parts produced, we have

$$\int_0^{+\infty} \left(1 - \Phi\left(E[\xi] + \sqrt{y}\right)\right) \mathrm{d}y = \int_{E[\xi]}^{+\infty} (1 - \Phi(x))\mathrm{d}(x - E[\xi])^2 = \int_{E[\xi]}^{+\infty} (x - E[\xi])^2 \mathrm{d}\Phi(x).$$

Similarly, substituting $e - \sqrt{y}$ with x and y with $(x - e)^2$, we obtain

$$\int_0^{+\infty} \Phi\left(E[\xi] - \sqrt{y}\right) \mathrm{d}y = \int_{E[\xi]}^{-\infty} \Phi(x)\, \mathrm{d}(x - E[\xi])^2 = \int_{-\infty}^{E[\xi]} (x - E[\xi])^2 \mathrm{d}\Phi(x).$$

It follows that the variance is

$$V[\xi] = \int_{E[\xi]}^{+\infty} (x - E[\xi])^2 \mathrm{d}\Phi(x) + \int_{-\infty}^{E[\xi]} (x - E[\xi])^2 \mathrm{d}\Phi(x) = \int_{-\infty}^{+\infty} (x - E[\xi])^2 \,\mathrm{d}\Phi(x).$$

The theorem is verified.

Theorem 9. *Let ξ be an uncertain random variable with a regular chance distribution Φ. If ξ has a finite expected value $E[\xi]$, then*

$$V[\xi]=\int_0^1 \left(\Phi^{-1}(r)-E[\xi]\right)^2 \mathrm{d}r.$$

Proof: Substituting $\Phi(x)$ with r and x with $\Phi^{-1}(r)$, it follows from the change of variables of integration and Theorem 8 that the variance is

$$V[\xi]=\int_{-\infty}^{+\infty} (x-E[\xi])^2\,\mathrm{d}\Phi(x)=\int_0^1 \left(\Phi^{-1}(r)-E[\xi]\right)^2 \mathrm{d}r.$$

The theorem is verified.

Example 1. *Let η be a random variable with probability distribution Ψ, and let τ be an uncertain variable with uncertainty distribution Υ. Then by Theorem 8, the sum $\xi=\eta+\tau$ has the variance*

$$V[\xi]=\int_{-\infty}^{+\infty} (x-E[\xi])^2\,\mathrm{d}\Phi(x)$$

where $\Phi(x)=\int_{-\infty}^{+\infty}\Upsilon(x-y)\mathrm{d}\Psi(y)$ and $E[\xi]=E[\eta]+E[\tau]$. By Theorem 9, the sum $\xi=\eta+\tau$ has the variance

$$V[\xi]=\int_0^1 \left(\Phi^{-1}(r)-(E[\eta]+E[\tau])\right)^2 \mathrm{d}r.$$

Example 2. *Let η be a random variable with probability distribution Ψ, and let τ be an uncertain variable with uncertainty distribution Υ. Then by Theorem 8, the product $\xi=\eta\tau$ has the variance*

$$V[\xi]=\int_{-\infty}^{+\infty} (x-E[\xi])^2\,\mathrm{d}\Phi(x)$$

where $\Phi(x)=\int_{-\infty}^{+\infty}\Upsilon(x/y)\mathrm{d}\Psi(y)$ and $E[\xi]=E[\eta]E[\tau]$. By Theorem 9, the product $\xi=\eta\tau$ has the variance

$$V[\xi]=\int_0^1 \left(\Phi^{-1}(r)-E[\eta]E[\tau]\right)^2 \mathrm{d}r.$$

Example 3. *Let η be a random variable with probability distribution Ψ, and let τ be an uncertain variable with uncertainty distribution Υ. Then by Theorem 8, the minimum $\xi=\eta\wedge\tau$ has the variance*

$$V[\xi]=\int_{-\infty}^{+\infty} (x-E[\xi])^2\,\mathrm{d}\Phi(x)$$

where $\Phi(x)=\Psi(x)+\Upsilon(x)-\Psi(x)\Upsilon(x)$ and $E[\xi]=E[\eta\wedge\tau]=\int_{\Re}\int_0^1(y\wedge\Upsilon^{-1}(\alpha))\mathrm{d}\alpha\mathrm{d}\Psi(y)$. By Theorem 9, the minimum $\xi=\eta\wedge\tau$ has the variance

$$V[\xi]=\int_0^1 \left(\Phi^{-1}(r)-E[\eta\wedge\tau]\right)^2 \mathrm{d}r.$$

Example 4. *Let η be a random variable with probability distribution Ψ, and let τ be an uncertain variable with uncertainty distribution Υ. Then by Theorem 8, the maximum $\xi = \eta \vee \tau$ has the variance*

$$V[\xi] = \int_{-\infty}^{+\infty} (x - E[\xi])^2 \, \mathrm{d}\Phi(x)$$

where $\Phi(x) = \Psi(x)\Upsilon(x)$ and $E[\xi] = E[\eta \vee \tau] = \int_{\Re} \int_0^1 (y \vee \Upsilon^{-1}(\alpha)) \mathrm{d}\alpha \mathrm{d}\Psi(y)$. By Theorem 9, the minimum $\xi = \eta \vee \tau$ has the variance

$$V[\xi] = \int_0^1 \left(\Phi^{-1}(r) - E[\eta \vee \tau]\right)^2 \mathrm{d}r.$$

Conclusions

In this paper, we mainly studied the variance of an uncertain random variable. Some formulas were derived to calculate the variance of an uncertain random variable using chance distribution and the inverse chance distribution. At the same time, several practical examples are also provided to calculate the variance of uncertain random variables with chance distribution and inverse chance distribution.

Acknowledgements
This work was supported by National Natural Science Foundation of China Grants No.61273044 and No.91224008.

Author details
[1]Department of Mathematical Sciences, Tsinghua University, Beijing 100084, China. [2]School of Management, University of Chinese Academy of Sciences, Beijing 100190, China.

References
1. Kolmogorov, AN: Grundbegriffe der Wahrscheinlichkeitsrechnung. Julius Springer, Berlin (1933)
2. Kahneman, D, Tversky, A: Prospect theory: an analysis of decision under risk. Econometrica. **47**(2), 263—292 (1979)
3. Liu, B: Uncertainty Theory: A Branch of Mathematics for Modeling Human Uncertainty. Springer-Verlag, Berlin (2010)
4. Liu, B: Why is there a need for uncertainty theory? J. Uncertain Syst. **6**(1), 3–10 (2012)
5. Liu, B: Uncertainty Theory. 2nd ed. Springer-Verlag, Berlin (2007)
6. Liu, B: Some research problems in uncertainty theory. J. Uncertain Syst. **3**(1), 3–10 (2009)
7. Gao, X: Some properties of continuous uncertain measure. Int. J. Uncertainty Fuzziness Knowledge-Based Syst. **17**(3), 419–426 (2009)
8. Peng, ZX, Iwamura, K: A sufficient and necessary condition of uncertainty distribution. J. Interdiscip. Math. **13**(3), 277–285 (2010)
9. Liu, YH, Ha, MH: Expected value of function of uncertain variables. J. Uncertain Syst. **4**(3), 181—186 (2010)
10. Yao, K: A formula to calculate the variance of uncertain variable. http://orsc.edu.cn/online/130831.pdf.
11. Sheng, YH, Samarjit, K: Some results of moments of uncertain variable through inverse uncertainty distribution. Fuzzy Optimization Decis. Making. (2014 in press)
12. Liu, YH: Uncertain random variables: a mixture of uncertainty and randomness. Soft Comput. **17**(4), 625–634 (2013)
13. Liu, YH: Uncertain random programming with applications. Fuzzy Optimization Decis. Making. **12**(2), 153—169 (2013)
14. Guo, HY, Wang, XS: Variance of uncertain random variables. J. Uncertainty Anal. Appl. (2014 in press)
15. Yao, K, Gao, JW: Law of large numbers for uncertain random variables. http://orsc.edu.cn-/online/120401.pdf.
16. Gao, JW, Yao, K: Some concepts and theorems of uncertain random process. Int. J. Intell. Syst. (2014 in press)
17. Hou, YC: Subadditivity of chance measure. http://orsc.edu.cn/online/130602.pdf.

Optimal dynamic production and price for reliability-dependent imperfect production with inventory-level-dependent demand in uncertain environment

Samar Hazari[1], Kalipada Maity[2*], Jayanta Kumar Dey[3] and Samarjit Kar[1]

*Correspondence:
kalipada_maity@yahoo.co.in
[2]Department of Mathematics, Mugberia Gangadhar Mahavidyalaya, Bhupatinagar Purba Medinipur, West Bengal 721425, India
Full list of author information is available at the end of the article

Abstract

An inventory system for reliability-dependent imperfect production is introduced in uncertain environment. The demand rate depends on the stock quantity displayed in the store as well as the sales price. With the goal to realize profit maximization, an optimization problem is addressed to seek for the optimal joint dynamic pricing and production policy which are obtained by solving the optimization problem by Euler-Lagrange equation of optimal control theory. Here, initial selling price, holding cost, and raw material cost are taken as uncertain variables, and using uncertain expectation mathematics, the uncertain variables are converted into crisp value. The numerical results demonstrate the advantages of the joint dynamic one and further show the effects of different system parameters on the optimal dynamic policy and the maximal total profit.

Keywords: Dynamic imperfect production; Optimum pricing; Reliability; Euler-Lagrangian equation; Uncertainty theory

Introduction

In classical inventory models, the demand whether deterministic or stochastic was often treated as an exogenous variable, without considering the effect of operating strategy. However, marketing practitioners and researchers have noticed that in many practical cases the demand rate depends on the stock quantity displayed in the store as well as the sales price. Generally, demand rate decreases with sales price yet increases with the stock quantity on display. In recent decades, extensive literature has focused on this phenomenon that demand rate depends on the stock quantity displayed in the store. Levin et al. [1] observed that large piles of consumer goods displayed in a supermarket would lead consumers to buy more. Gupta and Vrat [2] first set up a model to look into this phenomenon where it was assumed that the demand rate was a function of initial stock level. Baker and Urban [3] developed an economic order quantity (EOQ) model with a power-form inventory-level-dependent demand rate, decreasing with the inventory level throughout the sales period. Datta and Pal [4] extended the model developed by Baker and

Urban [3] on the basis of a hypothesis that the demand rate was dependent on the inventory level until a given level, beyond which it was a constant and the inventory level fell to zero at the end of the order period. Later, Urban [5] modified this model by relaxing the condition of zero-ending inventory. Ray et al. [6] developed an inventory model by assuming that the demand rate was stock dependent and two separate warehouses were used. When the stock level exceeded the capacity [5] gave a comprehensive review of the literature with inventory-level-dependent demand. Recently, Glock et al. [7] have published an optimal ordering policy for stock-dependent demand under progressive payment scheme. Some interesting results about the inventory-level-dependent demand were proposed in [8-15] and so on.

Reliability is closely related with the quality of a product and which is frequently considered a subset of quality. Quality can be defined qualitatively as the amount by which the product satisfies the users' (customers') requirements. Product quality is in part a function of design and conformance to design specifications. On the other hand, reliability is concerned with how long the product continuous to function once it becomes operational. A poor-quality product will likely have poor reliability and a high-quality product will likely have a high reliability. Reliability deals with reducing failures over a time interval and is a measure of the odds for failure-free operation during a given interval, improving reliability occurs at an increase development cost. Mettas [16] developed reliability allocation and optimization for complex systems and later by several researchers such as Panda and Maiti [17], Sarkar [18], and others.

One of the weaknesses of current production-inventory models is the unrealistic assumption that all items produced are of good quality. But production of defective units is a natural phenomenon due to different difficulties in a long-run production process. The production of defective items is increased with time and decreased with reliability parameters. The defective items as a result of imperfect quality production process were initially considered by Porteus [19] and later by several researchers such as Salameh and Jaber [20], Panda et al. [21], Maity et al. [22], Mandal et al. [23], Sana [24], Khan and Jaber [25], Yadav et al. [26], Krishnamoorthi and Panayappan [27], and others.

Uncertainty is common in real-life problems such as randomness, fuzziness, and roughness. Since Liu [28] introduced an uncertainty theory in 2007, it has been well developed and applied in a wide variety of real problems [29]. Recently, Liu [30] proposed an uncertain measure and developed an uncertainty theory which can be used to handle subjective imprecise quantity. Much research work has been done on the development of uncertainty theory and related theoretical work. You [31] proved some convergence theorems of uncertain sequences. Liu [29,32] has defined an uncertainty process and discussed an uncertainty theory. Recently, Jana et al. [33] developed a production-inventory model under permissible delay in payment in uncertain environment. It is often difficult to estimate the initial selling price, holding cost, and raw material cost. Depending upon different aspects, they fluctuate due to uncertainty in judgment, lack of evidence, insufficient information, etc. Sometimes it is not at all possible to get relevant precise data. So, these parameters are assumed to be flexible/imprecise in nature, i.e., uncertain sense and may be represented by uncertain variables.

In this paper, we consider an advertisement policy- and reliability-dependent imperfect production-inventory model in uncertain environment. Using expectation of uncertain number, the uncertain production inventory control problem is converted into equivalent crisp production inventory control problem. Then, the problem is solved by the Euler-Lagrangian equation [34] and GRG technique [35]. The optimum demand, production, and stock levels are determined. The model is illustrated through numerical examples, and results are presented in tabular and graphical form.

Necessary knowledge about uncertain variable

To better describe the subjective imprecise quantity, Liu [28,29] proposed an uncertain measure and further developed an uncertainty theory which is an axiomatic system of normality, duality, subadditivity, and product measure.

Uncertain measure

Definition 1. Let us define an uncertain measure $\mathcal{M}$ on the σ-algebra $\mathcal{L}$. That is, a number $\mathcal{M}\{\Lambda\}$ will be assigned to each event $\{\Lambda\}$ to indicate the belief degree with which we believe $\{\Lambda\}$ will happen. There is no doubt that the assignment is not arbitrary, and the uncertain measure $\mathcal{M}$ must have certain mathematical properties. In order to rationally deal with belief degrees, Liu [28] suggested the following three axioms:

Axiom 1. (Normality axiom) $\mathcal{M}\{\Gamma\}=1$

Axiom 2. (Duality axiom) $\mathcal{M}\{\Lambda\}+\mathcal{M}\{\Lambda^C\}=1$, for any event Λ

Axiom 3. (Subadditivity axiom) For every countable sequence of events $\Lambda_1, \Lambda_2, \ldots$, we have

$$\mathcal{M}\left\{\sum_{i=1}^{\infty} \Lambda_i\right\} \leq \sum_{i=1}^{\infty} \mathcal{M}\{\Lambda_i\}$$

Product uncertain measure

Product uncertain measure was defined by Liu [29], thus producing the fourth axiom of uncertainty theory. Let $(\Gamma_k, \mathcal{L}_k, \mathcal{M}_k)$ be uncertainty spaces for $k = 1, 2, \ldots$. Write $\Gamma = \Gamma_1 \times \Gamma_2 \times \ldots$ that is the set of all ordered tuples of the form $(\gamma_1, \gamma_2, \ldots)$ where $\gamma_k \in \Gamma_k$ for $k = 1, 2, \ldots$. A measurable rectangle in Γ is a set $\Lambda = \Lambda_1 \times \Lambda_2 \times \ldots$, where $\Lambda_k \in \mathcal{L}_k$ for $k = 1, 2, \ldots$. The smallest σ-algebra containing all measurable rectangles of Γ is called the product σ-algebra, denoted by $\mathcal{L} = \mathcal{L}_1 \times \mathcal{L}_2 \times \ldots$. Then, the product uncertain measure $\mathcal{M}$ on the product σ-algebra $\mathcal{L}$ is defined by the following product axiom [29]:

Axiom 4. (Product axiom) Let $(\Gamma_k, \mathcal{L}_k, \mathcal{M}_k)$ be uncertainty spaces for $k = 1, 2, \ldots$. The product uncertain measure $\mathcal{M}$ is an uncertain measure satisfying

$$\mathcal{M}\left\{\prod_{k=1}^{\infty} \Lambda_k\right\} \leq \prod_{k=1}^{\infty} \mathcal{M}\{\Lambda_k\}$$

where Λ_k are arbitrarily chosen events from $\mathcal{L}_k$ for $k = 1, 2 \ldots$, respectively.

Definition 2 ([28]). The uncertainty distribution $\Phi : R \to [0, 1]$ of an uncertain variable $\widehat{\xi}$ is defined by

$$\Phi(t) = \mathcal{M}\{\widehat{\xi} \leq t\}$$

Let $\widehat{\xi} \sim \mathcal{L}(a, b)$ be a linear uncertain variable. This linear uncertain variable is depicted in Figure 1.

Definition 3. Let $\widehat{\xi}$ be an uncertain variable with regular uncertainty distribution Φ. Then, the inverse function Φ^{-1} is called the inverse uncertainty distribution of $\widehat{\xi}$.

Definition 4 ([28])**.** Let $\widehat{\xi}$ be an uncertain variable. Then, the expected value of $\widehat{\xi}$ is defined by

$$E[\widehat{\xi}] = \int_0^{\infty} \mathcal{M}\{\widehat{\xi} \geq r\}dr - \int_{-\infty}^{0} \widehat{M}\{\widehat{\xi} \leq r\}dr$$

provided that at least one of the two integrals is finite.

Theorem 1 ([32])**.** Let $\widehat{\xi}$ be an uncertain variable with uncertainty distribution Φ. If the expected value exists, then $E[\widehat{\xi}] = \int_0^1 \Phi^{-1}(\alpha)d\alpha$

Theorem 2 ([30])**.** Let $\xi_1, \xi_2, \ldots, \xi_n$ be independent uncertain variables with uncertainty distributions $\phi_1, \phi_2, \ldots, \phi_n$, respectively. If f is a strictly increasing function, then $\xi = f(\xi_1, \xi_2, \ldots, \xi_n)$ is an uncertain variable with inverse uncertainty distribution

$$\Phi^{-1}(\alpha) = f\left(\Phi_1^{-1}(\alpha), \Phi_2^{-1}(\alpha), \ldots, \Phi_n^{-1}(\alpha)\right)$$

Theorem 3 ([32])**.** Let $\widehat{\xi}$ and $\widehat{\eta}$ be independent uncertain variables with finite expected values. Then, for any real numbers a_1 and a_2, we have

$$E[a_1\widehat{\xi} + a_2\widehat{\eta}] = a_1E[\widehat{\xi}] + a_2E[\widehat{\eta}] \tag{1}$$

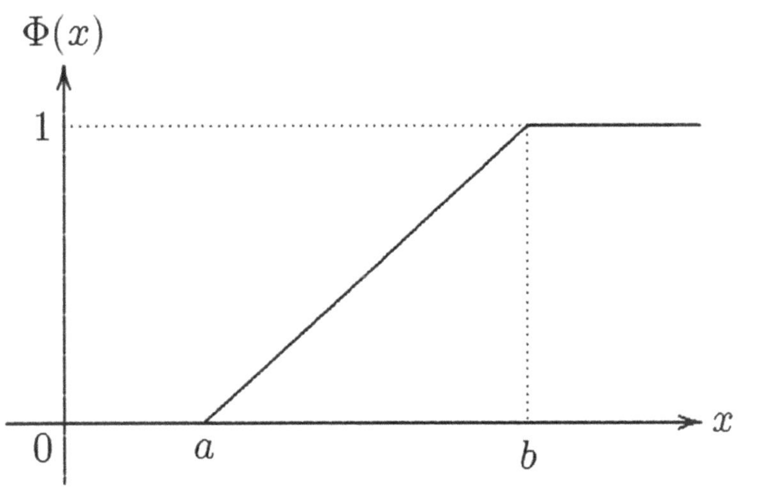

Figure 1 Linear uncertainty variable.

Lemma 1. Let $\widehat{\xi} \sim \mathcal{L}(a, b)$ be a linear uncertain variable. Then, its inverse uncertainty distribution $\Phi^{-1}(\alpha) = \frac{1}{2}[\,(1-\alpha)a + b\alpha]$, and it can be expressed as

$$E[\,\widehat{\xi}\,] = \int_0^1 \frac{1}{2}[\,(1-\alpha)a + b\alpha]\,d\alpha = \frac{a+b}{2} \tag{2}$$

Uncertain single-objective programming

Uncertain programming is a type of mathematical programming involving uncertain variables. Since an uncertain objective function $f(x, \widehat{\xi})$ cannot be directly maximized, we may maximize its expected value. Assume that x is a decision vector, $\widehat{\xi}$ is an uncertain vector, f is an objective function, and g_j are constraints functions for $j = 1, 2, \ldots, p$. Let us examine

$$\begin{cases} \max & f(x, \widehat{\xi}), \\ \text{subject to} & \begin{cases} g_j(x, \widehat{\xi}) \le 0, j = 1, 2, \ldots, p \\ x \in X \end{cases} \end{cases} \tag{3}$$

In order to obtain a decision with maximum expected objective value subject to a set of chance constraints, Liu [28] proposed that the above uncertain programming model (3) is equivalent to the crisp model (4) as follows.

Single-objective programming has been well developed and applied widely. For modelling single-objective decision-making problem with uncertain parameters, Liu [29] presented the following uncertain single-objective programming:

$$\begin{cases} \max_x & \left(E[f(x, \widehat{\xi}))]\right) \\ \text{subject to} & \\ & \mathcal{M}\left\{g_j(x, \widehat{\xi}) \le 0\right\} \ge \alpha_j, j = 1, 2, \ldots, p \end{cases} \tag{4}$$

where $f(x, \xi)$ are return functions and $g_j(x, \xi)$ are constraint functions for $j = 1, 2, \ldots, p$.

Optimal control framework

Assumption and notation

For an imperfect inventory control model, the following assumptions and notations are used.

Assumptions

- Selling price and production rates are a function of time which are also taken as control variables.
- Demand rate is dependent linearly on selling price and stock.
- Rate of defectiveness is dependent on reliability.
- This is a single-period inventory model with finite time horizon, T.
- Defective units occur only when the item is effectively produced and there is no repair or replacement of defective units over the period $[\,0, T]$,
- Unit production cost is produced quantity dependent.
- The salvage value price of the finish stock is the same as the selling price.
- The holding cost, raw material cost, and initial selling price are uncertain in nature.
- Shortage is not allowed.
- The inventory level is assumed to be a continuous function of time.

Notations

T = time length of the cycle
$U(t)$ = production rate at time t which is a control variable
$s(t)$ = selling price at time t which is a control variable
$\widehat{s_0} = \mathcal{L}(s_{01}, s_{02})$, linear uncertain selling price at time $t = 0$
$X(t)$ = inventory level at time t
$D(t)$ = demand rate at time t
$C_u(R) = M + Ne^{k\frac{(R-R_{\min})}{(R_{\max}-R)}}$ = development cost dependent on reliability parameter
$\widehat{C}_{u0} = \mathcal{L}(c_{u01}, c_{u02})$, linear uncertain raw material cost
C_{u2} = wear-tear cost
$C_p(R,t)$ = unit production cost which depends on production rate, raw material cost, development cost, and wear-tear cost
R = reliability parameter
δ = reliability parameter
$\delta\, e^{(1-R)}$ = reliability-dependent defective rate
$R_{\min}$ = minimum value of the reliability parameter
$R_{\max}$ = maximum value of the reliability parameter
M = the fixed cost like labor and energy costs which is independent of reliability factor R
N = the cost of technology, resource, and design complexity for production when $R = R_{\min}$
$\widehat{h} = \mathcal{L}(h_1, h_2)$, linear uncertain holding cost per unit item

Proposed optimal dynamic pricing and reliability-dependent imperfect production-inventory model in uncertain environment

In this model, single-item optimal dynamic pricing and reliability-dependent imperfect production-inventory problem is considered. Here, the items are produced at a variable rate $U(t)$ of which $\delta e^{(1-R)}$ (where $\delta e^{(1-R)} < 1$) is a reliability-dependent defective rate. The reliability R decreases the defective rate and improved the quality of the product. The defective item is treated as disposal part. The demand rate $D(t)$ of customers meet from the inventory. Moreover, $s(t)$ is the selling price of the item
Therefore, the differential equation for stock level $X(t)$ representing the above system during a fixed time horizon, T, is

$$\dot{X}(t) = (1 - \delta e^{(1-R)})U(t) - D(t) \tag{5}$$

$$\text{and} \quad D(t) = \{d_0 + d_1X(t) - d_2s(t)\} \tag{6}$$

where (.) denotes differentiation.

The unit production cost is considered as a function of produced quantity, raw material cost, wear-tear cost, and development cost. So, the unit production cost is

$$\widehat{C}_p(R,t) = \left(\widehat{C}_{u0} + \frac{C_{u_1}(R)}{U(t)} + C_{u2}U(t)\right) \tag{7}$$

Then, the profit function is

$$\text{Maximize } \widehat{J} = \int_0^T [\, s(t)\{d_0 + d_1x(t) - d_2s(t)\} - \widehat{h}X(t) - (\widehat{C}_{u_0}U(t) + C_{u_1}(R) + C_{u_2}U^2(t))]\, dt + s(t)X(T) \tag{8}$$

subject to $X(0) = 0$ and $X(T) = X_T$

From Equation 5, we get

$$U(t) = \frac{\dot{X}(t) + d_0 + d_1 x(t) - d_2 s(t)}{\left(1 - \delta e^{(1-R)}\right)} \tag{9}$$

Putting the value of $U(t)$ in (8), we get

$$\begin{aligned} \text{Maximize } \widehat{J} = \int_0^T \Bigg[& s(t)\{d_0 + d_1 x(t) - d_2 s(t)\} - \widehat{h}X(t) \\ & - \widehat{C}_{u0}\left\{\frac{\dot{X}(t) + d_0 + d_1 x(t) - d_2 s(t)}{\left(1 - \delta e^{(1-R)}\right)}\right\} - C_{u_1}(R) \\ & - C_{u_2}\left\{\frac{\dot{X}(t) + d_0 + d_1 x(t) - d_2 s(t)}{\left(1 - \delta e^{(1-R)}\right)}\right\}^2 \Bigg] dt + s(t)X(T) \end{aligned} \tag{10}$$

subject to $X(0) = 0$ and $X(T) = X_T$

The above problem is an optimal control problem with control variables $U(t)$ and $s(t)$ in uncertain environment.

The equivalent crisp production control problem

Using Theorem 3 and Lemma 1, we converted the uncertain production inventory control problem (10) into an equivalent crisp production control problem.

$$\begin{aligned} \text{Maximize } E(\widehat{J}) = & \int_0^T [s(t)D(t) - E(\widehat{h})X(t) - (E(\widehat{c}_{u_0})U(t) + C_{u_1}(R) + C_{u_2}U^2(t))]\, dt \\ & + s(T)X(T) \end{aligned} \tag{11}$$

subject to $X(0) = 0$ and $X(T) = X_T$

where $E(\widehat{c}_{u0}) = \frac{c_{u01}+c_{u02}}{2}$, $E(\widehat{h}) = \frac{h_1+h_2}{2}$ and $E(\widehat{s}_0) = \frac{s_{01}+s_{02}}{2}$ (using Lemma 1)

Solution method

The above control problem is solved using Euler-Lagrange equation [34]. Using this theory, we construct the Lagrangian function as

$$\begin{aligned} L = & s(t)\{d_0 + d_1 x(t) - d_2 s(t)\} - E(\widehat{h})X(t) - E(\widehat{C}_{u0})\left\{\frac{\dot{X}(t) + d_0 + d_1 x(t) - d_2 s(t)}{\left(1 - \delta e^{(1-R)}\right)}\right\} \\ & - C_{u_1}(R) - C_{u_2}\left\{\frac{\dot{X}(t) + d_0 + d_1 x(t) - d_2 s(t)}{\left(1 - \delta e^{(1-R)}\right)}\right\}^2 \end{aligned} \tag{12}$$

Therefore, our objective is to find out the optimal path of $U(t)$ and $s(t)$ such that L as well as J is maximum.

Lemma 2. The Lagrange function L is maximum in $[0, T]$ for optimum production rate $U(t) = Ce^{\frac{d1}{2d_2}t} - D$

$$\text{and selling price } s(t) = \left(E(\widehat{s}_0) + \frac{2d_2 E(\widehat{h})}{d_1}\right) e^{\frac{d1}{2d_2}t} - \frac{2d_2 E(\widehat{h})}{d_1} \tag{13}$$

Proof. For finding the optimal path $U(t)$ and $s(t)$, Euler-Lagrange equation is $U(t)$ and $s(t)$, we get

$$\frac{\partial L}{\partial X(t)} - \frac{d}{dt}\left(\frac{\partial L}{\partial \dot{X}(t)}\right) = 0$$

$$\Rightarrow d_1 s(t) - E(\widehat{h}) - \frac{d_1 E(\widehat{c}_{u0})}{1-\delta e^{(1-R)}}$$

$$- \frac{2c_{u_2} d_1}{\left(1-\delta e^{(1-R)}\right)^2}\left\{\dot{X}(t) + d_0 + d_1 X(t) - d_2 s(t)\right\}$$

$$+ \frac{d}{dt}\left[\frac{E(\widehat{c}_{u0})}{1-\delta e^{(1-R)}}\right.$$

$$\left. + \frac{2c_{u_2}}{\left(1-\delta e^{(1-R)}\right)^2}\left\{\dot{X}(t) + d_0 + d_1 X(t) - d_2 s(t)\right\}\right] = 0$$

and

$$\frac{\partial L}{\partial s(t)} - \frac{d}{dt}\left(\frac{\partial L}{\partial \dot{s}(t)}\right) = 0.$$

$$\Rightarrow -2d_2 s(t) + \frac{d_2 E(\widehat{c}_{u0})}{1-\delta e^{(1-R)}}$$

$$+ \frac{2c_{u_2} d_2}{\left(1-\delta e^{(1-R)}\right)^2}\left\{\dot{x}(t) + d_0 + d_1 x(t)\right\}$$

$$- \frac{2c_{u_2} d_2^2}{\left(1-\delta e^{(1-R)}\right)^2} s(t) - \frac{d}{dt}(0) = 0.$$

From Equation 12,

$$\dot{X}(t) + d_0 + d_1 x(t) - d_2 s(t) = \frac{\left(1-\delta e^{(1-R)}\right)^2}{2c_{u_2} d_2}\left[2d_2 s(t) - \frac{E(\widehat{c}_{u0}) d_2}{1-\delta e^{(1-R)}}\right]$$

Using in (13),

$$\dot{s}(t) - \frac{d_1}{2d_2} s(t) = E(\widehat{h})$$

which is a linear equation whose solution is

$$s(t) = \left(E(\widehat{s}_0) + \frac{2d_2 E(\widehat{h})}{d_1}\right) e^{\frac{d1}{2d_2}t} - \frac{2d_2 E(\widehat{h})}{d_1}$$

Solving the above equation, we get the optimal stock level

$$X(t) = F + Ge^{\left(\frac{d_1}{2d_2}\right)t} + He^{-d_1 t}$$

Using $X(t)$ in (9), we have

$$U(t) = Ce^{\frac{d1}{2d_2}t} - D$$

$$
\begin{aligned}
\text{where } A &= \left\{d_2 + \frac{(1-\delta e^{(1-R)})^2}{c_{u_2}}\right\}\left(E(\widehat{s}_0) + \frac{2d_2E(\widehat{h})}{d_1}\right),\\
B &= \frac{2d_2E(\widehat{h})}{d_1}\left\{d_2 + \frac{\left(1-\delta e^{(1-R)}\right)^2}{c_{u_2}}\right\} + d_0 + \frac{E(\widehat{c}_{u_0})\left(1-\delta e^{(1-R)}\right)}{2c_{u_2}},\\
C &= \left(E(\widehat{s}_0) + \frac{2d_2E(\widehat{h})}{d_1}\right)\frac{\left(1-\delta e^{(1-R)}\right)}{c_{u_2}},\\
D &= \frac{2d_2E(\widehat{h})}{d_1}\frac{\left(1-\delta e^{(1-R)}\right)}{c_{u_2}} + \frac{E(\widehat{c}_{u_0})}{2c_{u_2}},\\
F &= \frac{B}{d_1},\\
G &= \frac{2Ad_2}{d_1(1+2d_2)},\\
\text{and } H &= X_0 - \frac{2Ad_2}{d_1(1+2d_2)} - \frac{B}{d_1}.
\end{aligned}
$$

The maximum value of J is given by

$$
\begin{aligned}
J &= \int_0^T \left[s^*(t)D(t) - E(\widehat{h})X^*(t) - (E(\widehat{c}_{u_0})U^*(t) + C_{u_1}(R) + C_{u_2}(U^*(t))^2\right]dt + s(t)X^*(T)\\
&= d_0\left[\left(E(\widehat{s}_0) + \frac{2d_2E(\widehat{h})}{d_1}\right)\frac{d1}{2d_2}\left(e^{\frac{d1}{2d_2}T} - 1\right) - \frac{2d_2E(\widehat{h})}{d_1}T\right]\\
&\quad + d_1\left[\left(E(\widehat{s}_0) + \frac{2d_2E(\widehat{h})}{d_1}\right)\left\{F\frac{2d_2}{d_1}\left(e^{\frac{d1}{2d_2}T} - 1\right)\right.\right.\\
&\quad \left. + G\frac{d_2}{d_1}\left(e^{\frac{d1}{d_2}T} - 1\right) + \frac{2Hd_2}{d_1(1-2d_2)}\left(e^{(\frac{d1}{d_2}-d_1)T} - 1\right)\right\}\\
&\quad \left. - \frac{2d_2E(\widehat{h})}{d_1}\left\{FT + G\frac{2d_2}{d_1}\left(e^{\frac{d1}{2d_2}T} - 1\right) + \frac{H}{d_1}\left(1 - e^{d_1T}\right)\right\}\right]\\
&\quad - d_2\left[\left(E(\widehat{s}_0) + \frac{2d_2E(\widehat{h})}{d_1}\right)^2\frac{d_2}{d_1}\left(e^{\frac{d1}{d_2}T} - 1\right) - \frac{8d_2^2E(\widehat{h})}{d_1^2}\left(E(\widehat{s}_0) + \frac{2d_2E(\widehat{h})}{d_1}\right)\left(e^{\frac{d1}{2d_2}T} - 1\right)\right.\\
&\quad \left. + \frac{4d_2^2E(\widehat{h})^2}{d_1^2}T\right] - E(\widehat{h})\left[FT + G\frac{2d_2}{d_1}\left(e^{\frac{d1}{2d_2}T} - 1\right) + \frac{H}{d_1}\left(1 - e^{d_1T}\right)\right]\\
&\quad - E(\widehat{c}_{u_0})\left[\frac{2Cd_2}{d_1}\left(e^{\frac{d1}{2d_2}T} - 1\right) - DT\right] - \left\{M + Ne^{k\frac{(R-R_{\min})}{(R_{\max}-R)}}\right\}T - c_{u_2}\left[C^2\frac{d_2}{d_1}\left(e^{\frac{d1}{d_2}T} - 1\right)\right.\\
&\quad \left. - \frac{4CDd_2}{d_1}\left(e^{\frac{d1}{2d_2}T} - 1\right) + D^2T\right] + s(t)X(T) \qquad (14)
\end{aligned}
$$

□

Particular case when demand depends only on selling price in uncertain environment

In this case, the stock depends only on the selling price. The demand rate $D(t)$ of customers meets from the inventory.

Therefore, the differential equation for stock level $X(t)$ representing the above system during a fixed time horizon, T, is

$$\dot{X}(t) = (1 - \delta e^{(1-R)})U(t) - D(t) \qquad (15)$$

$$\text{and } \quad D(t) = \{d_0 - d_2s(t)\} \qquad (16)$$

where (.) denotes differentiation.

The unit production cost is considered as a function of produced quantity, raw material cost, wear-tear cost, and development cost. So, the unit production cost is

$$\widehat{C}_p(R,t) = \left(\widehat{C}_{u0} + \frac{C_{u_1}(R)}{U(t)} + C_{u2}U(t)\right) \tag{17}$$

Then, the profit function is

$$\begin{aligned} \text{Maximize } \widehat{J} = \int_0^T \Big[& s(t)\{d_0 - d_2 s(t)\} - \widehat{h}X(t) \\ & - (\widehat{C}_{u_0}U(t) + C_{u_1}(R) + C_{u_2}U^2(t))\Big]\, dt \end{aligned} \tag{18}$$

subject to $X(0) = 0$ and $X(T) = X_T$

From Equation 15, we get

$$U(t) = \frac{\dot{X}(t)+d_0-d_2 s(t)}{(1-\delta e^{(1-R)})} \tag{19}$$

Putting the value of $U(t)$ in (18), we get

$$\begin{aligned} \text{Maximize } \widehat{J} = \int_0^T \Bigg[& s(t)\{d_0 - d_2 s(t)\} - \widehat{h}X(t) \\ & - \widehat{C}_{u0}\left\{\frac{\dot{X}(t) + d_0 - d_2 s(t)}{\left(1-\delta e^{(1-R)}\right)}\right\} - C_{u_1}(R) \\ & - C_{u_2}\left\{\frac{\dot{X}(t) + d_0 - d_2 s(t)}{\left(1-\delta e^{(1-R)}\right)}\right\}^2\Bigg] dt + s(t)X(T) \end{aligned} \tag{20}$$

subject to $X(0) = 0$ and $X(T) = X_T$

The above problem is an optimal control problem with uncertain control variables $U(t)$, uncertain holding cost h, and uncertain initial selling price, i.e., $s_0 = s(0)$.

The equivalent crisp production control problem

Using Theorem 3 and Lemma 1, we converted the uncertain production inventory control problem (20) into an equivalent crisp production control problem.

$$\begin{aligned} \text{Maximize } E(\widehat{J}) = & \int_0^T \left[s(t)D(t) - E(\widehat{h})X(t) - (E(\widehat{c}_{u_0})U(t) + C_{u_1}(R) + C_{u_2}U^2(t))\right] dt \\ & + s(T)X(T) \end{aligned} \tag{21}$$

subject to $X(0) = 0$ and $X(T) = X_T$

where $E(\widehat{c}_{u0}) = \frac{c_{u01}+c_{u02}}{2}$, $E(\widehat{h}) = \frac{h_1+h_2}{2}$, and $E(\widehat{s}_0) = \frac{s_{01}+s_{02}}{2}$ (using Lemma 1).

Solution method

The above control problem is solved using Euler-Lagrange equation [34]. Using this theory, we construct the Lagrangian function as

$$\begin{aligned} L = {} & s(t)\{d_0 - d_2 s(t)\} - E(\widehat{h})X(t) - E(\widehat{C}_{u_0})\left\{\frac{\dot{X}(t) + d_0 - d_2 s(t)}{(1-\delta e^{(1-R)})}\right\} \\ & - C_{u_1}(R) - C_{u_2}\left\{\frac{\dot{X}(t) + d_0 - d_2 s(t)}{\left(1-\delta e^{(1-R)}\right)}\right\}^2 \end{aligned}$$

Therefore, our objective is to find out the optimal path of $U(t)$ and $s(t)$ such that L as well as J is maximum.

Lemma 3. The Lagrangian function L is maximum in $[0, T]$ for optimum production rate $U(t) = \frac{R_1 t+S}{1-\delta}$ and selling price $s(t) = \frac{E(\widehat{h})d_2 t+Q}{P}$

Proof. For finding the optimal path $U(t)$ and $s(t)$ Euler-Lagrange equation of $U(t)$ and $s(t)$, we get

$$\frac{\partial L}{\partial X(t)} - \frac{d}{dt}\left(\frac{\partial L}{\partial \dot{X}(t)}\right) = 0$$

$$\Rightarrow \; -E(\widehat{h}) - \frac{2c_{u_2}}{(1-\delta)^2}\left\{\dot{X}(t) + d_0 - d_2 s(t)\right\} = 0 \qquad (22)$$

and

$$\frac{\partial L}{\partial s(t)} - \frac{d}{dt}\left(\frac{\partial L}{\partial \dot{s}(t)}\right) = 0.$$

$$\Rightarrow \; d_0 - d_2 s(t) + \frac{d_2 E(\widehat{c}_{u_0})}{1-\delta} + \frac{2d_2 c_{u_2}}{(1-\delta)^2}\left\{\dot{X}(t) + d_0 - d_2 s(t)\right\} = 0 \qquad (23)$$

Solving Equations 22 and 23, we get

$$s(t) = \frac{E(\widehat{h})d_2 t + Q}{P}, \qquad (24)$$

$$U(t) = \frac{R_1 t + S}{1-\delta}, \qquad (25)$$

and

$$x(t) = Yt^2 + Zt,$$

where

$$P = 2d_2 + \frac{2c_{u_2}^2 d_2^2}{(1-\delta)^2},$$

$$Q = \frac{2c_{u_2}^2 d_2^2}{T(1-\delta)^2}\left\{X_T - \frac{E(\widehat{h})(1-\delta)^2 T^2}{4c_{u_2}}\right\} + d_0 + \frac{d_2 E(\widehat{c}_{u_0})}{1-\delta} + \frac{2c_{u_2}^2 d_2^2}{(1-\delta)^2},$$

$$R_1 = \frac{E(\widehat{h})(1-\delta)^2}{2c_{u_2}} - \frac{d_2^2 E(\widehat{h})}{P},$$

$$S = \frac{1}{T}\left\{X_T - \frac{E(\widehat{h})(1-\delta)^2 T^2}{4c_{u_2}}\right\} - d_0 - \frac{d_2 Q}{P(1-\delta)},$$

$$Y = \frac{E(\widehat{h})(1-\delta)^2}{4c_{u_2}},$$

and

$$Z = \frac{1}{T}\left\{X_T - \frac{E(\widehat{h})(1-\delta)^2 T^2}{4c_{u_2}}\right\}$$

Table 1 The uncertain input data

Uncertain parameters	Linear uncertain values (US$)	Expectation of uncertain parameters (US$)
Initial selling price ($\widehat{s_0}$)	(8, 12)	10
Holding cost ($\widehat{h}$)	(0.4, 0.8)	0.6
Raw material cost ($\widehat{C}_{u0}$)	(1, 3)	2

Table 2 The crisp input data

M	N (US$)	c_{u2} (US$)	k (US$)	$R_{\min}$	$R_{\max}$	δ	d_0	d_1	d_2	X_0	T
20	5	0.03	0.5	0.5	0.95	0.02	20	0.03	0.3	0	8

The maximum value of J is given by

$$J = \int_0^T \left[s^*(t)D(t) - E(\widehat{h})X^*(t) - (E(\widehat{c}_{u_0})U^*(t) + C_{u_1}(R) + C_{u_2}(U^*(t))^2 \right] dt + s^*(t)X^*(T)$$
$$= \frac{d_0}{P}\left[\frac{E(\widehat{h})d_2T^2}{2} + QT\right] + \frac{d_2}{P^2}\left[\frac{E(\widehat{h})^2 d_2^2 T^3}{3} + \frac{2QE(\widehat{h})d_2T^2}{2} + Q^2T\right] - E(\widehat{h})\left[\frac{YT^3}{3} + \frac{ZT^2}{2}\right]$$
$$- \frac{E(\widehat{c}_{u_0})}{1-\delta}\left[\frac{RT^2}{2} + ST\right] - c_{u_1}(R) - \frac{c_{u_2}}{(1-\delta)^2}\left[\frac{R^2T^3}{3} + RST^2 + S^2T\right] + s(t)X(T) \qquad (26)$$

□

Numerical illustration

To illustrate the above reliability-dependent uncertain production inventory model numerically, we consider uncertain and crisp input data given in Tables 1 and 2, respectively. For these uncertain and crisp input data, the optimum demand rate $D^*(t)$, production rate $u^*(t)$, and stock level $X^*(t)$ are given in Table 3 and shown in Figures 2,3,4, respectively.

Uncertain input data

The uncertain input data for advertisement cost, holding cost, and selling price and their expectation values by using Lemma 3 are given in Table 1.

Crisp input data

The other crisp inventory data are given in Table 2.

Optimum result

For the above uncertain and crisp input data, the optimum demand rate $D^*(t)$, production rate $u^*(t)$, and stock level $X^*(t)$ are given in Table 3 and shown in Figures 2,3,4, respectively. Also, the optimum results are $R^* = 0.7521$, $J^* =$ US$5,540.72, $X^*(t) = 3,095.54$

Table 3 Optimal values of demand, production, and stock level with relation to time

t	$J^*(t)$	$X^*(t)$	$U^*(t)$	$D^*(t)$	$s^*(t)$
0	0	0	291.46	17	10
1	−264.90	280.11	328.09	25.08	11.13
2	−386.49	589.54	366.61	34.00	12.31
3	−312.01	927.50	407.09	43.76	13.56
4	20	1,295.73	449.66	54.41	14.87
5	682.33	1,695.36	494.41	65.91	16.25
6	1,757.00	2,127.56	541.45	78.52	17.69
7	3,339.92	2,593.92	590.90	92.05	19.22
8	5,540.72	3,095.54	642.88	106.62	20.82

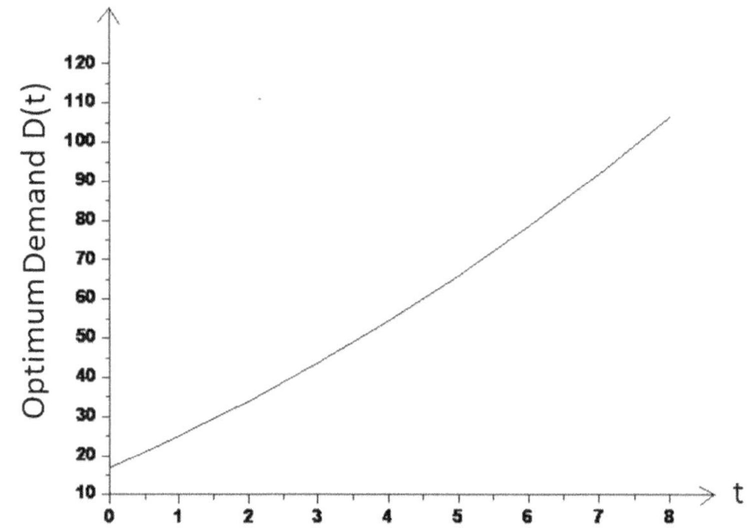

Figure 2 Optimum demand versus time.

units, $s^*(t) = 20.82$ units, $D^*(t) = 106.62$ units, and $U^*(t) = 642.88$ units. The optimum profit versus time are also shown in Figure 5. The optimum profit versus reliability are given in Figure 6. The optimum selling price versus time are given in Figure 7.

Numerical illustration for particular cases

To illustrate the above reliability-dependent production inventory model numerically, we consider the previous uncertain and crisp input data which are already given in Tables 1 and 2, respectively.

Optimum result

For the previous uncertain and crisp input data and using Lemma 3, the optimum result is obtained as $R^* = 0.5, J^* = \text{US\$}108,296.10, X^*(t) = 3,095.54$ units, $s^*(t) = 48.16$ units, $D^*(t) = 5.55$ units, and $U^*(t) = 402.71$ units.

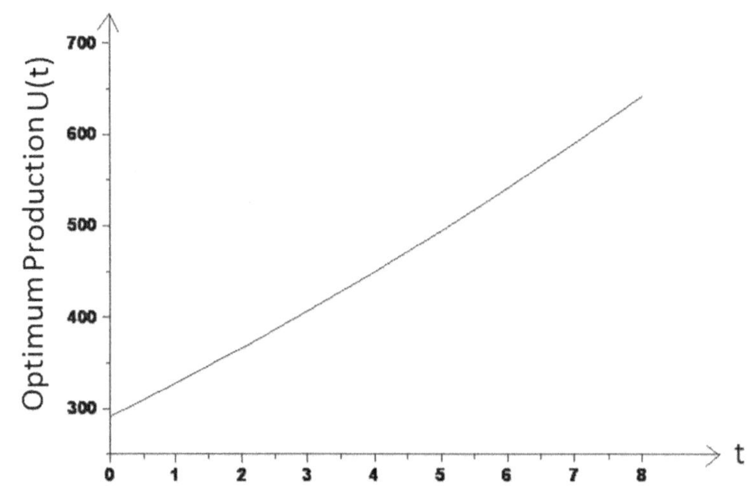

Figure 3 Optimum production rate versus time.

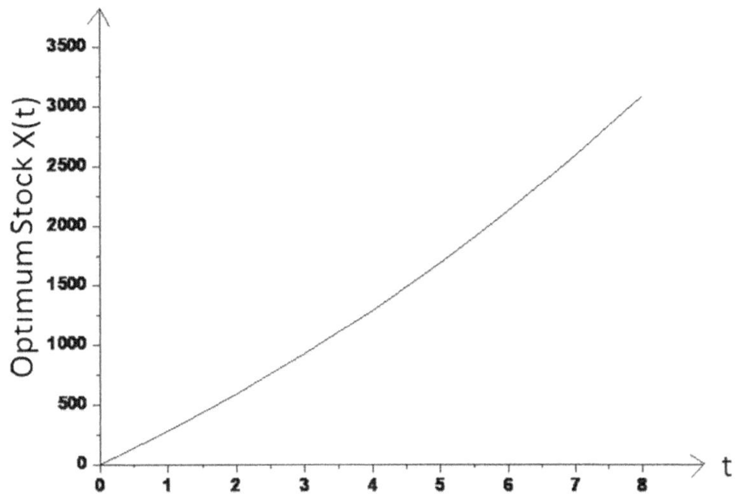

Figure 4 Optimum stock versus time.

Practical implications

A washing machine fails, a car battery goes dead, a toaster oven electrical plug burns, a water heater leaks, a floppy disk drive goes bad, a TV remote control quits functioning, a stergo amplifier quits, an automobile engine starter fails, and a house roof leaks are commonly experienced by the customers. These failures, however, are much more significant in both their economical and safety effects.

From the above examples, one can conclude that the impact of product and system failures varies from minor inconvenience and costs to personal injury, significant economical loss, and death. Causes of these failures include bad engineering design, faculty construction, human error, improper use, and lack of protection against excessive environmental stress. Under current laws and recent court decisions, the manufacturing can be held liable for failure to account properly for product safety and reliability. Moreover,

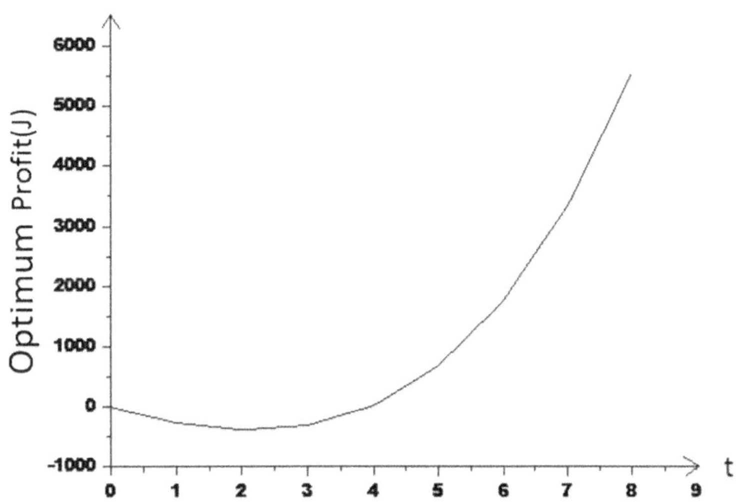

Figure 5 Optimum profit versus time.

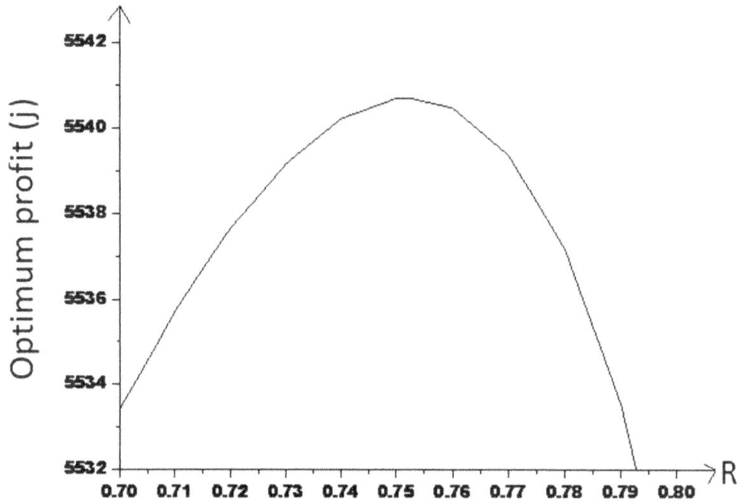

Figure 6 Optimum profit versus reliability.

the demand of these product are advertisement dependent. Also, the selling price fluctuated day by day, so it is taken as uncertain in nature. So, this model is practically applicable for such type of imperfect production inventory problems.

Conclusion

The present paper first time deals with the optimum production and advertising policy for a single-item production inventory system with reliability-dependent defective rate, reliability- and production rate-dependent unit production cost, depreciation rate of sales, marginal selling price of finish stock, and dynamic demand under uncertain environment. Also, some ideas such as (i) optimal control production problem for defective multi-items, (ii) advertisement-dependent demand, (iii) dynamic production function,

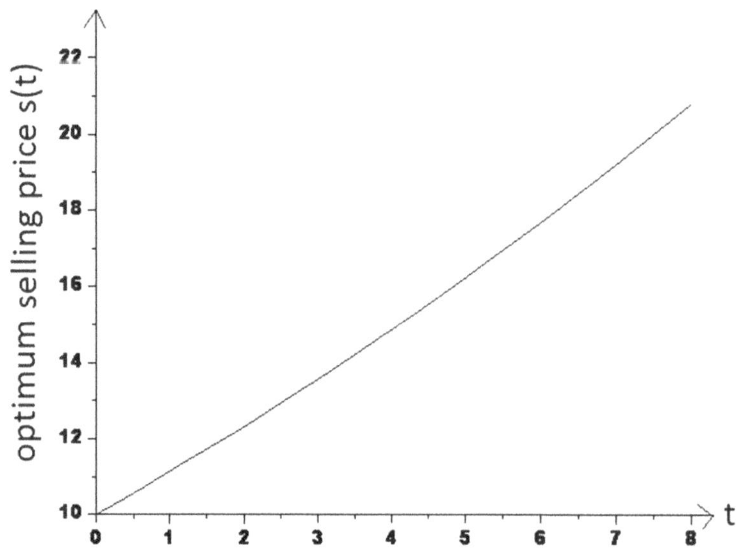

Figure 7 Optimum selling price versus time.

and (iv) production quantity-dependent unit cost. In the solution approach, the new ideas are reliability and uncertain expectation. In this connection, the theories 1, 2, and 3 and Lemma 1 for uncertain variable are established and used in optimal control problem. The formulation and analysis presented here can be extended to other production-inventory problems with different types of demand, advertisement, deterioration, defective, price discount, etc.

Acknowledgements
Dr. Jayanta Kumar Dey thanks the Minor Research Project (PSW-138/09-10, Government of India) for the financial support to do this research work.

Author details
[1]Department of Mathematics, National Institute of Technology, Durgapur, West Bengal 713209, India. [2]Department of Mathematics, Mugberia Gangadhar Mahavidyalaya, Bhupatinagar Purba Medinipur, West Bengal 721425, India. [3]Department of Mathematics, Mahisadal Raj College, Purba-Medinipur, West Bengal 721628, India.

References

1. Levin, RT, McLaughlin, CP, Lamone, RP, Kottas, JF: Production operations management: contemporary policy for managing operating systems. McGraw-Hill, New York (1972)
2. Gupta, R, Vrat, P: Inventory model for stock-dependent consumption rate. Opsearch. **23**, 19–24 (1986)
3. Baker, RC, Urban, TL: A deterministic inventory system with an inventory-level-dependent demand rate. J. Oper. Res. Soc. **39**, 823–831 (1988)
4. Datta, TK, Pal, AK: A note on an inventory model with inventory-level-dependent demand rate. J. Oper. Res. Soc. **41**, 971–975 (1990)
5. Urban, TL: Inventory models with inventory-level-dependent demand: a comprehensive review and unifying theory. Eur. J. Oper. Res. **162**, 792–804 (2005)
6. Ray, J, Goswami, A, Chaudhuri, KS: On an inventory model with two levels of storage and stock-dependent demand rate. Int. J. Syst. Sci. **29**, 249–254 (1998)
7. Glock, CH, Ries, JM, Schwindl, K: A note on optimal ordering policy for stock-dependent demand under progressive payment scheme. Eur J. Oper. Res. **232**, 423–426 (2014)
8. Zhou, YW, Min, J, Goyal, SK: Supply-chain coordination under an inventory-level-dependent demand rate. Int. J. Prod. Econ. **113**, 518–527 (2008)
9. Sajadieh, MS, Thorstenson, A, Jokar, MA: An integrated vendor-buyer model with stock-dependent demand. Transportation Res. Part E: Logistics Transportation Rev. **46**, 963–974 (2010)
10. Hsieh, TP, Dye, CY, Ouyang, LY: Optimal lot size for an item with partial backlogging rate when demand is stimulated by inventory above a certain stock level. Math. Comput. Model. **51**, 13–32 (2010)
11. Dey, JK, Mondal, SK, Maiti, M: Two storage inventory problem with dynamic demand and interval valued lead-time over finite time horizon under inflation and time-value of money. **185**, 170–194 (2008)
12. Pando, V, Garcΰşa-Laguna, J, San-Josqe, LA: Optimal policy for profit maximising in an EOQ model under non-linear holding cost and stock-dependent demand rate. Int. J. Syst. Sci. **43**, 2160–2171 (2012)
13. Kim, T, Glock, CH: A multi-stage joint economic lot size model with lead time penalty costs. Comput. Ind. Eng. **66**, 133–146 (2013)
14. Jana, DK, Maity, K, Das, B, Roy, TK: A fuzzy simulation via contractive mapping genetic algorithm approach to an imprecise production inventory model under volume flexibility. J. Simul. **7**, 90–100 (2013)
15. Maity, K: A supply-chain production inventory model with warehouse facilities under fuzzy environment. In: Kahraman, C, Oztaysi, B (eds.) Supply Chain Management Under Fuzziness, pp. 517–551. Springer, Heidelberg (2014)
16. Mettas, A: Reliability allocation and optimization for complex system, reliability and maintainability symposium. Relia Soft Corporation, Los Angeles (2000)
17. Panda, D, Maiti, M: Multi-item inventory models with price dependent demand under flexibility and reliability consideration and imprecise space constraint: a geometric programming approach. Math. Comput. Model. **49**, 1733–1749 (2009)
18. Sarkar, B: An inventory model with reliability in an imperfect production process. Appl. Math. Comput. **218**, 4881–4891 (2012)
19. Porteus, EL: Optimal lot sizing, process quality improvement and setup cost reduction. Oper. Res. **34**, 137–144 (1986)
20. Salameh, MK, Jaber, MY: Economic production quantity model for items with imperfect quality. Int. J. Prod. Econ. **64**, 59–64 (2000)
21. Panda, D, Kar, S, Maity, K, Maity, M: A single period inventory model with imperfect production and stochastic demand under chance and imprecise constraints. Eur. J. Oper. Res. **188**, 121–139 (2008)
22. Maity, AK, Maity, K, Maiti, M: Optimal remanufacturing control policy with defective items. International J. Oper. Res. **6**, 500–518 (2009)
23. Mandal, S, Maity, K, Mondal, S, Maiti, M: Optimal production inventory policy for defective items with fuzzy time period. Appl. Math. Model. **34**, 810–822 (2010)
24. Sana, SS: A production-inventory model of imperfect quality products in a three-layer supply chain. Decis. Support Syst. **50**, 539–547 (2011)

25. Khan, M, Jaber, MY: Optimal inventory cycle in a two-stage supply chain incorporating imperfect items from suppliers. Int. J. Oper. Res. **10**, 442–457 (2011)
26. Yadav, D, Singh, SR, Kumari, R: Three-stage supply chain coordination under fuzzy random demand and production rate with imperfect production process. Int. J. Oper. Res. **16**, 421–447 (2013)
27. Krishnamoorthi, C, Panayappan, S: An EPQ model for an imperfect production system with rework and shortages. Int. J. Oper. Res. **17**, 104–124 (2013)
28. Liu, B: Uncertainty Theory. 2nd edn. Springer-Verlag, Berlin (2007)
29. Liu, B: Some research problems in uncertainty theory. J. Uncertain Syst. **3**, 3–10 (2009)
30. Liu, B: Uncertainty Theory: a Branch of Mathematics for Modeling Human Uncertainty. Springer, Berlin (2010)
31. You, C: Some convergence theorems of uncertain sequences. Math. Comput. Model. **49**, 482–487 (2009)
32. Liu, B: Fuzzy process, hybrid process and uncertain process. J. Uncertain Syst. **2**, 3–16 (2008)
33. Jana, DK, Maity, K, Roy, TK, J. Uncertainty Anal. Appl: A three-layer supply chain integrated production-inventory model under permissible delay in payments in uncertain environments. **1**, 1–17 (2013)
34. Macki, J: Strauss, A: Introduction to Optimal Control. Springer, Berlin (1982)
35. Gabriel, GA, Ragsdell, KM: The generalized reduced gradient method. AMSE J. Eng. Ind. **99**, 384–400 (1977)

9

Towards uncertain network optimization

Jin Peng[1*], Bo Zhang[2] and Shengguo Li[1]

*Correspondence:
pengjin01@tsinghua.org.cn
[1] Institute of Uncertain Systems,
Huanggang Normal University,
Huanggang, Hubei 438000, China
Full list of author information is
available at the end of the article

Abstract

Uncertain network optimization is the study of network optimization with uncertain data which we often meet in decision making under the presence of uncertainties. The main purpose of this manuscript is to present a state-of-the-art review on the recent advances in uncertain network optimization and to show the general uncertain network optimization models based on an uncertainty theory. Some classical network optimization topics in an uncertain environment are revisited, and some challenging topics in future research are addressed in the field of uncertain network optimization. This paper divides uncertain network optimization into two main directions. One line of research deals with network optimization problems with an uncertain topology structure, and another line of research concerns the handling of network optimization with uncertain weights.

Keywords: Uncertain network; Uncertain graph; Network optimization; Uncertainty theory; Uncertain programming

Introduction

In real life, we are faced with so many networks such as road network, telecommunication network, television network, computer network, logistics network, social network, and so on. In practice, a lot of network optimization problems [1,2] are issued from a wide variety of the networks. Generally speaking, network optimization is to study the problems on how to design, manage, control, and optimize the network efficiently. As we can see, there are many variations of typical network optimization problems such as shortest path problem [3], minimum spanning tree [4], transportation problem [5], assignment problem [6], maximum flow problem [7], and so forth. These attractive problems have motivated many researchers to model the network optimization problems mathematically.

In the abovementioned research, the network optimization problems were usually investigated in a deterministic environment, in which the capacities of the arcs, the costs of the transportation, etc., are treated as positive crisp values. However, some indeterminacy factors might occur in the problems. In fact, due to the lack of adequate sample data, or the detail sample data are not easy to get because of economic reasons or technical difficulties, the capacities of the arcs and the costs of the transportation are not sharply known in advance. As a result, the capacity and the cost are described by some empirical data such as 'about 2 tons' and 'approximately 3 dollars,' respectively. In this situation, it is not suitable to employ the classical models and algorithms to study the network optimization problems directly.

Probabilistic approaches are borrowed in the analysis of network optimization problems. Moreover, some researchers employed probability theory to investigate the stochastic network optimization problems, the stochastic versions of the network optimization problems. Williams [8] proposed a stochastic transportation model, in which the demands were supposed to be random variables. Donath [9] investigated random assignment problems. Hall [10] concerned a problem of finding the least expected travel time path between two vertices in a network with travel times that are both random and time-dependent. Nawathe and Rao [11] provided a general approach for the maximum flow problem in probabilistic communication network. Mulvey and Vladimirou [12] described the stochastic network optimization models for investment planning. Focusing on communication and queueing systems, Neely [13] presented a modern theory of analysis, control, and optimization for stochastic networks. In [14], Peer and Sharma defined a probabilistic shortest path problem and formulated it as a linear programming problem. Ji et al. [15] proposed a new simulation-based multi-objective genetic algorithm approach to find a portfolio of reliable paths in stochastic networks.

In addition, some optimization problems were founded in the literature on fuzzy network, such as Chanas and Kolodziejczyk [16], Hanebeck and Schmidt [17], and Wu [18].

Essentially, the indeterminacy phenomena can be divided into two distinct types according to the present mathematical systems to deal with them. One is stochastic phenomena, and another is uncertain phenomena. When the sample size is too small or even no sample to estimate a probability distribution, we have to invite some domain experts to evaluate their belief degree that each event will occur. A lot of surveys showed that human beings usually estimate a much wider range of values than the object actually takes [19]. Probability theory is inappropriate in this case because it may lead to counterintuitive results [20]. In order to distinguish from randomness, this phenomenon was named uncertainty [20]. In order to deal with the uncertain phenomena, Liu [21] founded an uncertainty theory, which has become a branch of axiomatic mathematics for modeling human uncertainty.

In theoretical aspect, uncertain set (Liu [22,23]), uncertain calculus (Liu [24], Yao [25]), and uncertain differential equation (Liu [26], Chen and Liu [27], Yao [28]) have been established. From a practical aspect, uncertain finance (Liu [29], Peng [30], Peng and Yao [31]), uncertain inference (Liu [22], Gao et al. [32]), uncertain graph (Gao and Gao [33], Zhang and Peng [34,35]), etc., have also been developed. As a mathematical programming involving uncertain variables, uncertain programming was introduced by Liu [36] in 2009. After that, many scholars employed uncertain programming to some optimization problems in the real world. Yan [37] introduced portfolio selection in an uncertain environment and proposed two uncertain models. Rong [38] investigated economic order quantity for inventory based on uncertain programming. Gao [39] studied shortest path problem with uncertain lengths. Zhang and Chen [40] introduced uncertain programming to project scheduling problem. Furthermore, uncertain programming have also been used in supply chain design (Ding [41]), facility location problem (Gao [42]), parallel machine scheduling (Zhang and Meng [43]), and inverse minimum spanning tree problem (Zhang et al. [44,45]).

As an interdisciplinary subject of combinatorial optimization and uncertainty theory, we are interested in considering uncertain network optimization problem. There are two types of uncertain network optimization problems. One is network optimization with an uncertain topology structure. In order to deal with these kinds of uncertain factors, an uncertain graph was proposed by Gao and Gao [33]. After that, Euler tour [34], cycle index [46], Hamilton cycle [47], matching [48], connectivity of two vertices [35], and diameter [49] for an uncertain graph were considered by scholars. Another research field concerns network optimization problem with uncertain weights, including Chinese postman problem [50], network flow problem [51,52], optimal assignment problem [53,54], minimum spanning tree problem [55,56], and shortest path problem [39,57].

Instead of surveying all the researches on uncertain network optimization problems so far, this paper intends to provide an overview of an uncertain graph with uncertain edges and uncertain programming models which deal with uncertain network optimization problems based on an uncertain measure. The authors hope to provide the readers with some ideas to deal with uncertain factors in a graph and several basic types of uncertain network optimization models.

The rest of this paper is organized as follows. In the 'Preliminary' section, some necessary preliminary concepts and results extracted from uncertainty theory will be presented. After that, the classical network optimization problem and stochastic network optimization problems are introduced. Following that, network optimization with an uncertain topology structure is presented. In addition, we introduce network optimization with uncertain weights, including three kinds of uncertain programming models. What is more, we study some typical uncertain network optimization problems. Then a hybrid intelligent algorithm for solving the optimization models in general cases is introduced. After some topics on further research, some remarks are given in the last section.

Preliminary

As an efficient tool of modeling the behavior of uncertain phenomena, the uncertainty theory is employed to deal with uncertain network optimization problems. Uncertainty theory, proposed by Liu [21], is a branch of mathematics based on normality, duality, subadditivity, and product axioms. In this section, we are particularly interested in some related preliminaries from the uncertainty theory.

The first fundamental concept in the uncertainty theory is the uncertain measure that is interpreted as the personal belief degree (not frequency) of an uncertain event that may occur.

Let Γ be a nonempty set, and L a σ-algebra over Γ. For any $\Lambda \in L$, Liu [21] presented an axiomatic uncertain measure $\mathcal{M}\{\Lambda\}$ to express the chance that uncertain event Λ occurs. The set function $\mathcal{M}\{\cdot\}$ satisfies the following three axioms:

(*i*) (Normality) $\mathcal{M}\{\Gamma\} = 1$;

(*ii*) (Duality) $\mathcal{M}\{\Lambda\} + \mathcal{M}\{\Lambda^c\} = 1$ for any $\Lambda \in L$;

(*iii*) (Subadditivity) For every countable sequence of events $\{\Lambda_i\}$, we have $\mathcal{M}\{\cup_i \Lambda_i\} \leq \sum_i \mathcal{M}\{\Lambda_i\}$.

For the sake of convenience, the triplet $(\Gamma, L, \mathcal{M})$ is called an uncertainty space. Liu [24] defined a product of the uncertain measure by way of the fourth axiom of uncertainty theory, which makes major differences in operations between uncertainty theory and probability theory. Let $(\Gamma_k, L_k, \mathcal{M}_k)$ be uncertainty spaces for $k = 1, 2, \cdots$ Write

$$\Gamma = \Gamma_1 \times \Gamma_2 \times \cdots, \quad L = L_1 \times L_2 \times \cdots$$

Then the product uncertain measure $\mathcal{M}$ on the product σ-algebra L is defined by the following axiom:
(*iv*) (Product Axiom)

$$\mathcal{M}\left\{\prod_{k=1}^{\infty}\Lambda_k\right\} = \bigwedge_{k=1}^{\infty}\mathcal{M}_k\{\Lambda_k\}$$

where Λ_k are arbitrarily chosen events from L_k for $k = 1, 2, \cdots$, respectively.

An uncertain variable is defined by Liu [21] as a measurable function from an uncertainty space to the set of real numbers. An uncertain variable ξ can be characterized by its uncertainty distribution $\Phi : \Re \rightarrow [0, 1]$, which is defined by Liu [21] as follows

$$\Phi(x) = \mathcal{M}\left\{\gamma \in \Gamma \mid \xi(\gamma) \leq x\right\}.$$

Let ξ be an uncertain variable with uncertainty distribution Φ. Then the inverse function Φ^{-1} is naturally called the inverse uncertainty distribution of ξ.

The expected value of uncertain variable ξ is mathematically defined by Liu [21] as

$$E[\xi] = \int_0^{+\infty}\mathcal{M}\{\xi \geq r\}dr - \int_{-\infty}^{0}\mathcal{M}\{\xi \leq r\}dr$$

provided that at least one of the two integrals is finite.

Theorem 1. *(Liu [58]) Let ξ be an uncertain variable with uncertainty distribution Φ. If its expected value exists, then*

$$E[\xi] = \int_0^1 \Phi^{-1}(\alpha)d\alpha$$

where Φ^{-1} is the inverse uncertainty distribution of uncertain variable ξ.

Liu [21] introduced the independence concept of uncertain variables in the following way. The uncertain variables $\xi_1, \xi_2, \cdots, \xi_m$ are independent if and only if

$$\mathcal{M}\left\{\bigcap_{i=1}^{m}\{\xi_i \in B_i\}\right\} = \min_{1\leq i\leq m}\mathcal{M}\{\xi_i \in B_i\}$$

for any Borel sets $B_1, B_2, \cdots, B_m$ of $\Re$.

A real-valued function of multiple variables $f(x_1, x_2, \cdots, x_n)$ is said to be strictly increasing if

$$f(x_1, x_2, \cdots, x_n) \leq f(y_1, y_2, \cdots, y_n)$$

whenever $x_i \leq y_i$ for $i = 1, 2, \cdots, n$ and

$$f(x_1, x_2, \cdots, x_n) < f(y_1, y_2, \cdots, y_n)$$

whenever $x_i < y_i$ for $i = 1, 2, \cdots, n$.

Liu [58] provided the following useful theorem to determine the distribution function of the strictly increasing function of uncertain variables. Taking advantage of this theorem, we can transform an indeterminacy model into a deterministic one.

Theorem 2. *(Liu [58]) Let $\xi_1, \xi_2, \cdots, \xi_n$ be independent uncertain variables with uncertainty distributions Φ_1, Φ_2, $\cdots$, Φ_n, respectively. If f is a strictly increasing function, then $\xi = f(\xi_1, \xi_2, \cdots, \xi_n)$ is an uncertain variable with inverse uncertainty distribution*

$$\Phi^{-1}(\alpha) = f(\Phi_1^{-1}(\alpha), \Phi_2^{-1}(\alpha), \cdots, \Phi_n^{-1}(\alpha)).$$

Example 1. *Let ξ_1 and ξ_2 be independent uncertain variables with uncertainty distributions Φ_1 and Φ_2, respectively. Since $f(x_1, x_2) = x_1 + \exp(x_2)$ is a strictly increasing function, $\xi_1 + \exp(\xi_2)$ is an uncertain variable with inverse uncertainty distribution*

$$\Psi^{-1}(\alpha) = \Phi_1^{-1}(\alpha) + \exp(\Phi_2^{-1}(\alpha)).$$

Example 2. *Let ξ_1, ξ_2 and ξ_3 be independent and nonnegative uncertain variables with uncertainty distributions Φ_1, Φ_2 and Φ_3, respectively. Since $f(x_1, x_2, x_3) = x_1^2 + x_2 \times x_3$ is a strictly increasing function, $\xi_1^2 + \xi_2 \times \xi_3$ is an uncertain variable with inverse uncertainty distribution*

$$\Psi^{-1}(\alpha) = (\Phi_1^{-1}(\alpha))^2 + \Phi_2^{-1}(\alpha) \times \Phi_3^{-1}(\alpha).$$

Network optimization

Classical network optimization problem

Classical network optimization problems, including shortest path problem, maximum flow problem, transportation problem, matching problem, etc., have been researched for a long time and resulted in many gratifying achievements [59]. In a classical network optimization problem, the topology structure of the network is deterministic and the parameters in the network are crisp. With ignoring so complicated factors, the classical network optimization problems have permeated every aspect of our lives, and the models of these optimization problems have been widely used in the applications in the real world.

As we know, a network optimization problem is usually described as a special type of linear programming model. For simplicity, we use $\mathbf{x} = (x_1, x_2, \cdots, x_n)$ to represent the decision vector. Assume that $f(\mathbf{x})$ is the objective function of the network optimization, and the constrained conditions from the network can be expressed as $g_j(\mathbf{x}) \geq 0$ for $j = 1, 2, \cdots, m$. Then the classical network optimization model can be formulated as

$$\begin{cases} \min \quad f(\mathbf{x}) \\ \text{subject to :} \\ \qquad g_j(\mathbf{x}) \geq 0, \quad j = 1, 2, \cdots, m \\ \qquad \mathbf{x} \in D \end{cases}$$

where D emphatically denotes the restricted region for the sake of clarity.

Stochastic network optimization problem

Many real-life networks behave not deterministic but stochastic. Examples can be found in areas like communication systems, production, maintenance, and logistics systems. Construction and analysis of stochastic network optimization models rely on techniques from fundamental probability theory and mathematical statistics.

In stochastic network optimization, it is assumed that $\mathbf{x}$ is a decision vector, $\boldsymbol{\xi}$ is a random vector related to random variables, and $f(\mathbf{x}; \boldsymbol{\xi})$ is the objective function to be optimized stochastically. Suppose that we have a set of random constraints $g_j(\mathbf{x};\boldsymbol{\xi}) \geq 0$, $j = 1, 2, \cdots, m$.

Generally, a network optimization problem with random variables can be described by a stochastic programming model. In order to obtain a decision with minimum expected objective value subject to a set of chance constraints, the stochastic network optimization problem can be expressed as

$$\begin{cases} \min \quad E[f(\mathbf{x};\boldsymbol{\xi})] \\ \text{subject to :} \\ \qquad \Pr\{g_j(\mathbf{x};\boldsymbol{\xi}) \geq 0\} \geq \alpha_j, \quad j = 1, 2, \cdots, m \\ \qquad \mathbf{x} \in D \end{cases}$$

where $\alpha_j, j = 1, 2, \cdots, m$ are the predetermined confidence levels, and D is the restricted region.

As mentioned before, there are two kinds of indeterminacy. Probability theory was developed by Kolmogorov [60] for modeling frequencies, and uncertainty theory was founded by Liu [21] for modeling belief degrees. This paper aims at providing the general methodology of modeling uncertain network optimization problems rather than surveying the up-to-date advances in uncertain network optimization, in which the uncertain factors are dealt with by the uncertain measure. In this paper, we provide the uncertain network optimization problems in two main directions. One line of research concerns the handling of network with the uncertain topology structure, which is called an uncertain graph. Another line of research deals with network optimization with uncertain weights, such as an uncertain Chinese postman problem [50], uncertain assignment problem [53,54], and uncertain transportation problem [61-63].

Network optimization with an uncertain topology structure

In a classic graph theory, the vertices and edges are deterministic, either existing or not. However, with different scenarios from stochastic ones, very often we are lack of observed data due to economic reasons or technical difficulties. As we stated before, when there do not exist enough information for us to judge whether two nodes are joined by an edge, we usually use the belief degree to describe it. Since human beings usually overweigh unlikely events [64], the belief degree may have much larger variance than the real frequency; these belief degree data are no more suitable to be dealt with by probability theory. A counterexample was presented by Zhang and Peng [34].

Based on the analysis mentioned above, Gao and Gao [33] defined the uncertain graph, in which all edges are independent and exist with some belief degrees in an uncertain measure.

Definition 1. *(Gao and Gao [33]) A simple finite graph of order n is said to be uncertain if its adjacency matrix is*

$$A = \begin{pmatrix} 0 & \alpha_{12} & \cdots & \alpha_{1n} \\ \alpha_{21} & 0 & \cdots & \alpha_{2n} \\ \vdots & \vdots & \ddots & \vdots \\ \alpha_{n1} & \alpha_{n2} & \cdots & 0 \end{pmatrix}$$

where α_{ij} represents the truth values in the uncertain measure that the edges between vertices v_i and v_j exist, $i, j = 1, 2, \cdots, n$, respectively.

After that, Gao and Gao [33] discussed the connectivity of the uncertain graph. Briefly, a graph G is connected if there is a $u - v$ path whenever two vertices u, v of G. For the uncertain graph, in order to show how likely the uncertain graph is connected, Gao and Gao [33] defined connectedness index as below.

Definition 2. *(Gao and Gao [33]) A connectedness index of an uncertain graph is the uncertain measure that the uncertain graph is connected.*

In addition, a method for calculating connectedness index was given.

Theorem 3. *(Gao and Gao [33]) Let G be an uncertain graph of order n and its uncertain adjacency matrix is*

$$A = \begin{pmatrix} 0 & \alpha_{12} & \cdots & \alpha_{1n} \\ \alpha_{21} & 0 & \cdots & \alpha_{2n} \\ \vdots & \vdots & \ddots & \vdots \\ \alpha_{n1} & \alpha_{n2} & \cdots & 0 \end{pmatrix}.$$

If all edges are independent, then the connectedness index of G is

$$\rho(G) = \begin{cases} \sup\limits_{f(X)=1} \min\limits_{1\le i<j\le n} \nu_{ij}(X), & \text{if } \sup\limits_{f(X)=1} \min\limits_{1\le i<j\le n} \nu_{ij}(X) < 0.5 \\ 1 - \sup\limits_{f(X)=0} \min\limits_{1\le i<j\le n} \nu_{ij}(X), & \text{if } \sup\limits_{f(X)=1} \min\limits_{1\le i<j\le n} \nu_{ij}(X) \ge 0.5 \end{cases}$$

where

$$X = \begin{pmatrix} x_{11} & x_{12} & \cdots & x_{1n} \\ x_{21} & x_{22} & \cdots & x_{2n} \\ \vdots & \vdots & \ddots & \vdots \\ x_{n1} & x_{n2} & \cdots & x_{nn} \end{pmatrix},$$

$x_{ij} \in \{0, 1\}$, and ν_{ij} are defined by

$$\nu_{ij}(X) = \begin{cases} \alpha_{ij}, & \text{if } x_{ij} = 1 \\ 1 - \alpha_{ij}, & \text{if } x_{ij} = 0 \end{cases}$$

for $i, j = 1, 2, \cdots, n$, respectively,

$$f(X) = \begin{cases} 1, & \text{if each element of } I + X + X^2 + \cdots + X^{n-1} \text{ is positive} \\ 0, & \text{otherwise.} \end{cases}$$

For more research of the uncertain graph, we may consult Gao [46], Gao et al. [49], Zhang and Peng [47,48], and so on. Nevertheless, there are many new and fundamental topics of the uncertain graph needed to be discussed.

Network optimization with uncertain weights

Assume that the uncertain network is denoted by $N = (V, A, \boldsymbol{\xi})$, where V is the node set, A is the edge set, and the uncertain vector $\boldsymbol{\xi} = (\xi_1, \xi_2, \cdots, \xi_n)$ stands for the uncertain values corresponding to the edge vector $(e_1, e_2, \cdots, e_n)$. Without loss of generality, ξ_i are considered independent uncertain variables with uncertainty distributions Φ_i, $i = 1, 2, \cdots, n$, respectively.

For simplicity, we use $\mathbf{x} = (x_1, x_2, \cdots, x_n)$ to represent the decision vector, D to denote the restricted region. Assume the objective function is $f(\mathbf{x}; \boldsymbol{\xi})$. And the aim of the uncertain network optimization problem is to minimize $f(\mathbf{x}; \boldsymbol{\xi})$. The constrained conditions can be originally expressed as $g_j(\mathbf{x}; \boldsymbol{\xi}) \leq 0$ for $j = 1, 2, \cdots, m$. In accordance with the practical problems, it is also assumed that $g_j(\mathbf{x}; \boldsymbol{\xi})$ are strictly increasing with respect to $\xi_1, \xi_2, \cdots, \xi_n$.

Expected value model

In uncertain network optimization, it is difficult to optimize the uncertain objective value directly. Then an expected value model can be used to obtain a decision with an optimal expected objective value subject to the expected constraints.

Taking this point of view, we may minimize the expected value of the objective function, i.e.,

$$\min \quad E[f(\mathbf{x}; \boldsymbol{\xi})].$$

To make sense, the uncertain constraints $g_j(\mathbf{x}; \boldsymbol{\xi}) \leq 0$ for $j = 1, 2, \cdots, m$ should be expressed as:

$$E[g_j(\mathbf{x}; \boldsymbol{\xi})] \leq 0, \qquad j = 1, 2, \cdots, m.$$

Based on this modeling idea, the expected value model was constructed by Liu [36]:

$$\begin{cases} \min \quad E[f(\mathbf{x}; \boldsymbol{\xi})] \\ \text{subject to :} \\ \qquad E[g_j(\mathbf{x}; \boldsymbol{\xi})] \leq 0, \quad j = 1, 2, \cdots, m \\ \qquad \mathbf{x} \in D. \end{cases} \tag{1}$$

Taking advantage of properties of the operational law of the uncertain variable, the model (1) can be represented as

$$\begin{cases} \min \quad \displaystyle\int_0^1 \Phi_f^{-1}(\mathbf{x}; \alpha)\mathrm{d}\alpha \\ \text{subject to :} \\ \qquad \displaystyle\int_0^1 \Psi_{g_j}^{-1}(\mathbf{x}; \alpha)\mathrm{d}\alpha \leq 0, \quad j = 1, 2, \cdots, m \\ \qquad \mathbf{x} \in D \end{cases}$$

where Φ_f^{-1} and $\Psi_{g_j}^{-1}$ are the inverse uncertainty distributions of $f(\mathbf{x}; \boldsymbol{\xi})$ and $g_j(\mathbf{x}; \boldsymbol{\xi})$, respectively.

Belief degree-constrained programming model

Belief degree-constrained programming is another method to deal with optimization problem in an uncertain environment. In order to make a decision so that the objective function $f(\mathbf{x};\boldsymbol{\xi})$ is minimized in the sense of an uncertain measure, the α criteria can be taken as

$$\mathcal{M}\{f(\mathbf{x};\boldsymbol{\xi}) \leq \bar{f}\} \geq \alpha$$

where the critical value $\bar{f}$ will be minimized, and α is a predetermined confidence level.

In addition, it is naturally desired that the uncertain constraints $g_j(\mathbf{x};\xi) \leq 0$ for $j = 1, 2, \cdots, m$ hold with confidence levels $\alpha_1, \alpha_2, \cdots, \alpha_m$, respectively. Then the constrained conditions can be expressed by chance constraints

$$\mathcal{M}\{g_j(\mathbf{x};\boldsymbol{\xi}) \leq 0\} \geq \alpha_j, \quad j = 1, 2, \cdots, m.$$

Thus, Liu [36] presented the uncertain network optimization model as follows:

$$\begin{cases} \min \quad \bar{f} \\ \text{subject to}: \\ \qquad \mathcal{M}\{f(\mathbf{x};\boldsymbol{\xi}) \leq \bar{f}\} \geq \alpha \\ \qquad \mathcal{M}\{g_j(\mathbf{x};\boldsymbol{\xi}) \leq 0\} \geq \alpha_j, \quad j = 1, 2, \cdots, m \\ \qquad \mathbf{x} \in D \end{cases} \tag{2}$$

where α and α_j are the predetermined confidence levels.

Taking advantage of properties of the operational law of an uncertain variable, the model (2) can be equivalently represented as

$$\begin{cases} \min \quad \Phi_f^{-1}(\mathbf{x};\alpha) \\ \text{subject to}: \\ \qquad \Psi_{g_j}^{-1}(\mathbf{x};\alpha_j) \leq 0, \quad j = 1, 2, \cdots, m \\ \qquad \mathbf{x} \in D \end{cases}$$

where Φ_f^{-1} and $\Psi_{g_j}^{-1}$ are the inverse uncertainty distributions of $f(\mathbf{x};\boldsymbol{\xi})$ and $g_j(\mathbf{x};\boldsymbol{\xi})$, respectively.

Belief degree maximized-constrained programming model

Since there is no way to compare the uncertain objective functions directly, the decision maker may consider the problem from another point of view. The third criterion for the decision maker is a belief degree rule. The decision maker may firstly present a satisfying predetermined maximal value $\bar{f}$ and then maximize the belief degree that the objective function $f(\mathbf{x};\boldsymbol{\xi})$ is no more than the given value. In the sense of belief degree, the constrains can be expressed by chance constraints

$$\mathcal{M}\{g_j(\mathbf{x};\boldsymbol{\xi}) \leq 0\} \geq \alpha_j, \quad j = 1, 2, \cdots, m.$$

Taking this modeling idea, Liu [36] constructed belief degree maximized-constrained programming model as follows:

$$\begin{cases} \max \quad \mathcal{M}\{f(\mathbf{x};\boldsymbol{\xi}) \leq \bar{f}\} \\ \text{subject to}: \\ \qquad \mathcal{M}\{g_j(\mathbf{x};\boldsymbol{\xi}) \leq 0\} \geq \alpha_j, \quad j = 1, 2, \cdots, m \\ \qquad \mathbf{x} \in D \end{cases} \tag{3}$$

where α_j are the predetermined confidence levels.

Equivalently, it can be rewritten as

$$\begin{cases} \max \quad \mathcal{M}\{f(\mathbf{x};\boldsymbol{\xi}) \leq \bar{f}\} \\ \text{subject to :} \\ \qquad \Psi_{g_j}^{-1}(\mathbf{x};\alpha_j) \leq 0, \quad j = 1,2,\cdots,m \\ \qquad \mathbf{x} \in D \end{cases}$$

where $\Psi_{g_j}^{-1}$ are the inverse uncertainty distributions of $g_j(\mathbf{x};\boldsymbol{\xi})$.

Some uncertain network optimization problems

In this section, we focus on some typical uncertain network optimization problems to show the applications of the proposed models.

Uncertain Chinese postman problem

Assume $N = (V,A,\boldsymbol{\xi})$ is an uncertain network, where $V = \{1,2,\cdots,n\}$ is the set of nodes, $A = \{(i,j)|i,j \in V, i \neq j\}$ is the set of edges, and $\boldsymbol{\xi} = \{\xi_{ij}|(i,j) \in A, i \neq j\}$ is the set of the uncertain length of the edges. We also assume that ξ_{ij} are independent uncertain variables with uncertainty distributions Φ_{ij}, $i,j = 1,2,\cdots,n$, respectively. The Chinese postman problem is to find a shortest route R such that the postman must cover each street in his area at least once, and then end up at the same place where he begin his route. Without loss of generality, we assume x_{ij} is a decision variable on the edge (i,j). That is to say, $x_{ij} = 1$ means that there exists an edge from i to j in route R; Otherwise, $x_{ij} = 0$. In order to obtain a shortest route in the sense of expected value, Zhang and Peng [50] proposed the following expected value model,

$$\begin{cases} \min \quad E\left[\sum_{(i,j)\in A} x_{ij}\xi_{ij}\right] \\ \text{subject to :} \\ \qquad \sum_{j:(i,j)\in A} x_{ij} - \sum_{j:(j,i)\in A} x_{ji} = 0, \; i \in V \\ \qquad x_{ij} + x_{ji} \geq 1, \qquad (i,j) \in A \\ \qquad x_{ij} \in \{0,1\}, \qquad (i,j) \in A. \end{cases} \tag{4}$$

The first constraint requires that the route is a cycle, and the second constraint implies that the route traverses each edge at least once.

If the decision maker prefers treating the problem with a critical value, then the belief degree-constrained programming model was proposed by Zhang and Peng [50] as follows:

$$\begin{cases} \min \quad W \\ \text{subject to :} \\ \qquad \mathcal{M}\left\{\sum_{(i,j)\in A} x_{ij}\xi_{ij} \leq W\right\} \geq \alpha, \\ \qquad \sum_{j:(i,j)\in A} x_{ij} - \sum_{j:(j,i)\in A} x_{ji} = 0, \quad i \in V \\ \qquad x_{ij} + x_{ji} \geq 1, \qquad (i,j) \in A \\ \qquad x_{ij} \in \{0,1\}, \qquad (i,j) \in A \end{cases} \tag{5}$$

where W is the α-optimal shortest length defined as $\min\left\{W \mid \mathcal{M}\left\{\sum_{(i,j)\in A} x_{ij}\xi_{ij} \le W\right\} \ge \alpha\right\}$ which means the shortest length the postman can obtain at confidence level α.

Moreover, we can establish the belief degree maximized-constrained programming model as follows:

$$\begin{cases} \max \quad \mathcal{M}\left\{\sum_{(i,j)\in A} x_{ij}\xi_{ij} \le L\right\} \\ \text{subject to :} \\ \qquad \sum_{j:(i,j)\in A} x_{ij} - \sum_{j:(j,i)\in A} x_{ji} = 0, \; i \in V \\ \qquad x_{ij} + x_{ji} \ge 1, \qquad (i,j) \in A \\ \qquad x_{ij} \in \{0,1\}, \qquad (i,j) \in A \end{cases} \tag{6}$$

where L is a threshold value of all the route lengths.

Uncertain optimal assignment problem

In order to model an optimal assignment problem by a graph, some basic concepts of graph theory are introduced, which are from Bondy and Murty [65].

If the set of vertices of a graph can be partitioned into two subsets X and Y, such that each edge has one vertex in X and one vertex in Y, such a graph is called a bipartite graph. The partition (X, Y) is named a bipartition of the graph. A complete graph is one in which each pair of distinct vertices is joined by an edge. In addition, if a bipartite graph with bipartition (X, Y) in which each vertex of X is joined by exactly one edge to each vertex of Y, such a graph is called a complete bipartite graph.

Assume that n workers are available for n jobs in a company, and each worker is qualified for all of these jobs. Consider an uncertain weighted complete bipartite graph with bipartition $(X, Y, \boldsymbol{\xi})$, where $X = \{x_1, x_2, \cdots, x_n\}$, $Y = \{y_1, y_2, \cdots, y_n\}$, and $\boldsymbol{\xi} = \{\xi_{ij} \mid x_i \in X, y_j \in Y\}$ is the uncertain weight of the edge. Each ξ_{ij} indicates the uncertain profit made by worker x_i on job y_j. If the profits of the workers in different jobs are different, how can the decision maker design an assignment plan such that the total profit made by all workers is maximized? This is so-called optimal assignment problem. We also suppose that ξ_{ij} are independent uncertain variables with uncertainty distributions Φ_{ij}, $i, j = 1, 2, \cdots, n$, respectively.

In order to model the abovementioned uncertain optimal assignment problem, the following decision variables are employed

$$x_{ij} = \begin{cases} 1, & \text{if worker } x_i \text{ is assigned to job } y_j \\ 0, & \text{otherwise.} \end{cases}$$

As a simple choice, Zhang and Peng [66] built the following expected value model,

$$\left\{\begin{array}{l} \max \quad E\left[\sum\limits_{1\le i,j\le n} x_{ij}\xi_{ij}\right] \\ \text{subject to :} \\ \qquad \sum\limits_{1\le j\le n} x_{ij} = 1, \quad i = 1,2,\cdots,n \\ \qquad \sum\limits_{1\le i\le n} x_{ij} = 1, \quad j = 1,2,\cdots,n \\ \qquad x_{ij} \in \{0,1\}, \quad i,j = 1,2,\cdots,n \end{array}\right. \tag{7}$$

where the first constraint indicates that every worker is assigned to exactly one job, and the second constraint means that every job is assigned exactly one worker.

Sometimes the decision maker would like to set the specific value as the profit target for pursuit. In the uncertain environment, the target profit may not be obtained in some situations. It is natural that the decision maker would accept the inability to reach the target profit to some extent. However, the target profit must be achieved at a given confidence level. Based on this idea, Zhang and Peng [53] proposed the α-optimal assignment model pursing the maximal profit at the uncertain measure not less than a predetermined level. The α-optimal assignment model is as follows:

$$\left\{\begin{array}{l} \max \quad W \\ \text{subject to :} \\ \qquad \mathcal{M}\left\{\sum\limits_{1\le i,j\le n} x_{ij}\xi_{ij} \ge W\right\} \ge \alpha, \\ \qquad \sum\limits_{1\le j\le n} x_{ij} = 1, \quad i = 1,2,\cdots,n \\ \qquad \sum\limits_{1\le i\le n} x_{ij} = 1, \quad j = 1,2,\cdots,n \\ \qquad x_{ij} \in \{0,1\}, \quad i,j = 1,2,\cdots,n \end{array}\right. \tag{8}$$

where W is the α-optimal profit defined as $\max\left\{W|\mathcal{M}\left\{\sum\limits_{1\le i,j\le n} x_{ij}\xi_{ij} \ge W\right\} \ge \alpha\right\}$, the second constraint requires that every worker is assigned to exactly one job, and the third constraint requires that every job is assigned to exactly one worker.

If the decision maker gives the target profit T first, then he/she will want to maximize the belief degree that the uncertain profit is no less than the given target profit. According to this idea, the belief degree maximized-constrained programming model is built as follows:

$$\left\{\begin{array}{l} \max \quad \mathcal{M}\left\{\sum\limits_{1\le i,j\le n} x_{ij}\xi_{ij} \ge T\right\} \\ \text{subject to :} \\ \qquad \sum\limits_{1\le j\le n} x_{ij} = 1, \quad i = 1,2,\cdots,n \\ \qquad \sum\limits_{1\le i\le n} x_{ij} = 1, \quad j = 1,2,\cdots,n \\ \qquad x_{ij} \in \{0,1\}, \quad i,j = 1,2,\cdots,n \end{array}\right. \tag{9}$$

where T is the predetermined target profit.

Uncertain transportation problem

It is well-known that the transportation problem is a classical network optimization problem that has been researched for many years. As a problem in the uncertain environment, uncertain transportation problems have been studied by Cui and Sheng [67], Sheng and Yao [61,62] and Zhang and Peng [63] in recent years. Let us take the fixed charge solid transportation problem as an example.

As a generalization of traditional transportation problem, the solid transportation problem involves how to transport products from m sources to n destinations by l conveyances so that the total transportation cost is minimized. The fixed charge transportation consider two types of costs, including the direct cost and the fixed charge. The direct cost is the cost with respect to per unit transportation amount. When the transportation activity between a source and a destination by a conveyance occurs, the fixed charge will be paid. Denote x_{ijk} as the quantity transported from source i to destination j by conveyance k. The corresponding uncertain cost of unit amount and fixed charge are denoted as ξ_{ijk} and η_{ijk}, respectively. Let $\tilde{a}_i$ be the amount of products in source i which can be transported to n destinations, $\tilde{b}_j$ the minimal demand of products in destination j, and $\tilde{c}_k$ the capacity of conveyance k.

To structure the mathematical program, the following notations are employed:

$$y_{ijk} = \begin{cases} 1, & \text{if } x_{ijk} > 0 \\ 0, & \text{otherwise.} \end{cases}$$

This implies that if $x_{ijk} > 0$, we must add the fixed charge to the total transportation cost.

For the fixed charge solid transportation problem in the uncertain environment, we assume that $\tilde{a}_i, \tilde{b}_j, \tilde{c}_k, \xi_{ijk}$, and η_{ijk} are independent uncertain variables. Thus, Zhang and Peng [63] proposed the expected value model as follows:

$$\begin{cases} \min \quad E\left[\sum_{i=1}^{m}\sum_{j=1}^{n}\sum_{k=1}^{l}(\xi_{ijk}x_{ijk} + \eta_{ijk}y_{ijk})\right] \\ \text{subject to :} \\ \qquad E\left[\sum_{j=1}^{n}\sum_{k=1}^{l}x_{ijk} - \tilde{a}_i\right] \le 0, \quad i = 1, 2, \cdots, m \\ \qquad E\left[\sum_{i=1}^{m}\sum_{k=1}^{l}x_{ijk} - \tilde{b}_j\right] \ge 0, \quad j = 1, 2, \cdots, n \\ \qquad E\left[\sum_{i=1}^{m}\sum_{j=1}^{n}x_{ijk} - \tilde{c}_k\right] \le 0, \quad k = 1, 2, \cdots, l \\ \qquad x_{ijk} \ge 0, \quad y_{ijk} \in \{0, 1\} \\ \qquad i = 1, 2, \cdots, m; \quad j = 1, 2, \cdots, n; \quad k = 1, 2, \cdots, l. \end{cases} \tag{10}$$

Under other conditions, the decision maker prefers treating the problem under the belief degree-constraints. This fact provides a motivation for belief degree-constraint programming model founded by Zhang and Peng [63] as follows:

$$\begin{cases} \min \ \bar{f} \\ \text{subject to :} \\ \quad \mathcal{M}\left\{\sum_{i=1}^{m}\sum_{j=1}^{n}\sum_{k=1}^{l}(\xi_{ijk}x_{ijk}+\eta_{ijk}y_{ijk}) \le \bar{f}\right\} \ge \alpha \\ \quad \mathcal{M}\left\{\sum_{j=1}^{n}\sum_{k=1}^{l}x_{ijk} \le \tilde{a}_i\right\} \ge \alpha_i, \quad i=1,2,\cdots,m \\ \quad \mathcal{M}\left\{\sum_{i=1}^{m}\sum_{k=1}^{l}x_{ijk} \ge \tilde{b}_j\right\} \ge \beta_j, \quad j=1,2,\cdots,n \\ \quad \mathcal{M}\left\{\sum_{i=1}^{m}\sum_{j=1}^{n}x_{ijk} \le \tilde{c}_k\right\} \ge \gamma_k, \quad k=1,2,\cdots,l \\ \quad x_{ijk} \ge 0, \quad y_{ijk} \in \{0,1\} \\ \quad i=1,2,\cdots,m; \quad j=1,2,\cdots,n; \quad k=1,2,\cdots,l \end{cases} \tag{11}$$

where $\bar{f}$ is the α-optimal plan defined as $\min\left\{\bar{f}|\mathcal{M}\left\{\sum_{i=1}^{m}\sum_{j=1}^{n}\sum_{k=1}^{l}(\xi_{ijk}x_{ijk}+\eta_{ijk}y_{ijk}) \le \bar{f}\right\} \ge \alpha\right\}$.

In addition, the belief degree maximized-constrained programming model [63] can be built in the following way,

$$\begin{cases} \max \ \mathcal{M}\left\{\sum_{i=1}^{m}\sum_{j=1}^{n}\sum_{k=1}^{l}(\xi_{ijk}x_{ijk}+\eta_{ijk}y_{ijk}) \le f_0\right\} \\ \text{subject to :} \\ \quad \mathcal{M}\left\{\sum_{j=1}^{n}\sum_{k=1}^{l}x_{ijk} \le \tilde{a}_i\right\} \ge \alpha_i, \quad i=1,2,\cdots,m \\ \quad \mathcal{M}\left\{\sum_{i=1}^{m}\sum_{k=1}^{l}x_{ijk} \ge \tilde{b}_j\right\} \ge \beta_j, \quad j=1,2,\cdots,n \\ \quad \mathcal{M}\left\{\sum_{i=1}^{m}\sum_{j=1}^{n}x_{ijk} \le \tilde{c}_k\right\} \ge \gamma_k, \quad k=1,2,\cdots,l \\ \quad x_{ijk} \ge 0, \quad y_{ijk} \in \{0,1\} \\ \quad i=1,2,\cdots,m; \quad j=1,2,\cdots,n; \quad k=1,2,\cdots,l \end{cases} \tag{12}$$

where f_0 is the predetermined value.

Hybrid intelligent algorithm

In some special cases, we can convert the uncertain programming models into the corresponding crisp equivalents. But this is not the case in most situations due to the computational complexity of the distributions of uncertain objective function or constrained function. Nevertheless, if the uncertain functions are complex, then a suitable hybrid intelligent algorithm may be designed to do this work based on 99-method and genetic algorithm. In the algorithm, 99-method is used as uncertain simulation techniques for calculating the expected value, uncertain measure, and critical value, and genetic algorithm is employed for finding the optimal solution.

99-method

The 99-method was proposed by Liu [58]. It is suggested that an uncertain variable ξ_i with uncertainty distribution Φ_i can be represented by a 99-table (see Table 1). In the first row, $0.01, 0.02, \cdots, 0.99$ are the values of uncertainty distribution Φ_i of ξ_i. In the second row, $t_1^i, t_2^i, \cdots, t_{99}^i$ are the corresponding values of inverse uncertainty distribution Φ_i^{-1}. Let $\boldsymbol{\xi} = (\xi_1, \xi_2, \cdots, \xi_n)$, $\mathbf{x} = (x_1, x_2, \cdots, x_n)$, and $f(\mathbf{x}; \boldsymbol{\xi}) = f(\mathbf{x}; \xi_1, \xi_2, \cdots, \xi_n)$ be a strictly monotone increasing function with respect to $\xi_1, \xi_2, \cdots, \xi_n$. Assume that ξ_i are independent uncertain variables with uncertainty distributions Φ_i, $i = 1, 2, \cdots, n$, respectively. According to Theorem 2, the inverse uncertainty distribution of $f(\mathbf{x}; \boldsymbol{\xi})$ is $\Phi_f^{-1}(\mathbf{x}; \alpha) = f(x; \Phi_1^{-1}(\alpha), \Phi_2^{-1}(\alpha), \cdots, \Phi_n^{-1}(\alpha))$. In the following, we will introduce the methods for calculation expected value $E[f(\mathbf{x}; \boldsymbol{\xi})]$, uncertain measure $\mathcal{M}\{f(\mathbf{x}; \boldsymbol{\xi}) \le f_0\}$, and critical value $\bar{f}$ such that $\min\{\bar{f} | \mathcal{M}\{f(\mathbf{x}; \boldsymbol{\xi}) \le \bar{f}\} \ge \alpha\}$.

Expected value

It follows from Theorem 2 that the inverse uncertainty distribution of $f(\mathbf{x}; \boldsymbol{\xi})$ is

$$\Phi_f^{-1}(\mathbf{x}; \alpha) = f(\mathbf{x}; \Phi_1^{-1}(\alpha), \Phi_2^{-1}(\alpha), \cdots, \Phi_n^{-1}(\alpha)),$$

which can be expressed in the computer as Table 2.

By using Theorem 1, we have

$$E[f(\mathbf{x}; \boldsymbol{\xi})] = \int_0^1 f(\mathbf{x}; \Phi_1^{-1}(\alpha), \Phi_2^{-1}(\alpha), \cdots, \Phi_n^{-1}(\alpha))\mathrm{d}\alpha.$$

Thus we design the procedure for computing $E[f(\mathbf{x}; \boldsymbol{\xi})]$ as follows:

Step 1: Set $E = 0$ and $j = 1$.

Step 2: Calculate $y_j = 0.01 f(\mathbf{x}; t_j^1, t_j^2, \cdots, t_j^n)$, and let $E := E + y_j$.

Step 3: If $j < 99$, let $j := j + 1$. Turn back to Step 2.

Step 4: Report E as the estimation of $E[f(\mathbf{x}; \boldsymbol{\xi})]$.

Uncertain measure

As discussed above, the inverse uncertainty distribution $\Phi_f^{-1}(\mathbf{x}; \alpha)$ of $f(\mathbf{x}; \boldsymbol{\xi})$ can be expressed as Table 2. Now, the uncertain measure $\mathcal{M}\{f(\mathbf{x}; \boldsymbol{\xi}) \le f_0\}$ can be approximately obtained by the following method

Step 1: Set $i = 1$.

Step 2: Calculate $y_i = f(\mathbf{x}; t_i^1, t_i^2, \cdots, t_i^n)$.

Step 3: If $y_i < f_0$, then $i \leftarrow i + 1$, and go back to Step 2.

Step 4: Report $\alpha = 0.01i$ as the estimation of the uncertain measure $\mathcal{M}\{f(\mathbf{x}; \boldsymbol{\xi}) \le f_0\}$.

Critical value

Given a predetermined confidence level $0 < \alpha < 1$, we need to find the critical value $\bar{f}$ such that $\min\{\bar{f} | \mathcal{M}\{f(\mathbf{x}; \boldsymbol{\xi}) \le \bar{f}\} \ge \alpha\}$. Since $\mathcal{M}\{f(\mathbf{x}; \boldsymbol{\xi}) \le \bar{f}\} \ge \alpha$ can be transformed as

Table 1 The expression of uncertainty distribution of uncertain variable ξ_i

0.01	0.02	0.03	$\cdots$	0.99
t_1^i	t_2^i	t_3^i	$\cdots$	t_{99}^i

Table 2 The expression of inverse uncertainty distribution of uncertain variable f

α	$\Phi_f^{-1}(x;\alpha)$
0.01	$f(\mathbf{x};t_1^1,t_1^2,\cdots,t_1^n)$
0.02	$f(\mathbf{x};t_2^1,t_2^2,\cdots,t_2^n)$
0.03	$f(\mathbf{x};t_3^1,t_3^2,\cdots,t_3^n)$
...	...
0.99	$f(\mathbf{x};t_{99}^1,t_{99}^2,\cdots,t_{99}^n)$

$\Phi_f^{-1}(\mathbf{x};\alpha) \le \bar{f}$. Thus, for any $0 < \alpha < 1$, the critical value $\bar{f}$ can be approximately obtained by

$$\frac{\Phi_f^{-1}(\mathbf{x};0.01\lfloor 100\alpha \rfloor) + \Phi_f^{-1}(\mathbf{x};0.01\lceil 100\alpha \rceil)}{2}.$$

Hybrid intelligent algorithm

As we know, genetic algorithm was first introduced by Holland [68] and has been developed by many scholars. Many experiments show that genetic algorithm is effective to solve many optimization problems that are difficult to solve by traditional methods. In order to solve the proposed uncertain network models in a numerical method, it is feasible to employ the so-called hybrid intelligent algorithm, which combines 99-method with genetic algorithm. In the algorithm, the uncertain values are calculated by 99-method, and alternately the numerical results are embedded into the genetic algorithm for finding the optimal solution.

Now, the hybrid intelligent algorithm for solving the uncertain network optimization problems is summarized as follows.

Step 1: Initialize *pop_size* chromosomes at random. Calculate the constraint values and check the feasibility of the chromosomes by 99-method.

Step 2: Update the chromosomes by crossover and mutation operations.

Step 3: Calculate the objective values of the chromosomes by 99-method.

Step 4: Compute the fitness of each chromosome via the objective values, which can be obtained by 99-method.

Step 5: Select the chromosomes by spanning the roulette wheel.

Step 6: Repeat Step 2 to Step 5 for a given number of cycles.

Step 7: Take the best chromosome as the solution of the problems.

Further research problems

Generally speaking, network optimization problem can be classified into two categories: Network optimization with an uncertain topology structure (i.e., uncertain graph) and network optimization with uncertain weights.

Recently, much attention has been paid to the latter research field. The cited references on uncertain network optimization in this paper almost belong to this area. However, it is no doubt that there are many problems in the uncertain graph needed to be considered, such as tree, regular and vertex degrees, etc. In addition, the network optimization with uncertain weights can be further divided into the following three aspects: 1) Network optimization with uncertain edge weights; 2) Network optimization with uncertain node weights; and 3) Network optimization with uncertain both edge and node weights. As far as we know, the second and the third aspects have been seldom studied up to now.

Concluding remarks

The uncertain network optimization problem arises from various applications in real life, which is of both theoretical interest in mathematics and applied aspect in practice. Due to the existence of uncertainty almost everywhere and almost every time, uncertain network optimization will be widely applied in such various disciplines. In order to deal with the uncertainty in network optimization, uncertainty theory is employed to model an uncertain network optimization problem.

It is important to keep the following three things in mind: 1) Many network optimization problems appear indeed in an uncertain world; 2) Uncertainty theory provides an efficient tool for dealing with uncertain network optimization problems. 3) There are three types of models for uncertain network optimization with uncertain weights: Expected value model and belief degree-constrained programming model, as well as belief degree maximized-constrained programming model.

In a review manner, this paper presented a comprehensive summary on uncertain network optimization and exhibited the general methods for modeling uncertain network optimization problems. In summary, uncertain network optimization is a new field of interdisciplinary research with opportunities and challenges.

Competing interests
The authors declare that they have no competing interests.

Authors' contributions
JP proposed the framework of the study and drafted the manuscript. BZ conceived of the study and participated in its design and coordination. SL participated in the design of the study. All authors read and approved the final manuscript.

Acknowledgements
This work is supported by the Projects of the Humanity and Social Science Foundation of Ministry of Education of China (No. 13YJA630065), the Key Project of Hubei Provincial Natural Science Foundation (No. 2012FFA065), and the Scientific and Technological Innovation Team Project (No. T201110) of Hubei Provincial Department of Education, China.

Author details
[1]Institute of Uncertain Systems, Huanggang Normal University, Huanggang, Hubei 438000, China. [2]School of Statistics and Mathematics, Zhongnan University of Economics and Law, Wuhan, Hubei 430073, China.

References
1. Lawler, E: Combinatorial Optimization: Networks and Matroids. Holt, Reinhardt and Winston, New York (2001)
2. Lee, J: A first course in combinatorial optimization. Cambridge University Press, Cambridge (2004)
3. Dreyfus, S: An appraisal of some shortest path algorithms. Oper. Res. **17**(3), 395–412 (1969)
4. Graham, RL, Hell, P: On the history of the minimum spanning tree problem. IEEE Ann. Hist. Comput. **7**(1), 43–57 (1985)
5. Haley, KB: New methods in mathematical programming-the solid transportation problem. Oper. Res. **10**(4), 448–463 (1962)
6. Kuhn, HW: The Hungarian method for the assignment problem. Naval. Res. Logist. Quart. **2**(1–2), 83–97 (1955)
7. Mazzoni, G, Pallottino, S, Scutella, MG: The maximum flow problem: a max-preflow approach. Eur. J. Oper. Res. **53**(3), 257–278 (1991)
8. Williams, A: A stochastic transportation problem. Oper. Res. **11**(5), 759–770 (1963)
9. Donath, WE: Algorithm and average-value bounds for assignment problems. IBM J. Res. Dev. **13**(4), 380–386 (1969)
10. Hall, RW: The fastest path through a network with random time-dependent travel times. Transport. Sci. **20**(3), 182–188 (1986)
11. Nawathe, SP, Rao, BV: Maximum flow in probabilistic communication networks. Int. J. Circ. Theor. App. **8**(2), 167–177 (1980)
12. Mulvey, JM, Vladimirou, H: Stochastic network optimization models for investment planning. Ann. Oper. Res. **20**, 187–217 (1989)
13. Neely, MJ: Stochastic network optimization with application to communication and queueing systems. Morgan & Claypool Publishers, San Rafael (2010)
14. Peer, SK, Sharma, DK: Finding the shortest path in stochastic networks. Comput. Math. Appl. **53**(5), 729–740 (2007)

15. Ji, Z, Kim, Y, Chen, A: Multi-objective α-reliable path finding in stochastic networks with correlated link costs: a simulation-based multi-objective genetic algorithm approach (SMOGA). Expert Syst. Appl. **38**(3), 1515–1528 (2011)
16. Chanas, S, Kolodziejczyk, W: Real-valued flows in a network with fuzzy arc capacities. Fuzzy Sets Syst. **13**(2), 139–151 (1984)
17. Hanebeck, UD, Schmidt, GK: Genetic optimization of fuzzy networks. Fuzzy Sets Syst. **79**(1), 59–68 (1996)
18. Wu, H: Fuzzy optimization problems based on the embedding theorem and possibility and necessity measures. Math. Comput. Model. **40**(3–4), 329–336 (2004)
19. Liu, B: Uncertainty Theory. 4th edn. Springer, Berlin (2015)
20. Liu, B: Why is there a need for uncertainty theory. J. Uncertain Syst. **6**(1), 3–10 (2012)
21. Liu, B: Uncertainty Theory. 2nd edn. Springer, Berlin (2007)
22. Liu, B: Uncertain set theory and uncertain inference rule with application to uncertain control. J. Uncertain Syst. **4**(2), 83–98 (2010)
23. Liu, B: Uncertain risk analysis and uncertain reliability analysis. J. Uncertain Syst. **4**(3), 163–170 (2010)
24. Liu, B: Some research problems in uncertainty theory. J. Uncertain Syst. **3**(1), 3–10 (2009)
25. Yao, K: Uncertain calculus with renewal process. Fuzzy Optimization Decis. Mak. **11**(3), 258–297 (2012)
26. Liu, B: Fuzzy process, hybrid process and uncertain process. J. Uncertain Syst. **2**(1), 3–16 (2008)
27. Chen, X, Liu, B: Existence and uniqueness theorem for uncertain differential equations. Fuzzy Optimization Decis. Mak. **9**(1), 69–81 (2010)
28. Yao, K: Extreme values and integral of solution of uncertain differential equation. J. Uncertainty Anal. Appl. **1, Article 2** (2013)
29. Liu, B: Toward uncertain finance theory. J. Uncertainty Anal. Appl. **1, Article 1** (2013)
30. Peng, J: Risk metrics of loss function for uncertain system. Fuzzy Optimizational Decis. Mak. **12**(1), 53–64 (2013)
31. Peng, J, Yao, K: A new option pricing model for stocks in uncertainty markets. Int. J. Oper. Res. **8**(2), 18–26 (2011)
32. Gao, X, Gao, Y, Ralescu, DA: On Liu's inference rule for uncertain systems. Int. J. Uncertainty Fuzziness Knowl.-Based Syst. **18**(1), 1–11 (2010)
33. Gao, X, Gao, Y: Connectedness index of uncertain graph. Int. J. Uncertainty Fuzziness Knowledge-Based Syst. **21**(1), 127–137 (2013)
34. Zhang, B, Peng, J: Euler index in uncertain graph. Appl. Math. Comput. **218**(20), 10279–10288 (2012)
35. Zhang, B, Peng, J: Connectedness strength of two vertices in an uncertain graph. Int. J. Comput. Math. **90**(2), 246–257 (2013)
36. Liu, B: Theory and Practice of Uncertain Programming. 2nd edn. Springer, Berlin (2009)
37. Yan, L: Optimal portfolio selection models with uncertain returns. Modern Appl. Sci. **3**(8), 76–81 (2009)
38. Rong, L: Two new uncertainty programming models of inventory with uncertain costs. J. Inf. Comput. Sci. **8**(2), 280–288 (2011)
39. Gao, Y: Shortest path problem with uncertain arc lengths. Comput. Math. Appl. **62**(6), 2591–2600 (2011)
40. Zhang, X, Chen, X: A new uncertain programming model for project scheduling problem. Information. **15**(10), 3901–3910 (2012)
41. Ding, S: A new uncertain programming model for grain supply chain design. Information. **16**(2), 1069–1076 (2013)
42. Gao, Y: Uncertain models for single facility location problems on networks. Appl. Math. Model. **36**(6), 2592–2599 (2012)
43. Zhang, X, Meng, G: Expected-variance-entropy model for uncertain parallel machine scheduling. Information. **16**(2), 903–908 (2013)
44. Zhang, X, Wang, Q, Zhou, J: A chance-constrained programming model for inverse spanning tree problem with uncertain edge weights. Int. J. Adv. Comput. Technol. **5**(6), 76–83 (2013)
45. Zhang, X, Wang, Q, Zhou, J: Two uncertain programming models for inverse minimum spanning tree problem. Ind. Eng. Manag. Syst. **12**(1), 9–15 (2013)
46. Gao, X: Cycle index of uncertain graph. Information. **16**(2), 1131–1138 (2013)
47. Zhang, B, Peng, J: Hamilton index and its algorithm of uncertain graph (2014). http://orsc.edu.cn/online/120531.pdf
48. Zhang, B, Peng, J: Matching index of an uncertain graph: concept and algorithm. Appl. Comput. Math. **12**(3), 381–391 (2013)
49. Gao, Y, Yang, L, Li, S, Kar, S: On distribution function of the diameter in uncertain graph, http://orsc.edu.cn/online/131014.pdf (2014)
50. Zhang, B, Peng, J: Uncertain programming model for Chinese postman problem with uncertain weights. Ind. Eng. Manag. Syst. **11**(1), 18–25 (2012)
51. Han, S, Peng, Z, Wang, S: The maximum flow problem of uncertain network. Inform. Sci. **265**, 167–175 (2014)
52. Ding, S: Uncertain minimum cost flow problem. Soft Comput. **18**(11), 2201–2207 (2014)
53. Zhang, B, Peng, J: Uncertain programming model for uncertain optimal assignment problem. Appl. Math. Model. **37**(9), 6458–6468 (2013)
54. Zhang, B, Peng, J: Goal programming models for k-cardinality uncertain assignment problem, http://orsc.edu.cn/online/120420.pdf (2014)
55. Zhou, J, He, X, Wang, K: Uncertain quadratic minimum spanning tree problem. J. Commun. **9**(5), 385–390 (2014)
56. Zhou, J, Chen, L, Wang, K: Path optimality conditions for minimum spanning tree problem with uncertain edge weights, http://www.orsc.edu.cn/online/131223.pdf (2014)
57. Zhou, J, Yang, F, Wang, K: An inverse shortest path problem on an uncertain graph. J. Networks. **9**(9), 2353–2359 (2014)
58. Liu, B: Uncertainty theory: a branch of mathematics for modeling human uncertainty. Springer, Berlin (2010)
59. Newman, MEJ: Networks: An Introduction. Oxford University Press, Oxford (2010)
60. Kolmogorov, AN: Grundbegriffe der Wahrscheinlichkeitsrechnung. Julius Springer, Berlin (1933)
61. Sheng, Y, Yao, K: A transportation model with uncertain costs and demands. Information. **15**(8), 3179–3186 (2012)
62. Sheng, Y, Yao, K: Fixed charge transportation problem in uncertain environment. Ind. Eng. Manag. Syst. **11**(2), 183–187 (2012)

63. Zhang, B, Peng, J: Uncertain programming models for fixed charge solid transportation problem, http://orsc.edu.cn/online/120828.pdf (2014)
64. Kahneman, D, Tversky, A: Prospect theory: an analysis of decision under risk. Econometrica. **47**(2), 263–292 (1979)
65. Bondy, J, Murty, U: Graph theory with applications. Elsevier, New York (1976)
66. Zhang, B, Peng, J: Expected value model for optimal assignment problem with uncertain profits, http://orsc.edu.cn/online/111114.pdf (2014)
67. Cui, Q, Sheng, Y: Uncertain programming model for solid transportation problem. Information. **16**(2), 1207–1214 (2013)
68. Holland, J: Adaptation in Natural and Artificial Systems. University of Michigan Press, Ann Arbor (1975)

10

Chance distribution of the maximum flow of uncertain random network

Yuhong Sheng[1*] and Jinwu Gao[2]

*Correspondence:
sheng-yh12@mails.tsinghua.edu.cn
[1] Department of Mathematical Sciences, Tsinghua University, Beijing 100084, China
Full list of author information is available at the end of the article

Abstract

A network is called uncertain random network if some arc capacities of the network are uncertain variables and others are random variables. The main purpose of this paper is to study the maximum flow in an uncertain random network. Under the framework of chance theory, this paper obtains chance distribution of the maximum flow of an uncertain random network. At the same time, the expected value of maximum flow is given for an uncertain random network. A new method is derived to calculate chance distribution of the maximum flow for an uncertain random network.

Keywords: Network; Maximum flow problem; Uncertain random variable; Uncertain random network

Introduction

Quite a number of systems may be viewed as and are called networks although their physical appearance is quite different in nature. Most of these systems are results of human civilization and are of great importance for its functioning. In particular, transportation networks, like road, rail, and airline networks, canal networks for transporting water, waster water, oil and natural gas, and communication networks of telephones and computers networks and so on even natural systems such as caves and rivers can be viewed as networks. To see the common structure behind these different systems, they must be abstracted from their physical appearance. Then, their underlying structure may be seen as a collection of vertices, which might be road crossings, railway stations, airports, pumping stations and so on, and a collection of lines, which might be roads, canals, telephone cables, etc. connecting all or some of the vertices and every arc has weights or capacities. In deterministic network, the vertices, arcs, and capacities of arcs are deterministic. The deterministic network was developed and widely applied in the last century. However, in practice, different types of indeterminacy must be taken into account for a variety of reasons.

Random network was first investigated by Frank and Hakimi [1] in 1965 for modeling communication network with random capacities. From then on, the random network was well developed and widely applied. For example, Frank [2], Mirchandani [3], and Sigal et al. [4] studied probabilistic distribution of the shortest path when the network arc of weighs are random variables. Frank and Frisch [5] considered how to determine the maximum flow probability distribution in networks where each capacity is a continuous random variable. Doulliez [6] studied multi-terminal network with discrete probabilistic branch capacities. In addition, some researchers have tried to give lower and upper bounds on the expected maximum flow of network. Carey and Hendrickson [7] and Onaga [8] presented an efficient method to find a lower bound in general directed networks. Furthermore, Fishman [9], Goldberg and Tarjan [10], and Nawathe and Rao [11] mainly used stochastic optimization to solve the maximum flow problem in a random network. Other researchers, such as Fu and Rillet [12], Hall [13], and Lou [14], have done a lot of work in this field of random network.

Uncertain network was first explored by Liu [15] in 2009. Gao [16] investigates solutions to the α-shortest path and the most shortest path in an uncertain network in 2011. Besides, the maximum flow problem was discussed by Han et al. [17], the uncertain minimum cost flow problem was dealt with by Ding [18], and Chinese postman problem was explored by Zhang and Peng [19] for an uncertain network.

In many cases, uncertainty and randomness simultaneously appear in a complex system. In order describe this complex system, Liu [20] gave a concept of the uncertain random network in which some weights are random variables and others are uncertain variables, at the same time, studied the shortest path chance distribution of an uncertain random network.

In this paper, we will give the maximum flow of chance distribution of an uncertain random network. The remainder of this paper is organized as follows. In the section 'Preliminaries', some basic concepts and properties of uncertainty theory and chance theory used throughout this paper are introduced. In the section 'Uncertain random network', uncertain random network is recalled. In the section 'Maximum flow of uncertain random network', chance distribution of maximum flow is proved. The section 'Expected value of maximum flow' proposes the expected value the maximum flow in an uncertain random network. The section 'Conclusions' gives a brief summary to this paper.

Preliminaries

In this section, we first review some concepts of uncertainty theory, including uncertain measure, uncertain variable, uncertainty distribution, and operational law. Then, we introduce some useful definitions and properties about uncertain random variable, chance measure, chance distribution, and operational law.

Uncertainty theory

This subsection reviews some basic concepts of uncertainty theory. In reality, however, because of the lack of information that no samples are available to estimate a probability distribution, in this situation, we have to invite some domain experts to evaluate the belief degree that each event will happen, and this moment, belief degree are uncertain variables rather than random variables. If we insist on using probability theory to deal with indeterminacy, counterintuitive results will occur [21]. In order to deal with some indeterministic phenomena, Liu [21] founded uncertainty theory in 2007. And Liu [21] first presented uncertain measure as a set function satisfying four axioms. As a fundamental concept in uncertainty theory, the uncertain variable was presented by Liu [21] in 2007. Liu [22] proposed the concept of uncertainty distribution and inverse uncertainty distribution. After that, many researchers widely studied the uncertainty theory and made significative progress. Gao [23] studied the properties of continuous uncertain measure. Peng and

Iwamura [24] proved a sufficient and necessary condition for uncertainty distribution. In addition, the concept of independence was proposed by Liu [25]. After, Liu [22] presented the operational law of uncertain variables. In order to rank uncertain variables, Liu [21] proposed the concept of expected value of uncertain variables. The linearity of expected value operator was verified by Liu [22]. As an important contribution, Liu and Ha [26] derived a useful formula for calculating the expected values of strictly monotone functions of independent uncertain variables. Up to now, theory and practice have shown that uncertainty theory is an efficient tool to deal with indeterministic phenomena, especially expert belief degrees. Liu [27] presented a counterexample of truck-cross-over-bridge, it is inappropriate to model belief degrees by probability theory. It is a reason why there is a need for uncertainty theory. The similarities and differences between uncertainty concept and standard probabilistic concept, as well as other concepts of uncertainty could be found in [21].

Definition 1. (Liu [21]) Let $\mathcal{L}$ be a σ-algebra on a nonempty set Γ. A set function $\mathcal{M}$: $\mathcal{L} \to [0,1]$ is called an uncertain measure if it satisfies the following axioms:
Axiom 1: (Normality Axiom) $\mathcal{M}\{\Gamma\} = 1$ for the universal set Γ.
Axiom 2: (Duality Axiom) $\mathcal{M}\{\Lambda\} + \mathcal{M}\{\Lambda^c\} = 1$ for any event Λ.
Axiom 3: (Subadditivity Axiom) For every countable sequence of events $\Lambda_1, \Lambda_2, \cdots$, we have

$$\mathcal{M}\left\{\bigcup_{i=1}^{\infty} \Lambda_i\right\} \leq \sum_{i=1}^{\infty} \mathcal{M}\{\Lambda_i\}.$$

Besides, the product uncertain measure on the product σ-algebra $\mathcal{L}$ is defined by the following product axiom.

Axiom 4: (Product Axiom) (Liu [25]) Let $(\Gamma_k, \mathcal{L}_k, \mathcal{M}_k)$ be uncertainty spaces for $k = 1, 2, \cdots$ The product uncertain measure $\mathcal{M}$ is an uncertain measure satisfying

$$\mathcal{M}\left\{\prod_{i=1}^{\infty} \Lambda_k\right\} = \bigwedge_{k=1}^{\infty} \mathcal{M}_k\{\Lambda_k\}$$

where Λ_k are arbitrarily chosen events from $\mathcal{L}_k$ for $k = 1, 2, \cdots$, respectively.

Definition 2. (Liu [21]) An uncertain variable ξ is a measurable function from an uncertainty space $(\Gamma, \mathcal{L}, \mathcal{M})$ to the set of real numbers, i.e., for any Borel set B of real numbers, the set

$$\{\xi \in B\} = \{\gamma \in \Gamma | \xi(\gamma) \in B\}$$

is an event.

An uncertain variable is essentially a measurable function from an uncertain space to the set of real numbers. In order to describe an uncertain variable, a concept of uncertainty distribution is defined as follows.

Definition 3. (Liu [21]) The uncertainty distribution of an uncertain variable ξ is defined by

$$\Phi(x) = \mathcal{M}\{\xi \leq x\}$$

for any $x \in \Re$.

Definition 4. (Liu [21]) Let ξ be an uncertain variable with regular uncertainty distribution Φ. Then, the inverse function Φ^{-1} is called the inverse uncertainty distribution of ξ.

The distribution of a monotonous function of uncertain variables can be obtained by the following theorem.

Theorem 1. *(Liu [22]) Let $\xi_1, \xi_2, \cdots, \xi_n$ be independent uncertain variables with uncertainty distributions $\Phi_1, \Phi_2, \cdots, \Phi_n$, respectively. If $f(\xi_1, \xi_2, \cdots, \xi_n)$ is strictly increasing with respect to $\xi_1, \xi_2, \cdots, \xi_m$ and strictly decreasing with respect to $\xi_{m+1}, \xi_{m+2}, \cdots, \xi_n$, then $\xi = f(\xi_1, \xi_2, \cdots, \xi_n)$ is an uncertain variable with an inverse uncertainty distribution*

$$\Phi^{-1}(\alpha) = f\left(\Phi_1^{-1}(\alpha), \cdots, \Phi_m^{-1}(\alpha), \Phi_{m+1}^{-1}(1-\alpha), \cdots, \Phi_n^{-1}(1-\alpha)\right).$$

Definition 5. (Liu [21]) The expected value of an uncertain variable ξ is defined by

$$E[\xi] = \int_0^{+\infty} \mathcal{M}\{\xi \geq x\}\mathrm{d}x - \int_{-\infty}^0 \mathcal{M}\{\xi \leq x\}\mathrm{d}x$$

provided that at least one of the two integrals is finite.

Theorem 2. *(Liu [21]) Let ξ be an uncertain variable with uncertainty distribution Φ. If the expected value exists, then*

$$E[\xi] = \int_0^{+\infty} (1 - \Phi(x))\mathrm{d}x - \int_{-\infty}^0 \Phi(x)\mathrm{d}x.$$

Based on this result, Liu [25] proved the linearity property of the expected value operator. For two independent uncertain variables ξ and η, we have $E[a\xi + b\eta] = aE[\xi] + bE[\eta]$ where a and b are real numbers.

In 2010, Liu [22] first introduced a formula expected value by inverse uncertainty distribution, that is

$$E[\xi] = \int_0^1 \Phi^{-1}(\alpha)\mathrm{d}\alpha.$$

Liu and Ha [26] proposed a generalized formula for expected value by inverse uncertainty distribution.

Theorem 3. *(Liu and Ha [26]) Let $\xi_1, \xi_2, \cdots, \xi_n$ be independent uncertain variables with uncertainty distributions $\Phi_1, \Phi_2, \cdots, \Phi_n$, respectively. If $f(\xi_1, \xi_2, \cdots, \xi_n)$ is strictly increasing with respect to $\xi_1, \cdots, \xi_m$ and strictly decreasing with respect to $\xi_{m+1}, \cdots, \xi_n$, then the uncertain variable $\xi = f(\xi_1, \xi_2, \cdots, \xi_n)$ has an expected value*

$$E[\xi] = \int_0^1 f\left(\Phi_1^{-1}(\alpha), \cdots, \Phi_m^{-1}(\alpha), \Phi_{m+1}^{-1}(1-\alpha), \cdots, \Phi_n^{-1}(1-\alpha)\right)\mathrm{d}\alpha.$$

Meanwhile, Liu [21] presented the concept of variance of uncertain variables and also proposed some formulas to calculate the variance through uncertainty distribution. Recently, Yao [28] proposed a formula to calculate the variance using inverse uncertainty distribution. Sheng and Kar [29] verified some results of moment of uncertain variable through inverse uncertainty distribution.

Chance theory

Probability theory was developed by Kolmogorov [30] in 1933. Since then, probability theory has become an important branch of mathematics for modeling frequencies, while uncertainty theory was founded by Liu [21] in 2007 and subsequently studied by many researchers. Nowadays, uncertainty theory has become a branch of axiomatic mathematics for modeling belief degrees. However, in many cases, uncertainty and randomness simultaneously appear in a complex system. In order to describe this phenomenon, Liu [31] first proposed chance theory, which is a mathematical methodology for modeling complex systems with both uncertainty and randomness in 2013, including chance measure, uncertain random variable, chance distribution, operational law, expected value, and so on. As an important contribution to chance theory, Liu [32] presented an operational law of uncertain random variables. Meanwhile, Guo and Wang [33] proposed a formula to calculate the variance through chance distribution and Sheng and Yao [34] verified some results of variance through inverse chance distribution. Furthermore, in order to deal with uncertain random phenomenon evolving in time, Gao and Yao [35] presented an uncertain random process and an uncertain random renewal process in the light of chance theory.

Let $(\Gamma, \mathcal{L}, \mathcal{M})$ be an uncertainty space and $(\Omega, \mathcal{A}, \Pr)$ be a probability space. Then, the chance space refers to the product $(\Gamma, \mathcal{L}, \mathcal{M}) \times (\Omega, \mathcal{A}, \Pr)$.

Definition 6. (Liu [31]) Let $(\Gamma, \mathcal{L}, \mathcal{M}) \times (\Omega, \mathcal{A}, \Pr)$ be a chance space, and let $\Theta \in \mathcal{L} \times \mathcal{A}$ be an uncertain random event. Then, the chance measure of Θ is defined as

$$\mathrm{Ch}\{\Theta\} = \int_0^1 \Pr\{\omega \in \Omega \mid \mathcal{M}\{\gamma \in \Gamma | (\gamma, \omega) \in \Theta\} \geq r\} \mathrm{d}r.$$

Liu [31] proved that a chance measure satisfies normality, duality, and monotonicity properties, that is

(i) $\mathrm{Ch}\{\Gamma \times \Omega\} = 1$.

(ii) $\mathrm{Ch}\{\Theta\} + \mathrm{Ch}\{\Theta^c\} = 1$ for and event Θ.

(iii) $\mathrm{Ch}\{\Theta_1\} \leq \mathrm{Ch}\{\Theta_2\}$ for any real number set $\Theta_1 \subset \Theta_2$.

Besides, Hou [36] proved the subadditivity of chance measure, that is,

$$\mathrm{Ch}\left\{\bigcup_{i=1}^{\infty} \Theta_i\right\} \leq \sum_{i=1}^{\infty} \mathrm{Ch}\{\Theta_i\}$$

for a sequence of events $\Theta_1, \Theta_2, \cdots$.

Definition 7. (Liu [31]) An uncertain random variable is a measurable function ξ from a chance space $(\Gamma, \mathcal{L}, \mathcal{M}) \times (\Omega, \mathcal{A}, \Pr)$ to the set of real numbers, *i.e.*, $\{\xi \in B\}$ is an event for any Borel set B.

Random variables and uncertain variables can be regarded as special cases of uncertain random variables. Let η be a random variable, and let τ be an uncertain variable. Then, $\eta + \tau$ and $\eta \times \tau$ are both uncertain random variables.

To calculate the chance measure, Liu [31] presented a definition of chance distribution.

Definition 8. (Liu [31]) Let ξ be an uncertain random variable. Then, its chance distribution by

$$\Phi(x) = \mathrm{Ch}\{\xi \le x\}$$

for any $x \in \mathcal{R}$.

The chance distribution of a random variable is just its probability distribution, and the chance distribution of an uncertain variable is just its uncertainty distribution.

Theorem 4. *(Liu [32]) Let $\eta_1, \eta_2, \cdots, \eta_m$ be independent uncertain random variables with probability distributions $\Psi_1, \Psi_2, \cdots, \Psi_m$, respectively, and let $\tau_1, \tau_2, \cdots, \tau_n$ be uncertain variables. Then, the uncertain random variable*

$$\xi = f(\eta_1, \eta_2, \cdots, \eta_m, \tau_1, \tau_2, \cdots, \tau_n)$$

has a chance distribution

$$\Phi(x) = \int_{\mathcal{R}^m} F(x, y_1, \cdots, y_m)\mathrm{d}\Psi_1(y_1)\cdots\mathrm{d}\Psi_m(y_m)$$

where $F(x, y_1, \cdots, y_m)$ is the uncertainty distribution of uncertain variable

$$f(y_1, y_2, \cdots, y_m, \tau_1, \tau_2, \cdots, \tau_n)$$

for any real numbers $y_1, y_2, \cdots, y_m$.

Definition 9. (Liu [32]) Let ξ be an uncertain random variable. Then, its expected value is defined by

$$E[\xi] = \int_0^{+\infty} \mathrm{Ch}\{\xi \ge r\}\mathrm{d}r - \int_{-\infty}^0 \mathrm{Ch}\{\xi \le r\}\mathrm{d}r$$

provided that at least one of the two integrals is finite.

Let Φ denote the chance distribution of ξ. Liu [32] proved a formula to calculate the expected value of uncertain random variable with chance distribution if $E[\xi]$ exists, then

$$E[\xi] = \int_0^{+\infty} (1 - \Phi(x))\mathrm{d}x - \int_{-\infty}^0 \Phi(x)\mathrm{d}x.$$

Let Φ be an uncertain random variable with regular chance distribution Φ. Liu [32] proved a formula to calculate the expected value of uncertain random variable with inverse chance distribution if $E[\xi]$ exists, then

$$E[\xi] = \int_0^1 \Phi^{-1}(\alpha)\mathrm{d}\alpha.$$

Theorem 5. *(Liu [32]) Let $\eta_1, \eta_2, \cdots, \eta_m$ be independent uncertain random variables with probability distributions, $\Psi_1, \Psi_2, \cdots, \Psi_m$, respectively, and let $\tau_1, \tau_2, \cdots, \tau_n$ be uncertain variables, then the uncertain random variable*

$$\xi = f(\eta_1, \eta_2, \cdots, \eta_m, \tau_1, \tau_2, \cdots, \tau_n)$$

has an expected value

$$E[\xi] = \int_{\mathcal{R}^m} E\left[f(y_1, \cdots, y_m, \tau_1, \cdots, \tau_n)\right] \mathrm{d}\Psi_1(y_1) \cdots \mathrm{d}\Psi_m(y_m)$$

where $E[y_1, \cdots, y_m, \tau_1, \cdots, \tau_n]$ is the expected value of uncertain variable $f(y_1, \cdots, y_m, \tau_1, \cdots, \tau_n)$ for any real numbers $y_1, y_2, \cdots, y_m$.

Meanwhile, Liu [32] proved the linearity of expected value operator, that is

$$E[\eta + \tau] = E[\eta] + E[\tau]$$

where η is a random variable and is τ an uncertain variable.

In application, Liu [32] founded uncertain random programming in 2013. As extensions, Zhou et al. [37] proposed uncertain random multi-objective programming for optimizing multiple, incommensurable, and conflicting objectives. After that, uncertain random programming was developed steadily and applied widely; Qin [38] proposed uncertain random goal programming in order to satisfy as many goals as possible in the order specified, and Ke [39] proposed uncertain random multilevel programming for studying decentralized decision systems. In order to quantify the rise of uncertain random systems in which the leader and followers may have their own decision variables and objective functions, Liu and Ralescu [40] invented the tool of uncertain random risk analysis. Wen et al. [41] presented the tool of uncertain random reliability analysis for dealing with uncertain random systems.

Uncertain random network

In this section, we introduce a definition of an uncertain random network and the shortest path chance distribution of uncertain random network.

Definition 10. (Liu [20]) Assume $\mathcal{N}$ is the collection of nodes, $\mathcal{U}$ is the collection of uncertain arcs, $\mathcal{R}$ is the collection of random arcs, and $\mathcal{C}$ is the collection of uncertain and random arc capacities. Then, the quartette $(\mathcal{N}, \mathcal{U}, \mathcal{R}, \mathcal{C})$ is said to be an uncertain random network.

In this paper, we assume that the uncertain random network is of order n with a collection of nodes $\mathcal{N} = \{1, 2, \cdots, n\}$, where '1' is of the source node, and 'n' is the destination node. We defined two collections of arcs,

$$\mathcal{U} = \{(i,j) | (i,j) \text{are uncertain arcs}\},$$

$$\mathcal{R} = \{(i,j) | (i,j) \text{are random arcs}\}.$$

Note that all deterministic arcs are regarded as special uncertain ones. Let C_{ij} denote the capacities of arcs (i,j), $(i,j) \in \mathcal{U} \cup \mathcal{R}$, respectively. Then, C_{ij} are uncertain variables if $(i,j) \in \mathcal{U}$ and random variables if $(i,j) \in \mathcal{R}$. Write $\mathcal{C} = \{C_{ij} | (i,j) \in \mathcal{U} \cup \mathcal{R}\}$.

Figure 1 shows an uncertain random network $(\mathcal{N}, \mathcal{U}, \mathcal{R}, \mathcal{C})$ of order 6 in which

$$\mathcal{N} = \{1, 2, 3, 4, 5, 6\},$$

$$\mathcal{U} = \{(1, 2), (2, 5), (3, 5), (5, 6)\},$$

$$\mathcal{R} = \{(1, 3), (3, 4), (2, 4), (4, 6)\},$$

$$\mathcal{C} = \{C_{12}, C_{13}, C_{24}, C_{25}, C_{34}, C_{35}, C_{46}, C_{56}\}.$$

The uncertain random network degenerates to a random network (Frank and Hakimi [1]) if all capacities are random variables and degenerates to an uncertain network (Liu [22]) if all weights are uncertain variables.

Theorem 6. *(Shortest path chance distribution Liu [20]) Let $(\mathcal{N}, \mathcal{U}, \mathcal{R}, \mathcal{C})$ be an uncertain random network. Assume that the uncertain capacities ξ_{ij} have regular uncertainty distributions Υ_{ij} for $(i,j) \in \mathcal{U}$ and the random capacities ξ_{ij} have probability distributions Ψ_{ij} for $(i,j) \in \mathcal{R}$, respectively. Then, the shortest path from a source node to a destination node has a chance distribution*

$$\Phi(x) = \int_0^{+\infty} \cdots \int_0^{+\infty} F(x; y_{ij}, (i,j) \in \mathcal{R}) \prod_{(i,j)\in\mathcal{R}} d\Psi_{ij}(y_{ij})$$

where $F(x; y_{ij}, (i,j) \in \mathcal{R})$ is the uncertainty distribution of uncertain variable $f(y_{ij}, (i,j) \in \mathcal{R}, \xi_{ij}, (i,j) \in \mathcal{U})$ and it is determined by its inverse uncertainty distribution

$$F^{-1}(\alpha; y_{ij}, (i,j) \in \mathcal{R}) = f\left(\Upsilon_{ij}^{-1}(\alpha), (i,j) \in \mathcal{U}, y_{ij}, (i,j) \in \mathcal{R}\right),$$

and f can be calculated by the Dijkstra's algorithm.

Maximum flow of uncertain random network

In this section, we will introduce some algorithm of the maximum and apply chance theory to the maximum flow problem in an uncertain random network. This is a distinct contribution from other optimization methods. We will illustrate that chance theory can serve as a powerful tool to deal with the maximum flow in an uncertain random network.

Maximum flow problem

For a network, there is one of the important problems to study the maximum flow. In the past five decades, many efficient algorithms for the maximum flow problem have emerged

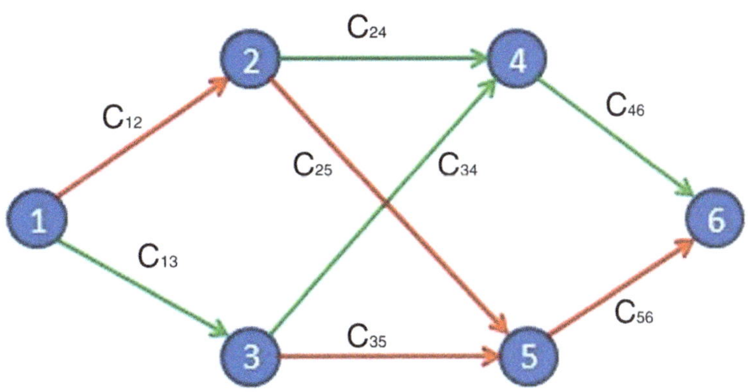

Figure 1 An uncertain random network.

for a fixed network. The maximum flow algorithm was first investigated by Fulkerson and Dantzig [42] in 1955. Then, representative methods in maximum flow algorithms are based on either augmenting paths or preflows. Augmenting path algorithms push flow along a path from the source to the sink in the residual network and include Ford-Fulkerson's labeling algorithm [43] and Dinic's blocking flow algorithm [44]. Edmonds and Karp [45] also independently proposed that the Ford and Fulkerson algorithm augments flow along shortest paths. In order to reduce the number of augmentations, preflow-based algorithms push flow along edges were investigated in the residual network and include Karzanov's blocking flow algorithms [46], which introduced the first preflow-push algorithm on layered networks, and Goldberg-Tarjan's push-relabeling algorithm [10]; it constructed distance labels instead of layered networks to improve the running time of preflow-push algorithm. They described a very flexible generic preflow-push algorithm that performs push and relabel operations at active nodes. In short, for a classical network, the maximum flow can be calculated by above for any algorithms.

How do we consider the maximum flow of a indeterminacy network? For a random network, Fishman [9], Goldberg and Tarjan [10], and Nawathe and Rao [11] mainly used stochastic optimization to solve the maximum flow problem in a random network. For an uncertain network, Han et al. [17] gave the inverse uncertain distribution of the maximum flow in an uncertain network. In this paper, according to chance theory, we will study the chance distribution, the maximum flow of an uncertain random network.

In here, we assume that the networks are directed with only one source and one sink. If the arc capacities of a network are given, then we can calculate the maximum flow f of the network using the above algorithm. For the different capacities, obtain different but unique maximum flow f. In other words, the maximum flow f is a function of arc capacities. In paper [17], the author proved that the maximum flow f is a continuous and strictly increasing function with respect to C_{ij}, where C_{ij} denote the capacities of the (i,j) arcs. For a random network, the maximum flow f is a random variable function of arc capacities. For an uncertain network, the maximum flow f is an uncertain variable function of arc capacities. Similarly, for an uncertain random network, the maximum flow f is an uncertain random variable function of arc capacities.

Chance distribution of maximum flow

Now, we employ chance theory to deal with this indeterministic factor. Define $\xi = \{\xi_{ij} | (i,j) \in \mathcal{U} \cup \mathcal{R}\}$. We can denote the network with uncertain random arc capacities as $(\mathcal{N}, \mathcal{U}, \mathcal{R}, \mathcal{C})$, its maximum flow is $f(\xi)$. Obviously, $f(\xi)$ is an uncertain random variable. Then, we can obtain the chance distribution of the maximum flow from a source node to a sink node by Theorem 4; we have the following theorem.

Theorem 7. *Let $(\mathcal{N}, \mathcal{U}, \mathcal{R}, \mathcal{C})$ be an uncertain random network. Assume that the uncertain capacities ξ_{ij} have regular uncertainty distributions Υ_{ij} for $(i,j) \in \mathcal{U}$ and the random capacities ξ_{ij} have probability distributions Ψ_{ij} for $(i,j) \in \mathcal{R}$, respectively. Then, the maximum flow $f(\xi_{ij}, (i,j) \in \mathcal{U} \cup \mathcal{R})$ has a chance distribution*

$$\Phi(x) = \int_0^{+\infty} \cdots \int_0^{+\infty} F(x; y_{ij}, (i,j) \in \mathcal{R}) \prod_{(i,j)\in\mathcal{R}} \mathrm{d}\Psi_{ij}(y_{ij})$$

where $F(x; y_{ij}, (i,j) \in \mathcal{R})$ is the uncertainty distribution of uncertain variable $f(y_{ij}, (i,j) \in \mathcal{R}, \xi_{ij}, (i,j) \in \mathcal{U})$, and it is determined by its inverse uncertainty distribution

$$F^{-1}\left(\alpha; y_{ij}, (i,j) \in \mathcal{R}\right) = f\left(\Upsilon_{ij}^{-1}(\alpha), (i,j) \in \mathcal{U}, y_{ij}, (i,j) \in \mathcal{R}\right),$$

and f may be calculated by the Ford-Fulkerson's algorithm.

Proof. By Definitions 6 and 8 and Theorem 4, we have

$$\begin{aligned}\Phi(x) &= \mathrm{Ch}\left\{f(\xi_{ij}, (i,j) \in \mathcal{U} \cup \mathcal{R}) \le x\right\} \\ &= \int_0^1 \Pr\left\{\omega \in \Omega | \mathcal{M}\left\{f(\xi_{ij}(\omega), (i,j) \in \mathcal{R}, \xi_{ij}, (i,j) \in \mathcal{U}) \le x\right\} \ge r\right\} \mathrm{d}r \\ &= \int_0^{+\infty} \cdots \int_0^{+\infty} \mathcal{M}\left\{f(\xi_{ij}(\omega), (i,j) \in \mathcal{R}, \xi_{ij}, (i,j) \in \mathcal{U}) \le x\right\} \prod_{(i,j)\in\mathcal{R}} \mathrm{d}\Psi_{ij}(y_{ij}) \\ &= \int_0^{+\infty} \cdots \int_0^{+\infty} F(x; y_{ij}, (i,j) \in \mathcal{R}) \prod_{(i,j)\in\mathcal{R}} \mathrm{d}\Psi_{ij}(y_{ij})\end{aligned}$$

where $F(x; y_{ij}, (i,j) \in \mathcal{R})$ is the uncertainty distribution of uncertain variable $f(y_{ij}, (i,j) \in \mathcal{R}, \xi_{ij}, (i,j) \in \mathcal{U})$ for any real numbers $y_{ij}, (i,j) \in \mathcal{R}$ and it is determined by its inverse uncertainty distribution

$$F^{-1}\left(\alpha; y_{ij}, (i,j) \in \mathcal{R}\right) = f\left(\Upsilon_{ij}^{-1}(\alpha), (i,j) \in \mathcal{U}, y_{ij}, (i,j) \in \mathcal{R}\right),$$

and $f\left(\Upsilon_{ij}^{-1}(\alpha), (i,j) \in \mathcal{U}, y_{ij}, (i,j) \in \mathcal{R}\right)$ is just the maximum flow of a determinacy network and f is a strictly increasing function with respect to ξ_{ij}, where ξ_{ij} denote the capacities of the (i,j) arcs. We may be calculated f by the Ford-Fulkerson's algorithm or Dinic's algorithm for each given α. The theorem is verified. □

Remark 1. *If the uncertain random network becomes a random network, then the probability distribution of maximum flow is*

$$\Phi(x) = \int_{f(y_{ij},(i,j)\in\mathcal{R})\le x} \prod_{(i,j)\in\mathcal{R}} \mathrm{d}\Psi_{ij}(y_{ij}).$$

Remark 2. *(Han el at. [17]) If the uncertain random network becomes an uncertain network, then the inverse uncertainty distribution of maximum flow is*

$$\Phi^{-1}(\alpha) = f\left(\Upsilon_{ij}^{-1}(\alpha), (i,j) \in \mathcal{U}\right).$$

Example 1. There is a series uncertain random network with n arcs defined by Figure 2. Assume that the uncertain capacities ξ_{ij} have regular uncertainty distributions Υ_{ij} for $(i,j) \in \mathcal{U}$, and the random capacities ξ_{ij} have probability distributions Ψ_{ij} for $(i,j) \in \mathcal{R}$, respectively. Then, by Theorem 6, we have that the maximum flow of the network

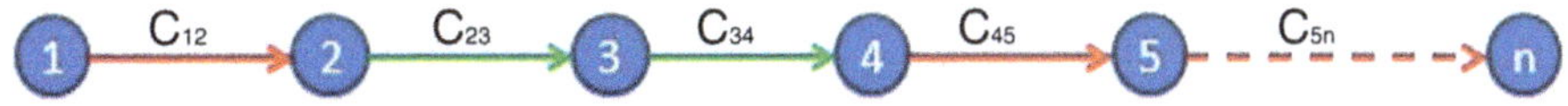

Figure 2 Network $(\mathcal{N}, \mathcal{U}, \mathcal{R}, \mathcal{C})$ for Example 1.

$(\mathcal{N}, \mathcal{U}, \mathcal{R}, \mathcal{C})$ is an uncertain random variable and its chance distribution is

$$\Phi(x) = \int_0^{+\infty} \cdots \int_0^{+\infty} F(x; y_{ij}, (i,j) \in \mathcal{R}) \prod_{(i,j)\in\mathcal{R}} \mathrm{d}\Psi_{ij}(y_{ij})$$

where $F(x; y_{ij}, (i,j) \in \mathcal{R})$ is determined by its inverse uncertainty distribution

$$F^{-1}(\alpha; y_{ij}, (i,j) \in \mathcal{R}) = \min_{(i,j)\in\mathcal{U}\cup\mathcal{R}} \left\{ \Upsilon_{ij}^{-1}(\alpha), y_{ij} \right\}.$$

Example 2. Assume an uncertain random network with four arcs defined by Figure 3. Assume that the uncertain capacities τ_1, τ_2 have regular uncertainty distributions Υ_1, Υ_2, and the random capacities ξ_1, ξ_2 have probability distributions Ψ_1, Ψ_2, respectively. Then, by Theorem 6, we have that the maximum flow of the network $(\mathcal{N}, \mathcal{U}, \mathcal{R}, \mathcal{C})$ is an uncertain random variable and its chance distribution is

$$\Phi(x) = \int_0^{+\infty} \int_0^{+\infty} F(x; y_1, y_2)\mathrm{d}\Psi_1(y_1)\mathrm{d}\Psi_2(y_2)$$

where $F(x; y_1, y_2)$ is determined by its inverse uncertainty distribution

$$F^{-1}(\alpha; y_1, y_2) = y_1 \wedge \Upsilon_1^{-1}(\alpha) + y_2 \wedge \Upsilon_2^{-1}(\alpha).$$

Example 3. Assume an uncertain random network with five arcs defined by Figure 4. Assume that the uncertain capacities τ_1, τ_2 have regular uncertainty distributions Υ_1, Υ_2, and the random capacities ξ_1, ξ_2, ξ_3 have probability distributions Ψ_1, Ψ_2, Ψ_3, respectively. Then, by Theorem 6, we have that the maximum flow of the network $(\mathcal{N}, \mathcal{U}, \mathcal{R}, \mathcal{C})$ is an uncertain random variable and its chance distribution is

$$\Phi(x) = \int_0^{+\infty} \int_0^{+\infty} \int_0^{+\infty} F(x; y_1, y_2, y_3)\mathrm{d}\Psi_1(y_1)\mathrm{d}\Psi_2(y_2)\mathrm{d}\Psi_3(y_3)$$

where $F(x; y_1, y_2, y_3)$ is determined by its inverse uncertainty distribution

$$F^{-1}(\alpha; y_1, y_2, y_3) = \left(\left(\left(y_1 - \Upsilon_1^{-1}(\alpha)\right) \vee 0\right) \wedge y_3 + y_2\right) \wedge \Upsilon_2^{-1}(\alpha) + \left(y_1 \wedge \Upsilon_1^{-1}(\alpha)\right).$$

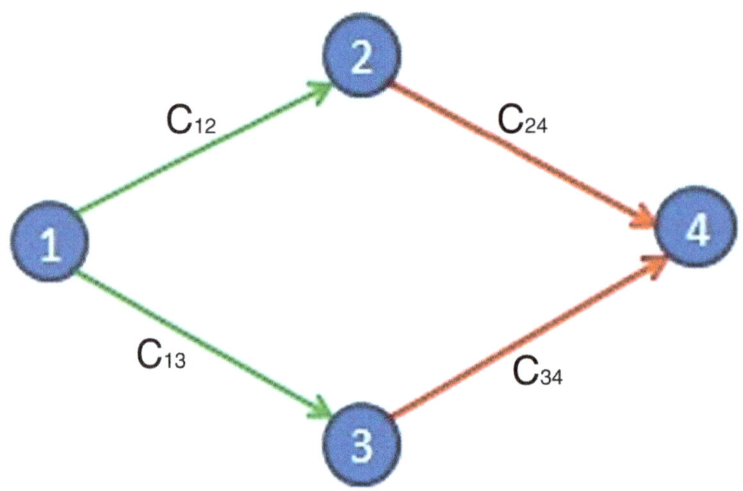

Figure 3 **Network $(\mathcal{N}, \mathcal{U}, \mathcal{R}, \mathcal{C})$ for Example 2.**

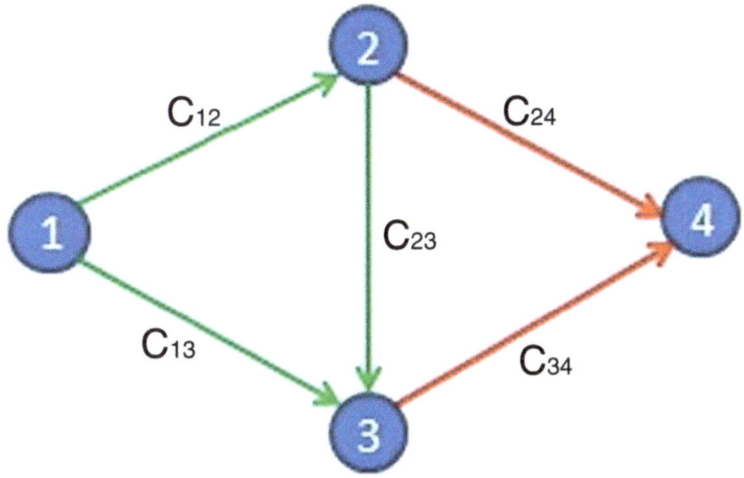

Figure 4 Network $(\mathcal{N}, \mathcal{U}, \mathcal{R}, \mathcal{C})$ for Example 3.

Expected value of maximum flow

In uncertain random network, we can obtain chance distribution of maximum flow, but it is difficult to calculate the maximum flow. Sometimes, we need not know the maximum flow for an uncertain random network, but only it is enough average value of flow, so we need only to calculate the expected value of the maximum flow. By Theorem 5, we have the following theorem.

Theorem 8. *Let* $(\mathcal{N}, \mathcal{U}, \mathcal{R}, \mathcal{C})$ *be an uncertain random network. Assume that the uncertain capacities* ξ_{ij} *have regular uncertainty distributions* Υ_{ij} *for* $(i,j) \in \mathcal{U}$ *and the random capacities* ξ_{ij} *have probability distributions* Ψ_{ij} *for* $(i,j) \in \mathcal{R}$*, respectively. Then, the maximum flow* $\xi = f(\xi_{ij}, (i,j) \in \mathcal{U} \cup \mathcal{R})$ *has an expected value*

$$E[\xi] = \int_0^{+\infty} \cdots \int_0^{+\infty} \int_0^1 f\left(\Upsilon_{ij}^{-1}(\alpha), (i,j) \in \mathcal{U}, y_{ij}, (i,j) \in \mathcal{R}\right) \mathrm{d}\alpha \prod_{(i,j)\in\mathcal{R}} \mathrm{d}\Psi_{ij}(y_{ij}).$$

Proof. Since the maximum flow $\xi = f(\xi_{ij}, (i,j) \in \mathcal{U} \cup \mathcal{R})$ is a strictly increasing function with respect to $\xi_{ij}, (i,j) \in \mathcal{U}$, we have □

$$E[y_{ij}, (i,j) \in \mathcal{R}, \xi_{ij}, (i,j) \in \mathcal{U}] = \int_0^1 f\left(\Upsilon_{ij}^{-1}(\alpha), (i,j) \in \mathcal{U}, y_{ij}, (i,j) \in \mathcal{R}\right) \mathrm{d}\alpha.$$

It follows from Theorem 5, we have

$$\begin{aligned} E[\xi] &= \int_0^{+\infty} \cdots \int_0^{+\infty} E[f(\xi_{ij}(\omega), (i,j) \in \mathcal{R}, \xi_{ij}, (i,j) \in \mathcal{U})] \prod_{(i,j)\in\mathcal{R}} \mathrm{d}\Psi_{ij}(y_{ij}) \\ &= \int_0^{+\infty} \cdots \int_0^{+\infty} \int_0^1 f\left(\Upsilon_{ij}^{-1}(\alpha), (i,j) \in \mathcal{U}, y_{ij}, (i,j) \in \mathcal{R}\right) \mathrm{d}\alpha \prod_{(i,j)\in\mathcal{R}} \mathrm{d}\Psi_{ij}(y_{ij}). \end{aligned}$$

The theorem is verified.

In Example 1, by Theorem 7, we can obtain the expected value of the maximum

$$E[\xi] = \int_0^{+\infty} \cdots \int_0^{+\infty} \int_0^1 \min_{(i,j)\in\mathcal{U}\cup\mathcal{R}} \left\{\Upsilon_{ij}^{-1}(\alpha), y_{ij}\right\} \mathrm{d}\alpha \prod_{(i,j)\in\mathcal{R}} \mathrm{d}\Psi_{ij}(y_{ij}).$$

In Example 2, by Theorem 7, we can obtain the expected value of the maximum

$$E[\xi] = \int_0^{+\infty}\int_0^{+\infty}\int_0^1 \left(y_1 \wedge \Upsilon_1^{-1}(\alpha) + y_2 \wedge \Upsilon_2^{-1}(\alpha)\right) \mathrm{d}\alpha \mathrm{d}\Psi_1(y_1)\mathrm{d}\Psi_2(y_2).$$

In Example 3, by Theorem 7, we can obtain the expected value of the maximum

$$E[\xi] = \int_0^{+\infty}\int_0^{+\infty}\int_0^{+\infty}\int_0^1 \left(\left(\left(\left(y_1 - \Upsilon_1^{-1}(\alpha)\right) \vee 0\right) \wedge y_3 + y_2\right) \wedge \Upsilon_2^{-1}(\alpha) + \left(y_1 \wedge \Upsilon_1^{-1}(\alpha)\right)\right)$$

$$\mathrm{d}\alpha \mathrm{d}\Psi_1(y_1)\mathrm{d}\Psi_2(y_2)\mathrm{d}\Psi_3(y_3).$$

Conclusions

Indeterministic factors often appear in network flow problems. In the past, probability theory and uncertainty theory have been employed to deal with these indeterministic factors. Chance theory provides a new approach to deal with indeterministic factors in a complex indeterministic network. In this paper, we investigated the maximum flow problem of network in an uncertain random environment. Under the framework of chance theory, we gave the chance distribution of the maximum flow and the expected value of the maximum flow of uncertain random network was derived. Some examples were derived to illustrate the theoretical considerations.

Acknowledgements
This work was supported by the National Natural Science Foundation of China Grants No.61273044, No.91224008, No.61374082 and No.61262023.

Author details
[1]Department of Mathematical Sciences, Tsinghua University, Beijing 100084, China. [2]School of Information, Renmin University of China, Beijing 100872, China.

References

1. Frank, H, Hakimi, SL: Probabilistic flows through a communication network. IEEE Trans. on Circuit Theory. **12**, 413–414 (1965)
2. Frank, H: Shortest paths in probability graphs. Oper. Res. **17**(4), 583–599 (1969)
3. Mirchandani, PB: Shortest distance and reliability of probabilistic networks. Comput. Oper. Res. **3**(4), 347–676 (1976)
4. Sigal, CL, Pritsker, AB, Solberg, JJ: The stochastic shortest route problem. Oper. Res. **5**(28), 1122–1129 (1980)
5. Frank, H, Frisch, IT: Communication, transmission, and transportation networks. Publishing Co., Reading, Mass., Addison-Wesley (1971)
6. Doulliez, P: Probability distribution function for the capacity of a multiterminal network. Revue Francaise d'Informatique et de Recherche Operationnelle. **1**(5), 39–49 (1971)
7. Carey, M, Henrickson, C: Bounds on expected performance of networks with links subject to failure. Networks. **14**(3), 439–456 (1984)
8. Onaga, K: Bounds on the average terminal capacity of probabilistic nets. IEEE Trans. on Inf. Theory. **14**(5), 766–768 (1968)
9. Fishman, GS: The distribution of maximum flow with applications to multistate reliability systems. Oper. Res. **35**(4), 607–618 (1987)
10. Goldberg, AV, Tarjan, RE: A new approach to the maximum flow problem. J. Assoc. Comput. Machinery. **35**(4), 921–940 (1988)
11. Nawathe, SP, Rao, BV: Maximum flow in probabilistic communication networks. J. Circuit Theory Appl. **8**, 167–177 (1980)
12. Fu, L, Rilett, LR: Expected shortest paths in dynamic and stochastic traffic networks. Transportation Res. Part B: Methodol. **32**(7), 499–516 (1998)
13. Hall, R: The fastest path through a network with random time-dependent travel time. Transportation Sci. **20**(3), 182–188 (1986)
14. Lou, P: Optimal paths in graphs with stochastic or multidimensional weights. Commun. Assoc. Comput. Machinery. **26**(9), 670–676 (1983)
15. Liu, B: Theory and practice of uncertain programming. 2nd ed. Springer-Verlag, Berlin (2009)
16. Gao, Y: Shortest path problem with uncertain arc lengths. Comput. Math. Appl. **62**, 2591–2600 (2011)
17. Han, SW, Peng, ZX, Wan, SQ: The maximum flow problem of uncertain network. Inf. Sci. **265**, 167–175 (2014)
18. Ding, SB: Uncertain minimum cost flow problem (2013). doi:10.1007/s00500-013-1194-4
19. Zhang, B, Peng, J: Uncertain programming model for Chinese postman problem with uncertain weights. Ind. Eng. Manag. Syst. **11**(1), 18–25 (2012)

20. Liu, B: Uncertain random graph and uncertain random network. J. Uncertain Syst. **8**(1), 3–12 (2014)
21. Liu, B: Uncertainty Theory. 2nd ed. Springer-Verlag, Berlin (2007)
22. Liu, B: Uncertainty Theory: A Branch of Mathematics for Modeling Human Uncertainty. Springer-Verlag, Berlin (2010)
23. Gao, X: Some properties of continuous uncertain measure. Int. J. Uncertainty Fuzziness Knowledge-Based Syst. **17**(3), 419–426 (2009)
24. Peng, ZX, Iwamura, K: A sufficient and necessary condition of uncertainty distribution. J. Interdiscip. Math. **13**(3), 277–285 (2010)
25. Liu, B: Some research problems in uncertainty theory. J. Uncertain Syst. **3**(1), 3–10 (2009)
26. Liu, YH, Ha, MH: Expected value of function of uncertain variables. J. Uncertain Syst. **4**(3), 181–186 (2010)
27. Liu, B: Why is there a need for uncertainty theory. J. Uncertain Syst. **6**(1), 3–10 (2012)
28. Yao, K: A formula to calculate the variance of uncertain variable. http://orsc.edu.cn/online/130831.pdf
29. Sheng, YH, Samarjit, K: Some results of moments of uncertain variable through inverse uncertainty distribution. Fuzzy Optimization Decis. Making (2014). in press
30. Kolmogorov, AN: Grundbegriffe der Wahrscheinlichkeitsrechnung. Julius Springer, Berlin (1933)
31. Liu, YH: Uncertain random variables: a mixture of uncertainty and randomness. Soft Comput. **17**(4), 625–634 (2013)
32. Liu, YH: Uncertain random programming with applications. Fuzzy Optimization Decis. Making. **12**(2), 153–169 (2013)
33. Guo, HY, Wang, XS: Variance of uncertain random variables. J. Uncertainty Anal. Appl. **2**, Article 6 (2014)
34. Sheng, YH, Yao, K: Some formulas of variance of uncertain random variable. J. Uncertainty Anal. Appl. **2**, Article 12 (2014)
35. Gao, JW, Yao, K: Some concepts and theorems of uncertain random process. Int. J. Intell. Syst (2014). in press
36. Hou, YC: Subadditivity of chance measure. J. Uncertainty Anal. Appl. **2**, Article 14 (2014)
37. Zhou, J, Yang, F, Wang, K: Multi-objective optimization in uncertain random environments. Fuzzy Optimization Decis. Making. doi:10.1007/s10700-014-9183-3
38. Qin, ZF: Uncertain random goal programming. http://orsc.edu.cn/online/130323.pdf
39. Ke, H: Uncertain random multilevel programming with application to product control problem. Soft Comput (2014). in press
40. Liu, YH, Ralescu, DA: Risk index in uncertain random risk analysis. Int. J. Uncertainty Fuzziness Knowledge-Based Syst. in press
41. Wen, ML, Kang, R: Reliability analysis in uncertain random system. http://orsc.edu.cn/online/120419.pdf
42. Fulkerson, DR, Dantzig, GB: Computations of maximum flow in networks. Naval Res. Logistics Q. **2**(4), 277–283 (1955)
43. Ford, LR, Fulkerson, DR: Maximal flow through a network. Can. J. Math. **8**(3), 399–404 (1956)
44. Dinic, EA: Algorithm for solution of a problem of maximum flow in networks with power estimation. Sov. Math. Doklady. **11**(8), 1277–1280 (1970)
45. Edmonds, J, Karp, RM: Theoretical improvements in algorithmic efficiency for network flow problems. J. Assoc. Comput. Machinery. **19**(2) (1972)
46. Karzanov, AV: Determining the maximum flow in a network by the method of preflows. Sov. Math. Doklady. **15**(3), 43–47 (1974)

11

Variance of uncertain random variables

Haiying Guo and Xiaosheng Wang*

*Correspondence:
xswang@hebeu.edu.cn
School of Sciences, Hebei University of Engineering, Handan 056038, China

Abstract

Uncertain random variable is developed to describe the phenomenon which mixes uncertainty with randomness. The variance of uncertain random variable may provide a degree of the spread of the distribution around its expected value. This paper presents a formula to obtain the variance of uncertain random variable.

Keywords: Uncertainty theory; Uncertain random variable; Chance measure; Variance

1 Introduction

Prior to today, probability theory and fuzzy set theory are two common mathematical tools to model indeterminacy phenomena and have been widely applied in information theory, engineering, management science, and so on. We know a fundamental premise of applying probability theory is that the estimated probability is closed enough to the real frequency. However, such as 'strength of bridge', 'about one million tons', 'tall', and 'most', due to lack of observed data and the complexity of environment, when making decisions, people have to consult with domain experts. In this case, information and knowledge cannot be described well by random variables. For fuzzy set theory, it was still challenged by many scholars after it was founded. Liu [1] presented several paradoxes to show that fuzzy variable and fuzzy set are not suitable for modeling uncertain quantities and unsharp concepts, respectively.

In order to deal with non-random phenomena, an uncertainty theory was founded by Liu [2] and refined by Liu [3] and then became a branch of mathematics for modeling belief degrees. One the hand, Liu presented some basic and important theoretical works of uncertainty theory [3]. On the other hand, as an application of uncertainty theory, Liu [4] proposed a spectrum of uncertain programming which is a type of mathematical programming involving uncertain variables and applied uncertain programming to system reliability design, facility location problem, vehicle routing problem, project scheduling problem, finance, and so on. In order to well interpret expert's experimental data by uncertainty theory, Liu first proposed uncertain statistics, and Wang et al. build several uncertain statistical methods such as the method of moments [5], uncertain Delphi method [6], and uncertain hypothesis testing for expert's empirical data [7] and discussed uncertain variance of sample and its application [8]. Nowadays, uncertainty theory was well developed on both theory section and practice section. For exploring the recent developments of uncertainty theory, the readers may consult the book [1].

Inspired by Kwakernaak [9,10], Puri and Ralescu [11], Kruse and Meyer [12], and Liu and Liu [13,14], uncertain random variable was first defined by Liu [15] to describe complex systems in which uncertainty and randomness frequently appear together. Otherwise, in order to model uncertain random phenomena well, Liu [15] also introduce the concepts of uncertain random arithmetic, chance measure, chance distribution, expected value, variance, and so on. As everyone knows, the variance of uncertain random variable will provide a degree of the spread of the distribution around its expected value. Although the concept of variance for uncertain random variable has been defined by expected value, we still do not know how to obtain the variance from chance distribution. This paper will build the formula for the convenience of obtaining the variance from chance distribution. The remainder of this paper is organized as follows. The next section is intended to introduce some concepts in uncertainty theory as they are needed. A formula for obtaining the variance of uncertain random variable and some examples are proposed in Section 3. Finally, a conclusion is drawn in Section 4.

2 Preliminary

Uncertainty theory was founded by Liu [2] in 2007 and refined by Liu [3] in 2010 and became a branch of mathematics based on normality, duality, subadditivity, and product axioms. For exploring the recent developments of uncertainty theory, the readers can consult the book by Liu [1]. Based on uncertainty theory and probability theory, chance theory was pioneered by Liu [15] in order to model the uncertain random compound systems. In this section, some useful concepts are introduced. Let $(\Gamma, \mathcal{L}, \mathcal{M})$ be an uncertainty space, where Γ is a non-empty set, $\mathcal{L}$ is a σ-algebra defined on Γ, and $\mathcal{M}$ is an uncertain measure, and let $(\Omega, \mathcal{A}, Pr)$ be a probability space, where Ω is a sample space, $\mathcal{A}$ is a σ-algebra defined on Ω, and Pr is a probability measure.

Definition 2.1. An uncertain random variable is a function ξ from a chance space $(\Gamma, \mathcal{L}, \mathcal{M}) \times (\Omega, \mathcal{A}, Pr)$ to the set of real numbers such that $\{\xi \in B\}$ is an event in $\mathcal{L} \times \mathcal{A}$ for any Borel set B [15].

Definition 2.2. Let ξ be an uncertain random variable on the chance space $(\Gamma, \mathcal{L}, \mathcal{M}) \times (\Omega, \mathcal{A}, Pr)$, and let B be a Borel set. Then, $\{\xi \in B\}$ is an uncertain random event with chance measure [15]

$$Ch\{\xi \in B\} = \int_0^1 Pr\{\omega \in \Omega \mid \mathcal{M}\{\gamma \in \Gamma \mid \xi(\gamma, \omega) \in B\} \geq x\} dx. \tag{2.1}$$

Note that the chance measure is in fact the expected value of the random variable $\mathcal{M}\{\xi(\cdot) \in B\}$, i.e.,

$$Ch\{\xi \in B\} = E[\mathcal{M}\{\xi(\cdot) \in B\}], \tag{2.2}$$

where E denotes the expected value operator for random variable under probability measure Pr.

Theorem 2.1. *Let ξ be an uncertain random variable. Then, the chance measure $Ch\{\xi \in B\}$ has normality, duality, and monotonicity properties, i.e., [15]*

(i) $Ch\{\xi \in \Re\} = 1$,

(ii) $Ch\{\xi \in B\} + Ch\{\xi \in B^c\} = 1$, *for any Borel set B,*

(iii) $Ch\{\xi \in B_1\} \leq Ch\{\xi \in B_2\}$, *for any Borel sets $B_1 \subset B_2$.*

Definition 2.3. Let ξ be an uncertain random variable. Then, its chance distribution of ξ is defined by [15]

$$\Phi(x) = Ch\{\xi \le x\}, \text{ for any } x \in \Re.$$

An operational law of uncertain random variables was given by Liu [16] as follows.

Theorem 2.2. *Let $\eta_1, \eta_2, \cdots, \eta_m$ be independent random variables with probability distributions $\Psi_1, \Psi_2, \cdots, \Psi_m$, and let $\tau_1, \tau_2, \cdots, \tau_n$ be independent uncertain variables with uncertainty distributions $\Upsilon_1, \Upsilon_2, \cdots, \Upsilon_n$, respectively. Then, the uncertain random variable $\xi = f(\eta_1, \eta_2, \cdots, \eta_m, \tau_1, \tau_2, \cdots, \tau_n)$ has a chance distribution [16]*

$$\Phi(x) = \int_{\Re^m} F(x; y_1, y_2, \cdots, y_m) d\Psi_1(y_1) d\Psi_2(y_2) \cdots d\Psi_m(y_m), \tag{2.3}$$

where $F(x; y_1, y_2, \cdots, y_m)$ is determined by its inverse function

$$F^{-1}(\alpha; y_1, y_2, \cdots, y_m) = f(y_1, y_2, \cdots, y_m, \Upsilon_1^{-1}(\alpha), \Upsilon_2^{-1}(\alpha), \cdots, \Upsilon_n^{-1}(\alpha)) \tag{2.4}$$

provided that $f(\eta_1, \eta_2, \cdots, \eta_m, \tau_1, \tau_2, \cdots, \tau_n)$ is a strictly increasing function with respect to $\tau_1, \tau_2, \cdots, \tau_n$.

Note that if $\eta_1, \eta_2, \cdots, \eta_m$ are non-independent random variables with joint distribution $\Psi(y_1, y_2, \cdots, y_m)$, Equation 2.3 can be modified as below,

$$\Phi(x) = \int_{\Re^m} F(x; y_1, y_2, \cdots, y_m) d\Psi(y_1, y_2, \cdots, y_m). \tag{2.5}$$

Definition 2.4. Let ξ be an uncertain random variable. Then its expected value is defined by [15]

$$E[\xi] = \int_0^{+\infty} Ch\{\xi \ge r\} dr - \int_{-\infty}^0 Ch\{\xi \le r\} dr \tag{2.6}$$

provided that at least one of the two integrals is finite.

Theorem 2.3. *Let $\eta_1, \eta_2, \cdots, \eta_m$ be independent random variables with probability distributions $\Psi_1, \Psi_2, \cdots, \Psi_m$, respectively, and let $\tau_1, \tau_2, \cdots, \tau_n$ be uncertain variables, then the uncertain random variable [16]*

$$\xi = f(\eta_1, \cdots, \eta_m, \tau_1, \cdots, \tau_n)$$

has an expected value

$$E[\xi] = \int_{\Re^m} E[f(y_1, \cdots, y_m, \tau_1, \cdots, \tau_n)] \, d\Psi_1(y_1) \cdots d\Psi_m(y_m)$$

where $E[f(y_1, \cdots, y_m, \tau_1, \cdots, \tau_n)]$ is the expected value of the uncertain variable $f(\eta_1, \cdots, \eta_m, \tau_1, \cdots, \tau_n)$ for any real numbers $y_1, \cdots, y_m$.

As a special case, Liu [16] proved the following theorem.

Theorem 2.4. *Let η be a random variable and let τ be an uncertain variable [16]. Then*

$$E[\eta + \tau] = E[\eta] + E[\tau] \tag{2.7}$$

and

$$E[\eta\tau] = E[\eta]\,E[\tau]\,. \tag{2.8}$$

3 Variance of uncertain random variables

A formal definition of variance for uncertain random variable was presented by Liu [15] as follows.

Definition 3.1. Let ξ be an uncertain random variable with expected value e. Then, its variance is defined by [15]

$$V[\xi] = E[(\xi - e)^2]\,. \tag{3.9}$$

The variance of uncertain random variable provides a degree of spread of the distribution around its expected value. The following theorem presents a formula to obtain the variance for a class of uncertain random variable.

Theorem 3.1. *Let $\eta_1, \eta_2, \cdots, \eta_m$ be independent random variables with probability distributions $\Psi_1, \Psi_2, \cdots, \Psi_m$, respectively, and let $\tau_1, \tau_2, \cdots, \tau_n$ be uncertain variables; then, the uncertain random variable*

$$\xi = f(\eta_1, \cdots, \eta_m, \tau_1, \cdots, \tau_n)$$

has variance

$$V[\xi] = 2\int_{R^m}\int_0^{\infty} x[\,1 - F(e+x; y_1, \cdots, y_m) + F(e-x; y_1, \cdots, y_m)]\,dx d\Psi_1(y_1)\cdots d\Psi_m(y_m) \tag{3.10}$$

where

$$F(x; y_1, \cdots, y_m) = \mathcal{M}\{f(y_1, \cdots, y_m, \tau_1, \cdots, \tau_n) \le x\}$$

and

$$e = \int_{\Re^m} E[f(y_1, \cdots, y_m, \tau_1, \cdots, \tau_n)]\,d\Psi_1(y_1)\cdots d\Psi_m(y_m)$$

are the uncertainty distribution and expected value of uncertain random variable ξ, respectively.

Proof. Suppose that the uncertain random variable $\xi = f(\eta_1, \cdots, \eta_m, \tau_1, \cdots, \tau_n)$ has the expected value e, by the definitions of variance and chance measure, we have

$$\begin{aligned}
V[\xi] &= \int_0^{+\infty} Ch\{(\xi - e)^2 \ge x\}dx \\
&= \int_0^{+\infty} E_{Pr}[\mathcal{M}\{\xi - e)^2 \ge x\}]\,dx \\
&= \int_0^{+\infty} E_{Pr}[\mathcal{M}\{(f(\eta_1, \cdots, \eta_m, \tau_1, \cdots, \tau_n) - e)^2 \ge x\}]\,dx \\
&= \int_0^{+\infty}\int_{R^m} \mathcal{M}\{(f(y_1, \cdots, y_m; \tau_1, \cdots, \tau_n) - e)^2 \ge x\}d\Psi_1(y_1)\cdots d\Psi_m(y_m)dx \\
&= \int_{R^m}\int_0^{+\infty} \mathcal{M}\{(f(y_1, \cdots, y_m; \tau_1, \cdots, \tau_n) - e)^2 \ge x\}dxd\Psi_1(y_1)\cdots d\Psi_m(y_m).
\end{aligned} \tag{3.11}$$

□

For the above equation, we have

$$\begin{aligned}
& \int_0^{+\infty} \mathcal{M}\{(f(y_1,\cdots,y_m;\tau_1,\cdots,\tau_n)-e)^2 \geq x\}dx \\
= & \int_0^{+\infty} \mathcal{M}\{f(y_1,\cdots,y_m;\tau_1,\cdots,\tau_n) \geq e+\sqrt{x} \cup f(y_1,\cdots,y_m;\tau_1,\cdots,\tau_n) \leq e-\sqrt{x}\}dx \\
\leq & \int_0^{+\infty} \mathcal{M}\{f(y_1,\cdots,y_m;\tau_1,\cdots,\tau_n) \geq e+\sqrt{x}\}dx \\
+ & \int_0^{+\infty} \mathcal{M}\{f(y_1,\cdots,y_m;\tau_1,\cdots,\tau_n) \leq e-\sqrt{x}\}dx \\
= & \int_0^{+\infty} [1-F(e+\sqrt{x};y_1,\cdots,y_m)+F(e-\sqrt{x};y_1,\cdots,y_m)]\,dx \\
= & \; 2\int_0^{+\infty} x[1-F(e+x;y_1,\cdots,y_m)+F(e-x;y_1,\cdots,y_m)]\,dx, \qquad (3.12)
\end{aligned}$$

where $F(x;y_1,\cdots,y_m)$ is the uncertainty distribution of uncertain variable $f(y_1,\cdots,y_m;\tau_1,\cdots,\tau_n)$ for any real numbers $y_1,y_2,\cdots,y_m$.

By Equations 3.11 and 3.12, this theorem is proved.

Example 1. Let η be a random variable with probability distribution Ψ, and let τ be an uncertain variable with uncertainty distribution Υ. Then, the sum

$$\xi = \eta + \tau$$

has the variance

$$V[\xi] = 2\int_{-\infty}^{+\infty}\int_0^{+\infty} x[1-\Upsilon(e+x-y)+\Upsilon(e-x-y)]\,dxd\Psi(y),$$

where e is the expected value of uncertain random variable ξ and the above integral is finite.

Especially, let η be a random variable having probability density function

$$\phi(x) = \begin{cases} \frac{1}{d-c}, & \text{if } c \leq x \leq d \\ 0, & \text{otherwise,} \end{cases} \qquad (3.13)$$

and let τ be an linear uncertain variable with uncertainty distribution

$$\Upsilon(x) = \begin{cases} 0, & \text{if } x < a \\ \frac{x-a}{b-a}, & \text{if } a \leq x \leq b \\ 1, & \text{otherwise.} \end{cases} \qquad (3.14)$$

By Theorem 2.3, the sum

$$\xi = \eta + \tau$$

has the variance

$$V[\xi] = \frac{(d-c)^2}{12} + \frac{(b-a)^2}{12} = V[\eta] + V[\tau].$$

Example 2. Let η be a random variable with probability distribution Ψ, and let τ be an uncertain variable with uncertainty distribution Υ. Then, the product

$$\xi = \eta\tau$$

has the variance

$$V[\xi] = 2\int_{-\infty}^{+\infty}\int_0^{+\infty} x\left[1-\Upsilon\left(\frac{e+x}{y}\right)+\Upsilon\left(\frac{e-x}{y}\right)\right]dxd\Psi(y),$$

where e is the expected value of uncertain random variable ξ and the above integral is finite.

Especially, let η be a random variable having probability density function

$$\phi(x) = \begin{cases} \frac{1}{d-c}, & \text{if } 0 \le c \le x \le d \\ 0, & \text{otherwise} \end{cases} \tag{3.15}$$

and let τ be an linear uncertain variable with uncertainty distribution

$$\Upsilon(x) = \begin{cases} 0, & \text{if } x < a \\ \frac{x-a}{b-a}, & \text{if } 0 \le a \le x \le b \\ 1, & \text{otherwise.} \end{cases} \tag{3.16}$$

Then, the product

$$\xi = \eta\tau$$

has the variance

$$V[\xi] = -\frac{1}{16}(a+b)^2(c+d)^2 + \frac{1}{9}(a^2+b^2+ab)(c^2+d^2+cd).$$

Remarks 3.1. The variance of uncertain random variable ξ provides a degree of spread of the distribution around its expected value e. If we only know its chance distribution Φ, then the variance

$$\begin{aligned} V[\xi] &= \int_0^{+\infty} Ch\{(\xi - e)^2 \ge x\}dx \\ &= \int_0^{+\infty} Ch\{(\xi \ge e + \sqrt{x}) \cup (\xi \le e - \sqrt{x})\}dx \\ &\le \int_0^{+\infty} (Ch\{\xi \ge e + \sqrt{x}\} + Ch\{\xi \le e - \sqrt{x}\})dx \\ &= \int_0^{+\infty} (1 - \Phi(e + \sqrt{x}) + \Phi(e - \sqrt{x}))dx. \end{aligned}$$

Then, we have the following stipulation

$$V[\xi] = \int_0^{+\infty} (1 - \Phi(e + \sqrt{x}) + \Phi(e - \sqrt{x}))dx.$$

4 Conclusions

Uncertain random variable and chance theory were introduced to model the uncertain random compound systems. As the same as random variable and uncertain variable, expected value is the average of uncertain random variable in the sense of chance measure and the variance of uncertain random variable provides a degree of spread of the distribution around its expected value. In this paper, a series of formulas were built to obtain the variance of uncertain random variable.

Acknowledgements

This work was supported by the National Natural Science Foundation of China (No. 61073121) and Hebei Natural Science Foundations (No. G2013402063, F2012402037).

References

1. Liu, B: Uncertainty theory, 4th ed. http://orsc.edu.cn/liu/ut.pdf. (2014)
2. Liu, B: Uncertainty Theory. 2nd ed. Springer-Verlag, Berlin (2007)
3. Liu, B: Uncertainty Theory: A Branch of Mathematics for Modeling Human Uncertainty. Springer-Verlag, Berlin (2010)

4. Liu, B: Theory and Practice of Uncertain Programming. 2nd ed. Springer-Verlag, Berlin (2009)
5. Wang, X, Peng, Z: Method of moments for estimating uncertainty distribution. J. Uncertainty Anal. Applications. **2**(5), 1–12 (2014)
6. Wang, X, Gao, Z, Guo, H: Delphi method for estimating uncertainty distributions. Inform. An Int. Interdiscip. J. **15**(2), 449–460 (2012)
7. Wang, X, Gao, Z, Guo, H: Uncertain hypothesis testing for two experts empirical data. Math. Comput. Model. **55**, 1478–1482 (2012)
8. Wang, X, Guo, H: Uncertain variance of sample and its application. Inform.: An Int. Interdiscip. J. **14**(1), 79–88 (2011)
9. Kwakernaak, H: Fuzzy random variables-1: definitions and theorems. Inform. Sci. **15**, 1–29 (1978)
10. Kwakernaak, H: Fuzzy random variables-2: algorithms and examples for the discrete case. Inform. Sci. **17**, 713–720 (1979)
11. Puri, M, Ralescu, D: Fuzzy random variables. J. Math. Appl. **114**, 409–422 (1986)
12. Kruse, R, Meyer, K: Statistics with Vague Data. Reidel Publishing Company, Dordrecht (1987)
13. Liu, YK, Liu, B: Fuzzy random variables: a scalar expected value operator. Fuzzy Optimization Decis. Making. **2**(2), 143–160 (2003)
14. Liu, YK, Liu, B: Fuzzy random programming with equilibrium chance constraints. Inform. Sci. **170**, 363–395 (2005)
15. Liu, YH: Uncertain random variables: a mixture of uncertainty and randomness. Soft Comput. **17**(4), 625–634 (2013)
16. Liu, YH: Uncertain random programming with applications. Fuzzy Optimization Decis. Making. **12**(2), 153–169 (2013)

12

Fuzzy cross-entropy

Xiang Li

Correspondence:
lixiang@mail.buct.edu.cn
School of Economics and Management, Beijing University of Chemical Technology, No. 15, Beisanhuan East Road, 100029 Beijing, China

Abstract

This paper deals with the divergence of fuzzy variables from *a priori* one. Within the framework of credibility theory, a fuzzy cross-entropy is defined to measure the divergence, and some mathematical properties are investigated. Furthermore, a minimum cross-entropy principle is proposed, which tells us that out of all membership functions satisfying given moment constraints, we should choose the one that is closest to the given *a priori* membership function.

Keywords: Fuzzy variable; Cross-entropy; Entropy; Credibility theory

Introduction

Fuzzy entropy provides a quantitative measure of the uncertainty associated with each fuzzy variable. Since Zadeh [1] introduced the fuzzy entropy as a weighted shannon entropy, researchers gave several definitions from different angles, such as De Luca and Termini [2], Yager [3], Kaufmann [4], Kosko [5], Pal and Pal [6]. The above definitions characterize the uncertainty resulting primarily from the linguistic vagueness rather than resulting from information deficiency and vanish when the fuzzy variable is an equipossible one. However, Liu [7] suggested that a fuzzy entropy should meet at least the following three basic requirements: the entropy of a crisp number is zero; the entropy of an equipossible fuzzy variable is maximum; and the entropy is applicable not only to finite and infinite cases but also to discrete and continuous cases. In order to meet these requirements, within the framework of credibility theory, Li and Liu [8] provided a new definition of fuzzy entropy to characterize the uncertainty resulting from information deficiency which is caused by the impossibility to predict the specified value that a fuzzy variable takes. Based on this definition, Li and Liu [9] proposed the fuzzy maximum entropy principle and proved some maximum entropy theorems.

This paper is devoted to formulate a fuzzy cross-entropy characterized by credibility measure. For this purpose, we organize this paper as follows. The 'Preliminaries' section recalls some useful definitions and properties about credibility theory. The 'Fuzzy cross-entropy' section defines the fuzzy cross-entropy and studies some useful properties. In the 'Minimum cross-entropy principle' section, the minimum cross-entropy principle is proposed. At the end of this paper, a brief summary is given.

Preliminaries

Credibility theory [10] is a branch of mathematics for studying the behavior of fuzzy phenomena. Let Θ be a nonempty set, and let $\mathcal{P}$ be the power set of Θ. Each element A of

$\mathcal{P}$ is called an event. In 2002, Liu and Liu [11] presented a credibility measure $\mathrm{Cr}\{A\}$ to express the chance that event A occurs. Furthermore, Li and Liu [12] proved that a set function is a credibility measure if and only if it satisfies the following axioms:

Axiom 1. (Normality) $\mathrm{Cr}\{\Theta\} = 1$;
Axiom 2. (Monotonicity) $\mathrm{Cr}\{A\} \leq \mathrm{Cr}\{B\}$ whenever $A \subset B$;
Axiom 3. (Self-duality) Cr is self-dual, i.e., $\mathrm{Cr}\{A\} + \mathrm{Cr}\{A^c\} = 1$ for any event A;
Axiom 4. (Countable subadditivity) $\mathrm{Cr}\{\cup_i A_i\} = \sup_i \mathrm{Cr}\{A_i\}$ for any events $\{A_i\}$ with $\sup_i \mathrm{Cr}\{A_i\} < 0.5$.

If Cr is a credibility measure, the triplet $(\Theta, \mathcal{P}, \mathrm{Cr})$ is called a credibility space. A fuzzy variable is defined as a function from a credibility space $(\Theta, \mathcal{P}, \mathrm{Cr})$ to the set of real numbers. Let ξ be a fuzzy variable. Then, its membership function is derived from the credibility measure by:

$$\mu(x) = (2\mathrm{Cr}\{\xi = x\}) \wedge 1, \ \forall x \in \Re.$$

Conversely, if ξ is a fuzzy variable with membership function μ, then, for any set $B \subseteq \Re$, we have:

$$\mathrm{Cr}\{\xi \in B\} = \frac{1}{2}\left(\sup_{x \in B} \mu(x) + 1 - \sup_{x \in B^c} \mu(x)\right).$$

This formula is also called the credibility inversion theorem.

Definition 2.1. Let ξ be a fuzzy variable taking values in $\{x_1, x_2, \cdots, x_n\}$ (Li and Liu [8]). Then, its fuzzy entropy is defined as:

$$H[\xi] = \sum_{i=1}^{n} S(\mathrm{Cr}\{\xi = x_i\})$$

where $S(t) = -t \ln t - (1-t)\ln(1-t)$.

Fuzzy entropy is used to quantify the uncertainty associated to fuzzy variables.

Theorem 2.1. Let ξ be a fuzzy variable taking values in $\{x_1, x_2, \cdots, x_n\}$ (Li and Liu [8]). Then, we have:

$$0 \leq H[\xi] \leq n \ln 2.$$

Especially, $H[\xi]$ attains its minimum value 0 if and only if ξ is a crisp number, and $H[\xi]$ attains its maximum value $n \ln 2$ if and only if ξ is an equipossible fuzzy variable.

Definition 2.2. Let ξ be a continuous fuzzy variable (Li and Liu [8]). Then, its fuzzy entropy is defined as:

$$H[\xi] = \int_{-\infty}^{+\infty} S(\mathrm{Cr}\{\xi = x\})\mathrm{d}x.$$

Theorem 2.2. Let ξ be a continuous fuzzy variable taking values in $[a,b]$ (Li and Liu [8]). Then, we have:

$$0 \leq H[\xi] \leq (b-a)\ln 2.$$

Especially, $H[\xi]$ attains its minimum value if and only if ξ is a crisp number, and $H[\xi]$ attains its maximum value if and only if ξ is an equipossible fuzzy variable.

In 2007, Li and Liu [9] proposed a fuzzy maximum entropy principle, which tells us that out of all the membership functions satisfying the given constraints, we should select the one that maximizes the entropy.

Fuzzy cross-entropy

In this section, we define a fuzzy cross-entropy for quantifying the divergence of fuzzy variables from an *a priori* one. The relation between fuzzy entropy and fuzzy cross-entropy is also discussed.

Definition 3.1. Let ξ and η be two discrete fuzzy variables taking values in $\{x_1, x_2, \cdots, x_n\}$. Then, the fuzzy cross-entropy of ξ from η is defined as:

$$D[\xi;\eta] = \sum_{i=1}^{n} T\left(\text{Cr}\{\xi = x_i\}, \text{Cr}\{\eta = x_i\}\right)$$

where $T(s,t) = s\ln(s/t) + (1-s)\ln((1-s)/(1-t))$.

It is easy to prove that $D[\xi;\eta]$ is permutationally symmetric, i.e., the value does not change if the outcomes are labeled differently.

Definition 3.2. Let ξ and η be two continuous fuzzy variables taking values in $[a,b]$. Then, the cross-entropy of ξ from η is defined as:

$$D[\xi;\eta] = \int_a^b T\left(\text{Cr}\{\xi = x\}, \text{Cr}\{\eta = x\}\right) \mathrm{d}x.$$

Let μ and ν be the membership functions of continuous fuzzy variables ξ and η, respectively. Since $\text{Cr}\{\xi = x\} = \mu(x)/2$ and $\text{Cr}\{\eta = x\} = \nu(x)/2$, the cross-entropy of ξ from η can be rewritten as:

$$D[\xi;\eta] = \int_a^b \mu(x)/2\ln\left(\mu(x)/\nu(x)\right) + (1-\mu(x)/2)\ln\left((2-\mu(x))/(2-\nu(x))\right) \mathrm{d}x.$$

Remark 3.1. It is easy to extend the concept of cross-entropy to fuzzy vectors. If $\boldsymbol{\xi} = (\xi_1, \xi_2, \cdots, \xi_m)$ and $\boldsymbol{\eta} = (\eta_1, \eta_2, \cdots, \eta_m)$ are discrete, we have:

$$\begin{aligned} D[\boldsymbol{\xi},\boldsymbol{\eta}] = \sum_{i_1=1}^{n_1}\sum_{i_2=1}^{n_2}\cdots\sum_{i_m=1}^{n_m} & T(\text{Cr}\{\xi_1 = x_{i_1}, \xi_2 = x_{i_2}, \cdots, \xi_m = x_{i_m}\}, \\ & \text{Cr}\{\eta_1 = x_{i_1}, \eta_2 = x_{i_2}, \cdots, \eta_m = x_{i_m}\}). \end{aligned}$$

If $\boldsymbol{\xi}$ and $\boldsymbol{\eta}$ are continuous variables, we have:

$$\begin{aligned} D[\boldsymbol{\xi},\boldsymbol{\eta}] = \int_{a_1}^{b_1}\int_{a_2}^{b_2}\cdots\int_{a_m}^{b_m} & T(\text{Cr}\{\xi_1 = x_1, \xi_2 = x_2, \cdots, \xi_m = x_m\}, \\ & \text{Cr}\{\eta_1 = x_1, \eta_2 = x_2, \cdots, \eta_m = x_m\})\mathrm{d}x_1\mathrm{d}x_2\cdots\mathrm{d}x_m. \end{aligned}$$

Remark 3.2. It is clear that $T(s,t)$ is a function from $[0,1] \times [0,1]$ to $[0,+\infty)$. Please also mention that:

$$T(s,0) = \begin{cases} 0, & \text{if } s = 0 \\ +\infty, & \text{if } s > 0, \end{cases} \qquad T(s,1) = \begin{cases} 0, & \text{if } s = 1 \\ +\infty, & \text{if } s < 1. \end{cases}$$

In addition, it is easy to prove that:

$$\frac{\partial T}{\partial s} = \ln\left(\frac{s}{t}\right) - \ln\left(\frac{1-s}{1-t}\right), \quad \frac{\partial T}{\partial t} = \frac{t-s}{t(1-t)},$$

$$\frac{\partial^2 T}{\partial s^2} = \frac{1}{s(1-s)}, \quad \frac{\partial^2 T}{\partial s \partial t} = \frac{\partial^2 T}{\partial t \partial s} = -\frac{1}{t(1-t)}, \quad \frac{\partial^2 T}{\partial t^2} = \frac{s}{t^2} + \frac{1-s}{(1-t)^2}.$$

Then, the following properties about $T(s,t)$ can be easily proved: (a) $T(s,t)$ is strictly convex with respect to (s,t) and attains its minimum value zero on the line $s = t$; and (b) for any $0 \le s \le 1$ and $0 \le t \le 1$, we have $T(s,t) = T(1-s, 1-t)$.

Theorem 3.1. For any fuzzy variables ξ and η, we have $D[\xi;\eta] \ge 0$, and the equality holds if and only if ξ and η have the same membership function.

Proof. Let μ and ν be the membership functions of discrete fuzzy variables ξ and η, respectively. Since $T(s,t)$ is strictly convex about (s,t) and attains its minimum value zero on the line $s = t$, we have $T(\text{Cr}\{\xi = x_i\}, \text{Cr}\{\eta = x_i\}) \ge 0$ for all i, which implies that:

$$D[\xi;\eta] = \sum_{i=1}^{\infty} T(\text{Cr}\{\xi = x_i\}, \text{Cr}\{\eta = x_i\}) \ge 0.$$

Furthermore, for any $0 \le s^* \le 1$, the unique minimum point of $T(s^*, t)$ is $t = s^*$. Thus, we have $D[\xi;\eta] = 0$ if and only if $T(\text{Cr}\{\xi = x_i\}, \text{Cr}\{\eta = x_i\}) = 0$, that is:

$$\mu(x_i) = (2\text{Cr}\{\xi = x_i\}) \wedge 1 = (2\text{Cr}\{\eta = x_i\}) \wedge 1 = \nu(x_i)$$

for all $i = 1, 2, \cdots, n$. If ξ and η are continuous fuzzy variables, the theorem can be proved in a similar way. The proof is complete. □

Theorem 3.2. Let τ be the equipossible fuzzy variable with membership function $\nu(x_i) = 1$ for all $i = 1, 2, \cdots, n$. Then, for any discrete fuzzy variable ξ taking values in $\{x_1, x_2, \cdots, x_n\}$, we have:

$$D[\xi, \tau] = n \ln 2 - H[\xi].$$

Proof. According to the credibility inversion theorem, it is easy to prove that $\text{Cr}\{\tau = x_i\} = 0.5$ for all $i = 1, 2, \cdots, n$. It follows from the definition of cross-entropy that $D[\xi, \tau]$ is:

$$\begin{aligned} & \sum_{i=1}^{n} \text{Cr}\{\xi = x_i\} \ln(2\text{Cr}\{\xi = x_i\}) + (1 - \text{Cr}\{\xi = x_i\}) \ln(2 - 2\text{Cr}\{\xi = x_i\}) \\ = & \sum_{i=1}^{n} \ln 2 + \text{Cr}\{\xi = x_i\} \ln \text{Cr}\{\xi = x_i\} + (1 - \text{Cr}\{\xi = x_i\}) \ln(1 - \text{Cr}\{\xi = x_i\}) \\ = & \; n \ln 2 - H[\xi]. \end{aligned}$$

The proof is complete. □

Theorem 3.3. Let τ be the equipossible fuzzy variable with membership function $\nu(x) = 1$ for all $x \in [a, b]$. Then, for any continuous fuzzy variable ξ taking values in $[a, b]$, we have:

$$D[\xi, \tau] = (b - a)\ln 2 - H[\xi].$$

Proof. It follows from the definition of cross-entropy that $D[\xi, \tau]$ is:

$$\begin{aligned} &\int_a^b (\mu(x)/2)\ln\mu(x) + (1 - \mu(x)/2)\ln(2 - \mu(x))\mathrm{d}x \\ = &\int_a^b \ln 2 + ((\mu(x)/2)\ln(\mu(x)/2) + (1 - \mu(x)/2)\ln(1 - \mu(x)/2))\,\mathrm{d}x \\ = &\ (b - a)\ln 2 - H[\xi]. \end{aligned}$$

The proof is complete. □

Minimum cross-entropy principle

In many real problems, the membership function of a fuzzy variable is unavailable except some partial information, for example, moment constraints, which may be based on observations. In this case, the maximum entropy principle (Li and Liu [9]) tells us that out of all the membership functions satisfying given constraints, choose the one that has maximum entropy. However, there may be another type of information, for example, *a priori* membership function, which may be based on intuition or experience with the problem. If both the *a priori* membership function and the moment constraints are given, which membership function should we choose? The following minimum cross-entropy principle tells us that out of all membership functions satisfying given moment constraints, choose the one that is closest to the given *a priori* membership function.

There is nothing mysterious about this principle. It is just based on common sense. Our membership function must be consistent with observations or given information, and if there are many membership functions consistent with the given information, we must choose the one that is nearest to our intuition and experience. On the other hand, if we have no *a priori* experience or intuition to guide us, we choose the membership function that is nearest to the equipossible one. In this sense, if the *a priori* membership function is not prescribed and the fuzzy variable is simple (bounded for continuous case), the maximum entropy principle and minimum cross-entropy principle are consistent because:

$$D[\xi; \upsilon] = \max_{\eta} D[\eta; \upsilon] - H[\xi]$$

where υ is the equipossible fuzzy variable.

Conclusion

Based on credibility measure, a definition of cross-entropy was proposed in this paper to measure the divergence of fuzzy variables from *a priori* one, and some properties were investigated. Furthermore, a minimum cross-entropy principle was proposed as an important entropy optimization principle.

Acknowledgements
This work was supported by the National Natural Science Foundation of China (No. 71371027) and Program for New Century Excellent Talents in University under Grant No. NCET-13-0649.

References
1. Zadeh, LA: Probability measures of fuzzy events. J. Math. Anal. Appl. **23**, 421–427 (1968)
2. De Luca, A, Termini, S: A definition of nonprobabilistic entropy in the setting of fuzzy sets theory. Inf. Control. **20**, 301–312 (1972)
3. Yager, RR: On measures of fuzziness and negation, part I: membership in the unit interval. Int. J. General Syst. **5**, 221–229 (1979)
4. Kaufmann, A: Introduction to the Theory of Fuzzy Subsets. Academic Press, New York (1975)
5. Kosko, B: Fuzzy entropy and conditioning. Inf. Sci. **40**, 165–174 (1986)
6. Pal, NR, Pal, SK: Higher order fuzzy entropy and hybrid entropy of a set. Inf. Sci. **61**, 211–231 (1992)
7. Liu, B: A survey of entropy of fuzzy variables. J. Uncertain Syst. **1**(1), 4–13 (2007)
8. Li, P, Liu, B: Entropy of credibility distributions for fuzzy variables. IEEE Trans. Fuzzy Syst. **16**(1), 123–129 (2008)
9. Li, X, Liu, B: Maximum entropy principle for fuzzy variables. Int. J. Uncertainty Fuzziness Knowledge-Based Syst. **15**(Supp 2), 40–48 (2007)
10. Liu, B: Uncertainty Theory. Springer-Verlag, Berlin (2004)
11. Liu, B, Liu, YK: Expected value of fuzzy variable and fuzzy expected value models. IEEE Trans. Fuzzy Syst. **10**(4), 445–450 (2002)
12. Li, X, Liu, B: A sufficient and necessary condition for credibility measures. Int. J. Uncertainty Fuzziness Knowledge-Based Syst. **14**(5), 527–535 (2006)

13

Analysis of order statistics of uncertain variables

Yuan Gao[1], Rong Gao[2*] and Lixing Yang[1]

*Correspondence:
gaor14@mails.tsinghua.edu.cn
[2]Department of Mathematical Sciences, Tsinghua University, Beijing 100084, China
Full list of author information is available at the end of the article

Abstract

In this paper, a concept of order uncertain variable is introduced as a main tool in analysis of k-out-of-n system with uncertain lifetimes. Then, by the operational law of uncertain variable, the uncertainty distribution of order uncertain variable is deduced. At last, the k-out-of-n system is studied, and the uncertainty distribution of k-out-of-n system lifetime is given.

Keywords: Uncertain variable; Uncertainty distribution; Order statistics of uncertain variables; Operational law; k-out-of-n system; System reliability

Introduction

The real world is filled with indeterminacy. In fact, randomness and fuzziness are two cases of indeterminacy, behind which probability theory and credibility theory have been proposed. However, some practical systems behave neither random nor fuzzy, so it cannot be modeled precisely by probability theory or credibility theory. In order to deal with this type of uncertainty, Liu [6] founded an uncertainty theory in 2007, which is a branch of mathematics based on normality, duality, subadditivity, and product axioms. Since then, based on uncertainty theory, significant work both in theory and engineering has been done by many researchers. As an application of uncertainty theory, Liu [7] proposed a spectrum of uncertain programming which is a type of mathematical programming involving uncertain variables and applied uncertain programming to system reliability design, facility location problem, vehicle routing problem, project scheduling problem, finance, and so on. In addition, Li and Liu [5] presented uncertain logic in which the truth value is defined as the uncertain measure that the proposition is true. Furthermore, uncertain inference was pioneered by Liu [8] as a process of deriving consequences from uncertain knowledge or evidence via the tool of conditional uncertainty. Other researchers also have done a lot of theoretical work in uncertainty theory, such as Chen [1], Chen and Liu [2], Gao [3], Gao et al. [4], Liu [9], Peng [11], Peng and Chen [12], Qin and Kar [13], Wang and Peng [14], Yang [15], Yao [16], You [17] and Zhu [18], etc.

Problems of k-out-of-n system play an important role in system reliability design. Up to now, k-out-of-n system with random lifetimes has been researched in many books related to system reliability design. This paper aims to give some analysis of k-out-of-n system with uncertain lifetimes. Based on the uncertainty theory, this paper gives a new concept of order uncertain variable, which is used as a tool to analyze k-out-of-n system with uncertain lifetimes in this paper. It is similar to the order statistics (Balakrishnan and Cohen [10]) in mathematical statistics. However, order uncertain variable just requires that the initial uncertain variables are independent. Besides, the uncertainty distribution of order statistics of uncertain variables has a much simpler form than the ones of random variables.

The remainder of this paper is organized as follows. In Section "Preliminary", basic concepts and properties regarding uncertain variables are reviewed. In Section "Order uncertain variable", the concept of order uncertain variable is given. In Section "Uncertainty distribution of order uncertain variable", the uncertainty distribution of order uncertain variable is deduced. In Section "k-out-of-n system with uncertain lifetimes", the order uncertain variable is used as a tool to analyze k-out-of-n system with uncertain lifetimes.

Preliminary

In this section, we introduce some foundational concepts and properties of uncertainty theory, which is used throughout this paper.

Let Γ be a nonempty set, and $\mathcal{L}$ a σ-algebra over Γ. Each element $\Lambda \in \mathcal{L}$ is assigned a number $\mathcal{M}\{\Lambda\}$. In order to ensure that the number $\mathcal{M}\{\Lambda\}$ has certain mathematical properties, Liu [6] presented the following four axioms:

Axiom 1. *(Normality Axiom)* $\mathcal{M}\{\Gamma\} = 1$.

Axiom 2. *(Duality Axiom)* $\mathcal{M}\{\Lambda\} + \mathcal{M}\{\Lambda^c\} = 1$ *for any event* Λ.

Axiom 3. *(Subadditivity Axiom) For every countable sequence of events* $\{\Lambda_i\}$, *we have*

$$\mathcal{M}\left\{\bigcup_{i=1}^{\infty} \Lambda_i\right\} \le \sum_{i=1}^{\infty} \mathcal{M}\{\Lambda_i\}.$$

Definition 1. *(Liu [6]) The set function* $\mathcal{M}$ *is called an uncertain measure if it satisfies the normality, duality, and subadditivity axioms.*

The triplet $(\Gamma, \mathcal{L}, \mathcal{M})$ is called an *uncertainty space*. Furthermore, the product uncertain measure on the product σ-algebra $\mathcal{L}$ is defined by Liu [6] as follows:

Axiom 4. *(Product Axiom) Let* $(\Gamma_i, \mathcal{L}_i, \mathcal{M}_i)$ *be uncertainty spaces for* $i = 1, 2, \cdots$ *The product uncertain measure* $\mathcal{M}$ *is an uncertain measure satisfying*

$$\mathcal{M}\left\{\prod_{i=1}^{\infty} \Lambda_i\right\} = \bigwedge_{i=1}^{\infty} \mathcal{M}_i\{\Lambda_i\}$$

where Λ_i *are arbitrarily chosen events from* $\mathcal{L}_i$ *for* $i = 1, 2, \cdots$, *respectively.*

Definition 2. *(Liu [6]) An uncertain variable is a measurable function* ξ *from an uncertainty space* $(\Gamma, \mathcal{L}, \mathcal{M})$ *to the set of real numbers, i.e., for any Borel set* B *of real numbers, the set*

$$\{\xi \in B\} = \{\gamma \in \Gamma \mid \xi(\gamma) \in B\}$$

is an event.

Definition 3. *(Liu [6]) The uncertainty distribution* $\Phi : \Re \to [0,1]$ *of an uncertain variable* ξ *is defined by*

$$\Phi(x) = \mathcal{M}\{\gamma \in \Gamma \mid \xi(\gamma) \le x\}. \tag{1}$$

Definition 4. *(Liu [8]) The uncertain variables* $\xi_1, \xi_2, \cdots, \xi_n$ *are said to be independent if*

$$\mathcal{M}\left\{\bigcap_{i=1}^{n}\{\xi_i \in B_i\}\right\} = \min_{1\le i\le n} \mathcal{M}\{\xi_i \in B_i\} \tag{2}$$

for any Borel sets $B_1, B_2, \cdots, B_n$ *of real numbers.*

More generally, the independence of uncertain variables was given by Liu [9].

Theorem 1. *(Liu [9]) The uncertain variables* $\xi_1, \xi_2, \cdots, \xi_n$ *are said to be independent if and only if*

$$\mathcal{M}\left\{\bigcup_{i=1}^{n}(\xi_i \in B_i)\right\} = \bigvee_{i=1}^{n} \mathcal{M}\{\xi_i \in B_i\} \tag{3}$$

Definition 5. *(Liu [6]) Let* ξ *be an uncertain variable. Then, the expected value of* ξ *is defined by*

$$E[\xi] = \int_0^{+\infty} \mathcal{M}\{\xi \ge r\}\mathrm{d}r - \int_{-\infty}^{0} \mathcal{M}\{\xi \le r\}\mathrm{d}r \tag{4}$$

provided that at least one of the two integrals is finite.

From Remark 1.6 in Liu ([6]), the following theorem is given without proof.

Theorem 2. *(Liu [8]) Let* $\xi_1, \xi_2, \cdots, \xi_n$ *be independent uncertain variables, and* $f : \Re^n \to \Re$ *a measurable function. Then,* $\xi = f(\xi_1, \xi_2, \cdots, \xi_n)$ *is an uncertain variable such that*

$$\mathcal{M}\{\xi \in B\} = \begin{cases} \sup\limits_{f(B_1,B_2,\cdots,B_n)\subset B} \min\limits_{1\le k\le n} \mathcal{M}_k\{\xi_k \in B_k\}, \\ \qquad \text{if} \sup\limits_{f(B_1,B_2,\cdots,B_n)\subset B} \min\limits_{1\le k\le n} \mathcal{M}_k\{\xi_k \in B_k\} > 0.5 \\ 1 - \sup\limits_{f(B_1,B_2,\cdots,B_n)\subset B^c} \min\limits_{1\le k\le n} \mathcal{M}_k\{\xi_k \in B_k\}, \\ \qquad \text{if} \sup\limits_{f(B_1,B_2,\cdots,B_n)\subset B^c} \min\limits_{1\le k\le n} \mathcal{M}_k\{\xi_k \in B_k\} > 0.5 \\ 0.5, \qquad \text{otherwise} \end{cases} \tag{5}$$

where $B, B_1, B_2, \cdots, B_n$ *are Borel sets of real numbers.*

Lemma 1. *Let* $\xi_1, \xi_2, \cdots, \xi_n$ *be independent uncertain variables taking values in* $\{0, 1\}$*, such that*

$$\mathcal{M}\{\xi_1 = 0\} = a_1,\ \mathcal{M}\{\xi_1 = 1\} = 1 - a_1,$$
$$\mathcal{M}\{\xi_2 = 0\} = a_2,\ \mathcal{M}\{\xi_2 = 1\} = 1 - a_2,$$
$$\cdots$$
$$\mathcal{M}\{\xi_n = 0\} = a_n,\ \mathcal{M}\{\xi_n = 1\} = 1 - a_n,$$

where $0 \leq a_i \leq 1, i = 1, 2, \cdots, n$, *and* $a_0 \leq a_1 \leq \cdots \leq a_n \leq a_{n+1} = 1$. *Then, for any* $0 \leq k \leq n, k \in \mathbb{N}$, *we have*

$$\mathcal{M}\left\{\sum_{i=1}^{n} \xi_i \leq k\right\} = a_{k+1}. \tag{6}$$

Proof. Because ξ_i takes values in $\{0, 1\}$, $i = 1, 2, \cdots, n$, according to Equation 5, we can get

$$\mathcal{M}\left\{\sum_{i=1}^{n} \xi_i \leq k\right\} = \begin{cases} \sup\limits_{B_1+B_2+\cdots+B_n \subset (-\infty,k]} \min\limits_{1\leq i\leq n} \mathcal{M}\{\xi_i \in B_i\}, \\ \quad \textit{if} \quad \sup\limits_{B_1+B_2+\cdots+B_n \subset (-\infty,k]} \min\limits_{1\leq i\leq n} \mathcal{M}\{\xi_i \in B_i\} \geq 0.5 \\ 1 - \sup\limits_{B_1+B_2+\cdots+B_n \subset [k+1,+\infty)} \min\limits_{1\leq k\leq n} \mathcal{M}\{\xi_k \in B_k\}, \\ \quad \textit{if} \quad \sup\limits_{B_1+B_2+\cdots+B_n \subset [k+1,+\infty)} \min\limits_{1\leq i\leq n} \mathcal{M}\{\xi_i \in B_i\} > 0.5 \end{cases}$$

It is easy to verify that

$$\begin{aligned} & \sup_{B_1+B_2+\cdots+B_n \subset (-\infty,k]} \min_{1\leq i\leq n} \mathcal{M}\{\xi_i \in B_i\} \\ &= \mathcal{M}\{\xi_1 \leq 1\} \wedge \mathcal{M}\{\xi_2 \leq 1\} \wedge \cdots \mathcal{M}\{\xi_k \leq 1\} \wedge \mathcal{M}\{\xi_{k+1} \leq 0\} \wedge \cdots \mathcal{M}\{\xi_n \leq 0\} \\ &= 1 \wedge 1 \wedge \cdots 1 \wedge a_{k+1} \wedge \cdots \wedge a_n \\ &= a_{k+1} \end{aligned}$$

and

$$\begin{aligned} & \sup_{B_1+B_2+\cdots+B_n \subset [k+1,+\infty)} \min_{1\leq i\leq n} \mathcal{M}\{\xi_i \in B_i\} \\ &= \mathcal{M}\{\xi_1 \geq 1\} \wedge \mathcal{M}\{\xi_2 \geq 1\} \wedge \cdots \mathcal{M}\{\xi_{k+1} \geq 1\} \wedge \mathcal{M}\{\xi_{k+2} \geq 0\} \wedge \cdots \mathcal{M}\{\xi_n \geq 0\} \\ &= (1 - a_1) \wedge (1 - a_2) \wedge \cdots \wedge (1 - a_{k+1}) \wedge 1 \cdots \wedge 1 \\ &= 1 - a_{k+1}. \end{aligned}$$

Then, we get

$$\mathcal{M}\left\{\sum_{i=1}^{n} \xi_i \leq k\right\} = a_{k+1}.$$

□

Order uncertain variable

Definition 6. *Let* $f_i : \Re^n \to \Re, i = 1, 2, \cdots, n$ *be a series of measurable functions. We call* $f_1, f_2, \cdots, f_n$ *the order functions of* $\Re^n$ *if for any* $(x_1, x_2, \cdots, x_n) \in \Re^n$,

$$f_i(x_1, x_2, \cdots, x_n) = x^{(i)}, \quad i = 1, 2, \cdots, n$$

where $\{x^{(1)}, x^{(2)}, \cdots, x^{(n)}\}$ *is the the rearrangement of* $\{x_1, x_2, \cdots, x_n\}$ *in ascending order of magnitude, that is,* $x^{(1)} \leq x^{(2)} \leq \cdots \leq x^{(n)}$.

Definition 7. *Let* $\xi_1, \xi_2, \cdots, \xi_n$ *be independent uncertain variables, and* $f_1, f_2, \cdots, f_n$ *order functions of* $\Re^n$. *Define*

$$\xi^{(i)} = f_i(\xi_1, \xi_2, \cdots, \xi_n), \quad i = 1, 2, \cdots, n.$$

Then, $\xi^{(1)}, \xi^{(2)}, \cdots, \xi^{(n)}$ *are called order uncertain variables of* $\xi_1, \xi_2, \cdots, \xi_n$, *and each* $\xi^{(i)}$ *is called the ith order uncertain variable of* $\xi_1, \xi_2, \cdots, \xi_n$.

To be more explicit, consider the case when $n = 2$ and the relationship between ξ_i and $\xi^{(i)}(i = 1, 2)$ is

$$\xi^{(1)} = \xi_1 \textit{ and } \xi^{(2)} = \xi_2 \textit{ when } \xi_1 < \xi_2,$$
$$\xi^{(1)} = \xi_2 \textit{ and } \xi^{(2)} = \xi_1 \textit{ when } \xi_2 < \xi_1.$$

Similarly, for $n = 3$ the relationship among the respective uncertain variables is

$$\xi^{(1)} = \xi_1,\ \xi^{(2)} = \xi_2,\textit{ and } \xi^{(3)} = \xi_3 \textit{ when } \xi_1 < \xi_2 < \xi_3,$$
$$\xi^{(1)} = \xi_1,\ \xi^{(2)} = \xi_3,\textit{ and } \xi^{(3)} = \xi_2 \textit{ when } \xi_1 < \xi_3 < \xi_2,$$
$$\cdots$$
$$\xi^{(1)} = \xi_3,\ \xi^{(2)} = \xi_2,\textit{ and } \xi^{(3)} = \xi_1 \textit{ when } \xi_3 < \xi_2 < \xi_1.$$

The most frequently encountered functions of order uncertain variable are $\xi^{(1)}$ and $\xi^{(n)}$. Obviously, $\xi^{(1)} = \min\limits_{1\le k\le n} \xi_k$, and $\xi^{(n)} = \max\limits_{1\le k\le n} \xi_k$.

Uncertainty distribution of order uncertain variable

In this section, the uncertainty distribution of the kth order uncertain variable for $k = 1, 2, \cdots, n$ will be given.

Theorem 3. *Let $\xi_1, \xi_2, \cdots, \xi_n$ be independent uncertain variables with distributions $\Phi_1(x), \Phi_2(x), \cdots, \Phi_n(x)$, respectively. The uncertainty distribution of the kth ($k = 1, 2, \cdots, n$) order uncertain variable $\xi^{(k)}$ is*

$$\Psi_k(x) = f_{n-k+1}(\Phi_1(x), \Phi_2(x), \cdots, \Phi_n(x)) \tag{7}$$

where $f_k : \Re^n \to \Re, k = 1, 2, \cdots, n$ are order functions of $\Re^n$.

Proof. For any fixed $x \in \Re$, define a series of new variables

$$\eta_i(x) = \begin{cases} 0, & \textit{if } \xi_i \le x \\ 1, & \textit{if } \xi_i > x \end{cases}$$

where $i = 1, 2, \cdots, n$. Then, $\eta_1(x), \eta_2(x), \cdots, \eta_n(x)$ are independent, and $\mathcal{M}\{\eta_i(x) = 0\} = \mathcal{M}\{\xi_i \le x\} = \Phi_i(x)$ for each i. According to the definitions of uncertain variable and order uncertain variable, by Lemma 1, we have

$$\begin{aligned}
\Psi_k(x) &= \mathcal{M}\left\{\xi^{(k)} \le x\right\} \\
&= \mathcal{M}\{\text{at most } (n-k) \text{ variables from } \xi_1, \xi_2, \cdots, \xi_n \text{ are greater than } x\} \\
&= \mathcal{M}\left\{\sum_{i=1}^{n} \eta_i(x) \le n-k\right\} \\
&= f_{n-k+1}\left(\Phi_1(x), \Phi_2(x), \cdots, \Phi_n(x)\right)
\end{aligned}$$

where $f_k : \Re^n \to \Re, k = 1, 2, \cdots, n$ are order functions of $\Re^n$. The theorem is proved. □

Corollary 1. *Let $\xi_1, \xi_2, \cdots, \xi_n$ be independent uncertain variables with the same uncertainty distribution $\Phi(x)$, and let $\xi^{(1)}, \xi^{(2)}, \cdots, \xi^{(n)}$ be the corresponding order uncertain variables. Then, for any $k = 1, 2, \cdots, n$, the kth order uncertain variable has uncertainty distribution $\Phi(x)$.*

Proof. It follows from Theorem 2 immediately. □

Corollary 2. *Let $\xi_1, \xi_2, \cdots, \xi_n$ be independent uncertain variables with the same uncertainty distribution $\Phi(x)$, and let $\xi^{(1)}, \xi^{(2)}, \cdots, \xi^{(n)}$ be the corresponding order uncertain variables. Then, for any $k = 1, 2, \cdots, n$, the kth order uncertain variable has expected value $E[\xi_1]$ provided that $E[\xi_1]$ exists.*

Proof. It follows from Corollary 1 and Definition 4 immediately. □

k-out-of-n system with uncertain lifetimes

System reliability design with uncertain lifetimes was first studied by Liu [7]. The k-out-of-n system is a significant kind of problem in system reliability design. In this section, order uncertain variable is used as a tool in analysis of k-out-of-n system with uncertain lifetimes.

A system of n components is called a k-out-of-n system if it remains operational only if at least k components continue to function. Assume that the ith element of this component has independent lifetime ξ_i with uncertainty distribution $\Phi_i(x)$, $i = 1, 2, \cdots, n$, respectively. Then, what is the distribution $\Phi(x)$ of this system lifetime η?

Let $\xi^{(1)}, \xi^{(2)}, \cdots, \xi^{(n)}$ be the order uncertain variables of $\xi_1, \xi_2, \cdots, \xi_n$. Then, for any $x > 0, x \in \Re$, by Theorem 2

$$\begin{aligned} \Phi(x) &= \mathcal{M}\{\eta \leq x\} \\ &= \mathcal{M}\{\text{at most } (k-1) \text{ components function at time } x\} \\ &= \mathcal{M}\left\{\xi^{(n-k+1)} \leq x\right\} \\ &= f_k(\Phi_1(x), \Phi_2(x), \cdots, \Phi_n(x)) \end{aligned} \tag{8}$$

where $f_k : \Re^n \to \Re, k = 1, 2, \cdots, n$ are order functions of $\Re^n$. Thus, we get the uncertainty distribution of η.

The special cases $k = 1$ and $k = n$ correspond respectively to parallel system and series system (Figure 1). According to the formula (Equation 8), the uncertainty distribution $\Phi_p(x)$ of parallel system lifetime η_p is

$$\begin{aligned} \Phi_p(x) &= \mathcal{M}\{\eta_p \leq x\} \\ &= f_1(\Phi_1(x), \Phi_2(x), \cdots, \Phi_n(x)) \\ &= \Phi_1(x) \wedge \Phi_2(x) \wedge \cdots \wedge \Phi_n(x) \end{aligned}$$

and the uncertainty distribution $\Phi_s(x)$ of series system lifetime η_s is

$$\begin{aligned} \Phi_s(x) &= \mathcal{M}\{\eta_s \leq x\} \\ &= f_n(\Phi_1(x), \Phi_2(x), \cdots, \Phi_n(x)) \\ &= \Phi_1(x) \vee \Phi_2(x) \vee \cdots \vee \Phi_n(x). \end{aligned}$$

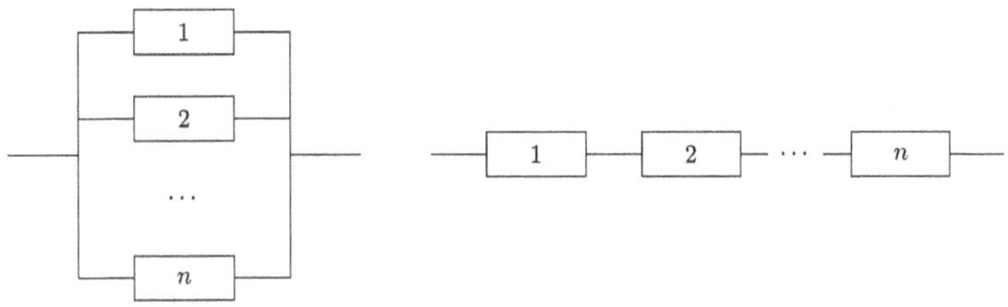

Figure 1 Parallel system and series system.

Conclusion

In order to analyze k-out-of-n system with uncertain lifetimes, we presented a new concept of order uncertain variable and gave uncertainty distribution of order uncertain variable in this paper. By using order uncertain variable, the uncertainty distribution of k-out-of-n system lifetime was deduced.

Acknowledgments

The research was supported by the Fundamental Research Funds for the central universities (No. 2014JBM155) and the State Key Laboratory of Rail Traffic Control and Safety (No. RCS2014ZQ001 and No. RCS2014ZT22, Beijing Jiaotong University).

Author details

[1]State Key Laboratory of Rail Traffic Control and Safety, Beijing Jiaotong University, Beijing 100044, China. [2]Department of Mathematical Sciences, Tsinghua University, Beijing 100084, China.

References

1. Chen, X, Dai, W: Maximum entropy principle for uncertain variables. Int. J. Fuzzy Syst. **13**(3), 232–236 (2011)
2. Chen, X, Liu, B: Existence and uniqueness theorem for uncertain differential equations. Fuzzy Optimization Decis. Making. **9**(1), 69–81 (2010)
3. Gao, X: Some properties of continuous uncertain measure. Int. J. Uncertainty Fuzziness Knowledge-Based Syst. **17**(3), 419–426 (2009)
4. Gao, Y, Yang, L, Li, S, Kar, S: On distribution function of the diameter in uncertain graph. Inf. Sci. **296**, 61–74 (2015)
5. Li, X, Liu, B: Hybrid logic and uncertain logic. J. Uncertain Syst. **3**(2), 83–94 (2009)
6. Liu, B: Uncertainty Theory. 2nd ed. Springer-Verlag, Berlin (2007)
7. Liu, B: Theory and Practice of Uncertain Programming. 2nd ed. Springer-Verlag, Berlin (2009)
8. Liu, B: Some research problem in uncertainty theory. J. Uncertain Syst. **3**(1), 3–10 (2009)
9. Liu, B: Polyrectangular theorem and independence of uncertain vectors. J. Uncertainty Anal. Appl. **1**, 9 (2013). doi:10.1186/2195-5468-1-9
10. Balakrishnan, N, Cohen, AC: Order statistics and inference: estimation methods. Academic Press, Inc., San Diego (1991). ISBN 0120769484, 9780120769483
11. Peng, Z, Iwamura, K: Some properties of product uncertain measure. J. Uncertain Syst. **6**(4), 263–269 (2012)
12. Peng, Z, Chen, X: Uncertain systems are universal approximators. J. Uncertainty Anal. Appl. **2**, 13 (2014). doi:10.1186/2195-5468-2-13
13. Qin, Z, Kar, S: Single-period inventory problem under uncertain environment. Appl. Math. Comput. **219**(18), 9630–9638 (2013)
14. Wang, X, Peng, Z: Method of moments for estimating uncertainty distributions. J. Uncertainty Anal. Appl. **2**, 5 (2014). doi:10.1186/2195-5468-2-5
15. Yang, L, Liu, P, Li, S, Gao, Y, Ralescu, D: Reduction methods of type-2 uncertain variables and their applications to solid transportation problem. Inf. Sci. **291**, 204–237 (2015)
16. Yao, K: Extreme values and integral of solution of uncertain differencial equation. J. Uncertainty Anal. Appl. 1, Article, 2 (2013)
17. You, C: Some convergence theorems of uncertain sequences. Math. Comput. Modell. **49**(3-4), 482–487 (2009)
18. Zhu, Y: Uncertain optimal control with application to a portfolio selection model. Cybernet. Syst. **41**(7), 535–547 (2010)

14

Uncertain random time-cost trade-off problem

Hua Ke

Correspondence:
hke@tongji.edu.cn
School of Economics and Management, Tongji University, Shanghai 200092, China

Abstract

Since for most projects, 'new' parts (with little or no historical data) and 'old' parts (with sufficient historical data) always coexist in the same project, uncertainty and randomness should be considered simultaneously in time-cost trade-off problem. In this paper, combined with uncertainty theory and dependent-chance programming, an uncertain random time-cost trade-off model is built. A crisp equivalent model is also given for some special case. Besides, uncertain random simulation and genetic algorithm are integrated for solving the proposed model.

Keywords: Project scheduling; Time-cost trade-off; Uncertain variable; Uncertain random programming; Dependent-chance programming

Introduction

Time-cost trade-off problem (TCTP), a specific type of project scheduling problem, desires to balance two important goals in real-life projects, project completion times, and project total costs to achieve decision-making requirements. For real-life projects, project managers always consider the trade-offs among the performance goals (e.g., project cost, project completion time, project quality, etc.) for project scheduling and control. The time-cost trade-off problem takes into account the project time-cost trade-off by crashing or prolonging project activity durations. In 1961, Kelly [1] first did research on the TCTP. In the following years, many researches have been done on the TCTP with assumption of deterministic activity durations [2-5].

In real-life projects, activity durations may be variational for many external factors, such as technological advance, weather change, labor force shortage, etc. For some projects, since many similar projects have ever been fulfilled before, there are enough historical data for deducing probability distributions for describing imprecise quantities in projects. Hence, probability theory was introduced into the TCTP for estimating nondeterministic factors. In 1985, Wollmer [6] discussed a stochastic version of the deterministic linear TCTP. Following that, many authors discussed the TCTP with stochastic environmental factors [7-9].

In project management, one of the most important characteristics of the concept 'project' is that there is always something new in a project. That is, there are always some parts which are never performed before or only similarly performed. For such parts in projects, historical data for estimating probability distributions of some imprecise quantities are always not enough when facing external indeterminacy, which means that probability theory is no longer valid for describing imprecise quantities (e.g., activity durations) in such project parts. For this case, project activity durations were originally described by fuzzy variables, and many researchers studied the TCTP with fuzzy parameters [10-13].

In the past years, subjective probability theory and fuzzy set theory were applied to estimate imprecise quantities (or indeterminacy) with no or little statistical data in decision systems, including project scheduling and control. However, actually both these two theories were proved not suitable for the above case, illustrated by some examples [14]. Since such a type of indeterminacy cannot be depicted via randomness or fuzziness, a new theory is needed. On this occasion, Liu founded uncertainty theory in 2007 [15] and refined it in 2010 [16], which is a branch of axiomatic mathematics for modeling human uncertainty. Since its foundation, uncertainty theory has been widely applied in many decision systems, such as portfolio selection [17], inventory problem [18], facility location [19], supply chain integrated production-inventory problem [20], investment problem [21], differential games [22], and coalitional game [23]. Especially, Ke [24] applied uncertain measure for estimating activity durations in time-cost trade-off problem.

For most projects, though they always include some 'new' parts, meanwhile they may contain some common parts with many other finished projects. With enough statistical data from other projects, probability theory may be introduced to estimate imprecise quantities in these common parts. Then for many projects, integrated by some 'new' parts and some other 'old' (common) parts, randomness and uncertainty may coexist. For such cases, the concepts of uncertain random variable and chance measure and the philosophy of uncertain random programming were presented by Liu [25] and Liu [26], respectively. Uncertain random programming has been further studied in some aspects, e.g., uncertain random graph and uncertain random network [27], uncertain random risk analysis [28], uncertain random multi-objective optimization [29], and uncertain random multilevel programming [30]. In this paper, we introduce chance measure and uncertain random programming to model the TCTP with coexisted randomness and uncertainty. As Huang and Ding [31] illustrated that the standard path algorithms (e.g., the well-known Dijkstra method) were not able to arrive at solutions for searching critical path of this problem, genetic algorithm (GA) is applied in this paper.

The remainder of this paper is organized as follows: in Section 'Preliminaries of uncertainty theory', some concepts and useful theorems of uncertainty theory are presented. Section 'Description of time-cost trade-off problem' briefly describes the problem with uncertain random parameters. In Section 'Uncertain random time-cost trade-off model with dependent-chance programming', with dependent-chance programming, an uncertain random time-cost trade-off model is built and a crisp TCTP model is also given for some special case of uncertain random variable. Section 'Algorithm description with numerical experiment' conducts a numerical experiment to illustrate the proposed model. Finally, Section 'Conclusions' draws some concluding statements.

Preliminaries of uncertainty theory

In this section, we introduce some basic concepts and theorems for building uncertain random TCTP model. Let Γ be a nonempty set, $\mathcal{L}$ a σ-algebra over Γ, and each element Λ in $\mathcal{L}$ is called an event.

Definition 1. *(Liu [15]) The set function $\mathcal{M}$ is called an uncertain measure if it satisfies:*

Axiom 1 (*Normality Axiom*). *$\mathcal{M}\{\Gamma\} = 1$ for the universal set Γ.*

Axiom 2 (*Duality Axiom*). *$\mathcal{M}\{\Lambda\} + \mathcal{M}\{\Lambda^c\} = 1$ for any event Λ.*

Axiom 3 (*Subadditivity Axiom*). *For every countable sequence of events $\Lambda_1, \Lambda_2, \cdots$, we have*

$$\mathcal{M}\left\{\bigcup_{i=1}^{\infty} \Lambda_i\right\} \leq \sum_{i=1}^{\infty} \mathcal{M}\{\Lambda_i\}.$$

Besides, the product uncertain measure on the product σ-algebra $\mathcal{L}$ was defined by Liu [32] as follows:

Axiom 4 (*Product Axiom*). Let $(\Gamma_k, \mathcal{L}_k, \mathcal{M}_k)$ be uncertainty spaces for $k = 1, 2, \cdots$. The product uncertain measure $\mathcal{M}$ is an uncertain measure satisfying

$$\mathcal{M}\left\{\prod_{k=1}^{\infty} \Lambda_k\right\} = \bigwedge_{k=1}^{\infty} \mathcal{M}_k\{\Lambda_k\}$$

where Λ_k are arbitrarily chosen events from $\mathcal{L}_k$ for $k = 1, 2, \cdots$, respectively.

Definition 2. *(Liu [15]) An uncertain variable is a measurable function ξ from an uncertainty space $(\Gamma, \mathcal{L}, \mathcal{M})$ to the set of real numbers, i.e., for any Borel set B of real numbers, the set*

$$\{\xi \in B\} = \{\gamma \in \Gamma \mid \xi(\gamma) \in B\}$$

is an event.

Definition 3. *(Liu [32]) The uncertain variables $\xi_1, \xi_2, \cdots, \xi_n$ are said to be independent if*

$$\mathcal{M}\left\{\bigcap_{i=1}^{n} (\xi_i \in B_i)\right\} = \bigwedge_{i=1}^{n} \mathcal{M}\{\xi_i \in B_i\}$$

for any Borel sets $B_1, B_2, \cdots, B_n$.

Sometimes, to describe real-world optimization problems with uncertain parameters, it is sufficient to know the uncertainty distribution rather than the uncertain variable itself.

Definition 4. *(Liu [15]) The uncertainty distribution Φ of an uncertain variable ξ is defined by*

$$\Phi(x) = \mathcal{M}\{\xi \leq x\}$$

for any real number x.

Definition 5. *(Liu [16]) An uncertainty distribution $\Phi(x)$ is said to be regular if it is a continuous and strictly increasing function with respect to x at which $0 < \Phi(x) < 1$, and*

$$\lim_{x \to -\infty} \Phi(x) = 0, \qquad \lim_{x \to +\infty} \Phi(x) = 1.$$

Liu [16] also defined the inverse function Φ^{-1} as the inverse uncertainty distribution of uncertain variable ξ.

Theorem 1. *(Liu [16]) Let $\xi_1, \xi_2, \cdots, \xi_n$ be independent uncertain variables with regular uncertainty distributions $\Phi_1, \Phi_2, \cdots, \Phi_n$, respectively. If the function $f(x_1, x_2, \cdots, x_n)$ is strictly increasing with respect to $x_1, x_2, \cdots, x_m$ and strictly decreasing with respect to $x_{m+1}, x_{m+2}, \cdots, x_n$, then*

$$\xi = f(\xi_1, \xi_2, \cdots, \xi_n)$$

is an uncertain variable with inverse uncertainty distribution

$$\Phi^{-1}(\alpha) = f\left(\Phi_1^{-1}(\alpha), \cdots, \Phi_m^{-1}(\alpha), \Phi_{m+1}^{-1}(1-\alpha), \cdots, \Phi_n^{-1}(1-\alpha)\right).$$

Based on the definitions of uncertain variable and random variable, the concept of uncertain random variable is given as follows:

Definition 6. *(Liu [25]) An uncertain random variable is a function ξ from a chance space $(\Gamma, \mathcal{L}, \mathcal{M}) \times (\Omega, \mathcal{A}, \Pr)$ to the set of real numbers such that $\{\xi \in B\}$ is an event in $\mathcal{L} \times \mathcal{A}$ for any Borel set B.*

Example 1 A random variable is a special uncertain random variable since any real value is a special uncertain variable.

Example 2 The sum of a random variable and an uncertain variable is an uncertain random variable.

Definition 7. *(Liu [25]) Let $(\Gamma, \mathcal{L}, \mathcal{M}) \times (\Omega, \mathcal{A}, \Pr)$ be a chance space, and $\Theta \in \mathcal{L} \times \mathcal{A}$ be an event. Then, the chance measure of Θ is defined as*

$$Ch\{\Theta\} = \int_0^1 \Pr\{\omega \in \Omega \,|\, \mathcal{M}\{\gamma \in \Gamma \big| (\gamma, \omega) \in \Theta\} \geq x\} \mathrm{d}x.$$

Liu [25] proved that the chance measure is self-dual.

Definition 8. *(Liu [25]) Let ξ be an uncertain random variable. Then, the chance distribution of ξ is defined by*

$$\Phi(x) = Ch\{\xi \leq x\}$$

for any $x \in R$.

For some special case, some operational laws are presented as follows:

Theorem 2. *(Liu [26]) Let $\eta_1, \eta_2, \cdots, \eta_m$ be independent random variables with probability distributions $\Psi_1, \Psi_2, \cdots, \Psi_m$, respectively, and $\tau_1, \tau_2, \cdots, \tau_n$ be uncertain variables (not necessarily independent). Then, the uncertain random variable $\xi = f(\eta_1, \eta_2, \cdots, \eta_m, \tau_1, \tau_2, \cdots, \tau_n)$ has a chance distribution*

$$\Phi(x) = \int_{R^m} F(x; y_1, \cdots, y_m) \mathrm{d}\Psi_1(y_1) \cdots \mathrm{d}\Psi_m(y_m)$$

where $F(x; y_1, \cdots, y_m)$ *is the uncertainty distribution of the uncertain variable* $f(y_1, \cdots, y_m, \tau_1, \cdots, \tau_n)$ *for any real numbers* $y_1, \cdots, y_m$.

Theorem 3. *(Liu [33]) Let* $\eta_1, \eta_2, \cdots, \eta_m$ *be independent random variables with probability distributions* $\Psi_1, \Psi_2, \cdots, \Psi_m$, *and* $\tau_1, \tau_2, \cdots, \tau_n$ *be independent uncertain variables with regular uncertainty distributions* $\Upsilon_1, \Upsilon_2, \cdots, \Upsilon_n$, *respectively. If* $f(\eta_1, \cdots, \eta_m, \tau_1, \cdots, \tau_n)$ *is strictly increasing with respect to* $\tau_1, \cdots, \tau_k$ *and strictly decreasing with respect to* $\tau_{k+1}, \cdots, \tau_n$, *then*

$$Ch\{f(\eta_1, \cdots, \eta_m, \tau_1, \cdots, \tau_n) \leq 0\} = \int_{R^m} G(y_1, \cdots, y_m) \mathrm{d}\Psi_1(y_1) \cdots \mathrm{d}\Psi_m(y_m)$$

where $G(y_1, \cdots, y_m)$ *is the root* α *of the equation*

$$f\Big(y_1, \cdots, y_m, \Upsilon_1^{-1}(\alpha), \cdots, \Upsilon_k^{-1}(\alpha), \Upsilon_{k+1}^{-1}(1-\alpha), \cdots, \Upsilon_n^{-1}(1-\alpha)\Big) = 0.$$

Description of time-cost trade-off problem

Whether a project is successful or not depends on the opinion of the responsible managers. Some of them may focus more on whether the project is on time or before. Some others may pay more attentions to the project budget. For some projects, e.g., China's manned space program, the highest priority is technical performance. Generally, the success of a project is measured by three dimensions, i.e., time, cost, and performance (quality or agreed-upon specifications). For many projects, since project managers only need to meet some quality standards, the dimension of quality is not so important for project success. In this case, project managers are required to make trade-off between project cost and project completion time with the project progress. Furthermore, project cost and project completion time are always in some relation. Sometimes, project managers may make decisions to finish projects sooner with project cost augment by accelerating project schedule, which is also named as project crashing in project management. In other cases, motivated by reducing project costs, managers may be conscripted to sacrifice with prolonging project completion times. To sum up, it is natural for project managers to find some schedules to complete projects with the balance of project costs and project completion times.

Generally, a project can be represented by an activity-on-arc (AoA) network $G = (V, A)$, where $V = \{1, 2, \cdots, n\}$ is the set of nodes representing the milestones and A is the set of arcs representing the activities, shown as Figure 1. The normal duration of activity (i, j) denoted as ξ_{ij} represents the duration without the influence of decision made by the project manager. Correspondingly, the normal cost per time unit of activity (i, j) is denoted by c_{ij}. The decision variable x_{ij} indicates the duration change of activity (i, j) controlled by the manager, which may be realized by determining work force change, instrument change, etc. The variable x_{ij}, supposed to be an integer for simplicity, is bounded by some interval $[l_{ij}, u_{ij}]$ owing to practical conditions, where l_{ij} and u_{ij} are also assumed to be integers. Similarly, the decision associated cost d_{ij} is regarded as the additional cost of per unit change of x_{ij}. For simplicity, c_{ij} and d_{ij} are both assumed to be constants. Due to the changeable and indeterminate external environment, the normal durations ξ_{ij} are always nondeterministic for all activities $(i, j) \in A$. For many projects, since the 'new' parts and ever performed parts coexist, the normal activity durations here are assumed to be uncertain random variables, concisely written as

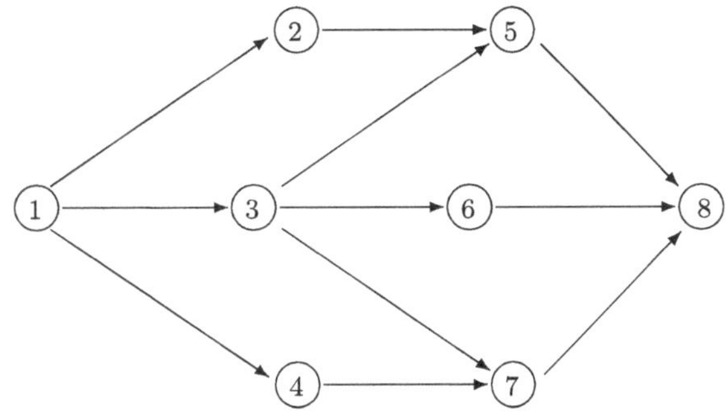

Figure 1 An illustration of a project.

$\boldsymbol{\xi} = \{\xi_{ij} : (i,j) \in A\}$. For the trade-off between project completion time and project cost, the goal is to decide the optimal vector $\boldsymbol{x} = \left\{x_{ij} : (i,j) \in A\right\}$ to meet different scheduling requirements.

The starting time of activity (i,j) is denoted by $T_{ij}(\boldsymbol{x}, \boldsymbol{\xi})$, and the starting time of activity $(1,j) \in A$ is defined as $T_{1j}(\boldsymbol{x}, \boldsymbol{\xi}) = 0$, which means that the starting time of the total project is assumed to be 0. To simplify the problem, it is assumed that each activity can be processed only if all the foregoing activities are finished and should be processed without interruption, lead times, and lag times. With the assumptions, the starting time of activity (i,j), $i = 2, 3, \cdots, n-1$ can be calculated by

$$T_{ij}(\boldsymbol{x}, \boldsymbol{\xi}) = \max_{(k,i)\in A} \{T_{ki}(\boldsymbol{x}, \boldsymbol{\xi}) + \xi_{ki} + x_{ki}\}.$$

And the project completion time can be written by

$$T(\boldsymbol{x}, \boldsymbol{\xi}) = \max_{(k,n)\in A} \{T_{kn}(\boldsymbol{x}, \boldsymbol{\xi}) + \xi_{kn} + x_{kn}\}. \tag{1}$$

The project cost, composed of the normal cost and the additional cost, can be calculated by

$$C(\boldsymbol{x}, \boldsymbol{\xi}) = \sum_{(i,j)\in A} (c_{ij}\xi_{ij} - d_{ij}x_{ij}). \tag{2}$$

Uncertain random time-cost trade-off model with dependent-chance programming

Due to the complex project environment, achieving project goals faces many types of indeterminacy or risks, e.g., the weather variation and the government policy adjustment, leading to the fact that many project objectives cannot be obtained completely. Such a case may be more vital when project objectives are in conflict, as this is mostly the truth. Hence, being sure to maximize the opportunity of achieving project goals is as important as the trade-off between project goals in project management. In this section, dependent-chance programming (DCP), initiated by Liu [34], is applied in uncertain random TCTP. Generally, DCP is applied for risk-averse project managers to ensure the achievement of project goals for possibly extra expenses. Based on the DCP philosophy, the uncertain random chance maximization model is built as follows:

$$\begin{cases} \max \mathrm{Ch}\left\{C(\boldsymbol{x}, \boldsymbol{\xi}) \leq C^0\right\} \\ \text{subject to:} \\ \qquad \mathrm{Ch}\{T(\boldsymbol{x}, \boldsymbol{\xi}) \leq T^0\} \geq \alpha_0 \\ \qquad x_{ij} \in [l_{ij}, u_{ij}], \ \forall(i,j) \in A \\ \qquad x_{ij}, \ \forall(i,j) \in A, \ \text{integers} \end{cases}$$

where α_0 is the confidence level given in advance, T^0 is the due date of the project, C^0 is the budget, l_{ij} and u_{ij} are integers given in advance, and $T(\boldsymbol{x}, \boldsymbol{\xi})$ and $C(\boldsymbol{x}, \boldsymbol{\xi})$ are defined by Equations 1 and 2, respectively.

As we mentioned above, in most projects, 'new' parts and 'old' parts integrate into the whole project. That means the uncertain random environment for many projects can be divided into sole randomness and uncertainty. In this case, the uncertain random vector $\boldsymbol{\xi}$ can be represented with the form $(\eta_1, \eta_2, \cdots, \eta_k, \tau_1, \tau_2, \cdots, \tau_l)$, where $\eta_1, \eta_2, \cdots, \eta_k$ are independent random variables with probability distributions $\Psi_1, \Psi_2, \cdots, \Psi_k$, and $\tau_1, \tau_2, \cdots, \tau_l$ are independent uncertain variables with uncertainty distributions Υ_1, $\Upsilon_2, \cdots, \Upsilon_l$, respectively. Since $C(\boldsymbol{x}, \boldsymbol{\xi}) - C^0$ and $T(\boldsymbol{x}, \boldsymbol{\xi}) - T^0$ are strictly increasing with respect to all the uncertain random variables, applying Theorem 3, the above model can be transformed into the following crisp model:

$$\begin{cases} \max \int_{\mathcal{R}^k} G(\boldsymbol{x}, z_1, \cdots, z_k) \mathrm{d}\Psi_1(z_1) \cdots \mathrm{d}\Psi_k(z_k) \\ \text{subject to :} \\ \qquad \int_{\mathcal{R}^k} H(\boldsymbol{x}, z_1, \cdots, z_k) \mathrm{d}\Psi_1(z_1) \cdots \mathrm{d}\Psi_k(z_k) \geq \alpha_0 \\ \qquad x_{ij} \in [l_{ij}, u_{ij}], \ \forall(i,j) \in A \\ \qquad x_{ij}, \ \forall(i,j) \in A, \ \text{integers} \end{cases}$$

where $G(\boldsymbol{x}, z_1, \cdots, z_k)$ is the root β of

$$C\Big(\boldsymbol{x}, z_1, \cdots, z_k, \Upsilon_1^{-1}(\beta), \cdots, \Upsilon_l^{-1}(\beta)\Big) - C^0 = 0,$$

and $H(\boldsymbol{x}, z_1, \cdots, z_k)$ is the root β of

$$T\Big(\boldsymbol{x}, z_1, \cdots, z_k, \Upsilon_1^{-1}(\beta), \cdots, \Upsilon_l^{-1}(\beta)\Big) - T^0 = 0.$$

Algorithm description with numerical experiment

Since Huang and Ding [31] demonstrated that the standard path algorithms were not able to approach the critical path for project scheduling problem with random activity durations, it is more difficult to solve uncertain random TCTP. Hence, uncertain random simulation and GA are integrated for solving the proposed chance maximization model in this section. An uncertain random simulation algorithm for chance measure (e.g., $\mathrm{Ch}\left\{C(\boldsymbol{x}, \boldsymbol{\xi}) \leq C^0\right\}$) can be obtained as follows:

Uncertain random simulation for chance measure:

Step 1. Set $e = 0$.

Step 2. Randomly generate ω from the probability space according to the probability distribution.

Step 3. $e \leftarrow e + \mathcal{M}\{C(\boldsymbol{x}, \boldsymbol{\xi}(\cdot, \omega)) \leq C^0\}$.

Step 4. Repeat the second and third steps N times, where N is a sufficiently large number.

Step 5. Return e/N as the chance measure.

The above designed uncertain random simulation approach can be embedded in GA simply for solving the proposed model.

Now, consider a project illustrated in Figure 1. The activity durations, the normal costs, and the additional costs of the activities are presented in Table 1, respectively. Note that the uncertain random activity durations are partly linear uncertain variables with the form of $\mathcal{L}(a, b)$ and partly uniformly distributed random variables denoted by $\mathcal{U}(c, d)$, where $a < b$ and $c < d$ are crisp numbers. The chance maximization model is given as follows:

$$\begin{cases} \max \text{Ch}\{C(\boldsymbol{x}, \boldsymbol{\xi}) \leq 24100\} \\ \text{subject to:} \\ \qquad \text{Ch}\{T(\boldsymbol{x}, \boldsymbol{\xi}) \leq 35\} \geq 0.9 \\ \qquad x_{ij} \in [-4, 3], \ \forall (i, j) \in A \\ \qquad x_{ij}, \ \forall (i, j) \in A, \ \text{integers.} \end{cases}$$

Uncertain random simulation and GA are employed to search for the optimal solution. For the above project, the parameters in GA are set as the population size 70, the mutation probability 0.8, and the crossover probability 0.5. After a run of 1,000 generations, the optimal solution and value are obtained: $\boldsymbol{x}^* = (1, -1, 0, 3, 0, -3, 1, 3, -1, 3, -1)$ and $\text{Ch}\{C(\boldsymbol{x}, \boldsymbol{\xi}) \leq 24100\} = 0.938$, respectively.

Conclusions

In project management, one of the characteristics of the concept 'project' is that there is always something new in a project, i.e., a project always has some 'new' parts with little or no historical data. Meanwhile, most projects may contain some common or 'old' parts (with enough historical data) ever performed by many other finished projects. Hence, uncertainty and randomness should be considered simultaneously in project scheduling

Table 1 Activity durations and costs of project

Activity (i, j)	Normal duration ξ_{ij}	Normal cost c_{ij}	Additional cost d_{ij}
(1,2)	$\mathcal{L}(7, 10)$	170	200
(1,3)	$\mathcal{U}(9, 12)$	300	280
(1,4)	$\mathcal{L}(8, 10)$	65	70
(2,5)	$\mathcal{U}(8, 12)$	270	300
(3,5)	$\mathcal{L}(10, 15)$	135	150
(3,6)	$\mathcal{L}(9, 13)$	75	90
(3,7)	$\mathcal{U}(11, 15)$	150	100
(4,7)	$\mathcal{L}(10, 13)$	600	400
(5,8)	$\mathcal{U}(10, 14)$	85	100
(6,8)	$\mathcal{L}(12, 14)$	300	400
(7,8)	$\mathcal{U}(9, 14)$	95	90

problem. In this paper, an uncertain random time-cost trade-off model with dependent-chance programming was built with a crisp equivalent model for the case that uncertain random parameters in the problem are partly random variables and partly uncertain variables. Furthermore, uncertain random simulation and GA were integrated for solving the proposed model.

The TCTP with uncertain random parameters can be regarded as the extension of the fuzzy TCTP, random TCTP, uncertain TCTP, and fuzzy random TCTP. Besides, since in this paper continuous time-cost trade-off relationship was assumed, in future research, discrete or mixed time-cost trade-off relationships can be introduced for modeling other project optimization problems.

Acknowledgements

The work was partly supported by the National Natural Science Foundation of China (71371141, 71001080).

References

1. Kelley, JE: Critical path planning and scheduling mathematical basis. Oper. Res. **9**(3), 296–320 (1961)
2. Heberta, JE, Deckro, RF: Combining contemporary and traditional project management tools to resolve a project scheduling problem. Comput. Oper. Res. **38**, 21–32 (2011)
3. Weglarz, J, Jozefowska, J, Mika, M, Waligora, G: Project scheduling with finite or infinite number of activity processing modes - A survey. Eur. J. Oper. Res. **208**(3), 177–205 (2011)
4. Salmasnia, A, Mokhtari, H, Abadi, INK: A robust scheduling of projects with time, cost, and quality considerations. Int. J. Adv. Manuf. Tech. **60**(5–8), 631–642 (2012)
5. Ghoddousi, P, Eshtehardian, E, Jooybanpour, S, Javanmardi, A: Multi-mode resource-constrained discrete time-cost-resource optimization in project scheduling using non-dominated sorting genetic algorithm. Automat. Constr. **30**, 216–227 (2013)
6. Wollmer, RD: Critical path planning under uncertainty. Math. Program. Stud. **25**, 164–171 (1985)
7. Ke, H, Ma, W, Ni, Y: Optimization models and a GA-based algorithm for stochastic time-cost trade-off problem. Appl. Math. Comput. **215**(1), 308–313 (2009)
8. Mokhtari, H, Aghaie, A, Rahimi, J, Mozdgir, A: Project time-cost trade-off scheduling: a hybrid optimization approach. Int. J. Adv. Manuf. Tech. **50**, 811–822 (2010)
9. Ke, H, Ma, W, Chen, X: Modeling stochastic project time-cost trade-offs with time-dependent activity durations. Appl. Math. Comput. **218**(18), 9462–9469 (2012)
10. Leu, SS, Chen, AT, Yang, CH: A GA-based fuzzy optimal model for construction time-cost trade-off. Int. J. Proj. Manage. **19**, 47–58 (2001)
11. Ghazanfari, M, Yousefli, A, Ameli, MSJ, Bozorgi-Amiri, A: A new approach to solve time-cost trade-off problem with fuzzy decision variables. Int. J. Adv. Manuf. Tech. **42**, 408–414 (2009)
12. Ke, H, Ma, W, Gao, X, Xu, W: New fuzzy models for time-cost trade-off problem. Fuzzy Optim. Decis. Ma. **9**(2), 219–231 (2010)
13. Chen, S, Tsai, M: Time-cost trade-off analysis of project networks in fuzzy environments. Eur. J. Oper. Res. **212**(2), 386–397 (2011)
14. Liu, B: Why is there a need for uncertainty theory. J. Uncertain Syst. **6**, 3–10 (2012)
15. Liu, B: Uncertainty Theory. 2nd ed. Springer-Verlag, Berlin (2007)
16. Liu, B: Uncertainty Theory: A Branch of Mathematics for Modeling Human Uncertainty. Springer-Verlag, Berlin (2010)
17. Bhattacharyya, R, Chatterjee, A, Kar, S: Uncertainty theory based multiple objective mean-entropy-skewness stock portfolio selection model with transaction costs. J. Uncertain. Anal. Appl. **1**, Article 16 (2013)
18. Ding, S: Uncertain random newsboy problem. J. Intell. Fuzzy Syst. **26**, 483–490 (2014)
19. Gao, Y: Uncertain models for single facility location problems on networks. Appl. Math. Model. **36**, 2592–2599 (2012)
20. Jana, DK, Maity, K, Roy, TK: A three-layer supply chain integrated production-inventory model under permissible delay in payments in uncertain environments. J. Uncertain. Anal. Appl. **1**, Article 6 (2013)
21. Lu, Y, Huang, Z, Gao, J: Under-investment problem with adverse selection in uncertain environment. J. Uncertain Syst. **8**, 83–89 (2014)
22. Yang, X, Gao, J: Uncertain differential games with application to capitalism. J. Uncertain. Anal. Appl. **1**, Article 17 (2013)
23. Yang, X, Gao, J: Uncertain core for coalitional game with uncertain payoffs. J. Uncertain Syst. **8**, 13–21 (2014)
24. Ke, H: A genetic algorithm-based optimizing approach for project time-cost trade-off with uncertain measure. J. Uncertain. Anal. Appl. **2**, Article 8 (2014)
25. Liu, YH: Uncertain random variables: A mixture of uncertainty and randomness. Soft. Comput. **17**, 625–634 (2013)
26. Liu, YH: Uncertain random programming with applications. Fuzzy Optim. Decis. Ma. **12**, 153–169 (2013)
27. Liu, B: Uncertain Random Graph and Uncertain Random Network. J. Uncertain Syst. **8**, 3–12 (2014)
28. Liu, YH, Ralescu, DA: Risk index in uncertain random risk analysis. Int. J. Uncertain. Fuzz. **22**, 491–504 (2014)
29. Zhou, J, Yang, F, Wang, K: Multi-objective optimization in uncertain random environments. Fuzzy Optim. Decis. Ma. **13**, 397–413 (2014)

30. Ke, H, Su, T, Ni, Y: Uncertain random multilevel programming with application to production control problem (2014). doi:10.1007/s00500-014-1361-2
31. Huang, W, Ding, L: Project-scheduling problem with random time-dependent activity duration times. IEEE T. Eng. Manage. **58**, 377–387 (2011)
32. Liu, B: Some research problems in uncertainty theory. J. Uncertain Syst. **3**, 3–10 (2009)
33. Liu, YH, Ralescu, DA: Risk index in uncertain random risk analysis. Int. J. Uncertain. Fuzz. **22**, 491–504 (2014)
34. Liu, B: Dependent-chance programming: a class of stochastic programming. Comput. Math. Appl. **34**, 89–104 (1997)

15

Maximum-mean uncertain measure

Xiang Li

Correspondence:
lixiang@mail.buct.edu.cn
School of Economics and Management, Beijing University of Chemical Technology, No. 15, Beisanhuan East Road, 100029 Beijing, China

Abstract

Uncertain measure, which is used to indicate the degree of belief that an uncertain event may occur, is a set function satisfying the normality, duality, and subadditivity axioms. Although the tools of uncertain measure, uncertainty space, and uncertain variable have been well developed and became the foundations of uncertainty theory, there is no study on how to construct the uncertainty space for an uncertain variable, which seriously hinders the applications of uncertainty theory in practice. In this paper, we propose an expert knowledge-based construction approach for uncertainty space with maximum-mean uncertain measure, which takes the maximum and mean operations on confidence function.

Keywords: Maximum-mean uncertain measure; Confidence function; Expert knowledge

Introduction

Uncertainty theory introduced by Liu [1] is a branch of nonadditive measure theory for describing human uncertainty. Compared with randomness, Liu's uncertainty amounts to subject indeterminacy arising from information loss on some exact value of a certain quantity. The main difference between randomness and Liu's uncertainty is that the former is used to model unfixed values, while the latter is used to model fixed but ill-known values. For example, if we toss a coin in the air and choose between head and tail before it lands on the ground, the outcome is random since both head and tail are possible. However, if we make the choice after the coin has landed on the ground, the outcome is uncertain since it has been fixed although we do not know the result.

Nowadays, uncertainty theory has become a branch of axiomatic mathematics for modeling human uncertainty, and it has been well developed and applied to uncertain programming [2], uncertain statistics [3], uncertain risk analysis [4], uncertain set [5], uncertain logic [6], uncertain inference [7,8], uncertain process [9], uncertain renewal process [10], uncertain calculus [11], uncertain differential equation [12], uncertain finance [13], and so on. The latest results related to uncertainty theory may be found in [14].

Uncertain measure and uncertain variable are two basic concepts in uncertainty theory. Uncertain measure, which is used to indicate the degree of belief that an uncertain event may occur, was defined in [1] as a set function satisfying the normality, duality, and subadditivity axioms. The normality is used to ensure that degree of belief is consistent with human thinking, and the duality is an extension of the law of contradiction. For a sequence of countable events, subadditivity means that the entire degree of belief must be less than or equal to the sum of their individual degrees of belief. Liu [15] proved the monotonicity theorem, null-additivity theorem, and asymptotic theorem, and Gao [16] studied some properties of continuous uncertain measures. The product uncertain measure of compound event was defined by Liu [17] as the minimum of uncertain measures of individual events, and the product uncertain measure theorem was proved by Peng and Iwamura [18]. In order to study the dependence of uncertain events, Liu [1] defined the conditional uncertain measure of an event given that some other event has occurred. Uncertain variable is used to represent quantity with uncertainty, which was defined in [1] as a measurable function from uncertainty space to the set of real numbers.

Possibility theory is another type of nonadditive measure theory for dealing with the subjective uncertainty. Possibility measure was defined by Zadeh [19] as a set function satisfying the normality, nonnegativity, and maximality. Necessity measure was also defined by Zadeh [20] as the dual part of possibility measure. It is proved that both possibility measure and necessity measure satisfy the properties of normality, nonnegativity, and monotonicity. However, neither of them are self-dual. Since the duality is intuitive and important in real problems, Liu and Liu [21] defined a credibility measure as the average of possibility measure and necessity measure, which was lately proved to be a set function satisfying normality, monotonicity, duality, and partial maximality [22]. As an extension of credibility measure, Liu [1] proposed the uncertain measure by slacking the partial maximality to subadditivity, which generalizes the uncertain model to describe ill-known but fixed value.

Although the tools of uncertain measure, uncertainty space, and uncertain variable have been well developed and became the foundations for uncertainty theory, there is no study on how to construct the uncertainty space for an uncertain variable. This seriously hinders the applications of uncertainty theory in practice. The purpose of this paper is to provide a general construction approach for uncertainty space with maximum-mean uncertain measure based on the expert knowledge. The rest of this paper is organized as follows. The 'Preliminaries' section reviews some basic definitions and properties on uncertainty theory. The 'Confidence function' section introduces a concept of confidence function. In the 'Maximum-mean uncertain measure' section, we propose the concept of maximum-mean uncertain measure based on the maximum and mean operations on confidence function, and we provide a general procedure on generating an uncertainty space with the maximum-mean uncertain measure. The 'An illustrative example' section provides a practical example to illustrate how to construct an uncertainty space for an uncertain quantity. Finally, a brief summary is given.

Preliminaries

This section will briefly review some basic concepts and properties of uncertainty theory. Let Γ be a nonempty set, and let Λ be a σ-algebra over Γ. Each element of Λ is called an event. In order to present an axiomatic definition of uncertain measure, it is necessary to assign to each event a number which indicates the degree of belief that the event will occur. In order to ensure that this assignment has certain meaningful mathematical properties, Liu [1] proposed the following three axioms:

Axiom 1. (Normality) $\mathrm{M}\{\Gamma\} = 1$;

Axiom 2. (Duality) $\mathrm{M}\{A\}+\mathrm{M}\{A^c\}=1$ for any $A \in \Lambda$;

Axiom 3. (Subadditivity) For every countable sequence of events $\{A_i\}$, we have:

$$\mathrm{M}\left\{\bigcup_{i=1}^{\infty} A_i\right\} \le \sum_{i=1}^{\infty} \mathrm{M}\{A_i\}.$$

The normality is used to ensure that degree of belief is consistent with human thinking. The duality is an extension of the law of contradiction, which ensures that an event and its complement cannot occur simultaneously. For a sequence of countable events, subadditivity means that the entire degree of belief must be less than or equal to the sum of their individual degrees of belief.

Definition 2.1. Let Γ be a nonempty set, and let Λ be a σ-algebra (Liu [1]). The set function M is called an uncertain measure if it satisfies the normality, duality, and subadditivity axioms.

Liu [15] proved the monotonicity theorem for uncertain measure. That is, for any events $A_1 \subseteq A_2$, we have $\mathrm{M}\{A_1\} \le \mathrm{M}\{A_2\}$. In addition, the null-additivity theorem and asymptotic theorem were also proved. Liu [17] defined the product uncertain measure for compound event, and Liu [1] defined a conditional uncertain measure of an event after it has been learned that some other event has occurred.

Definition 2.2. Suppose that Γ is a nonempty set, Λ is a σ-algebra, and M is an uncertain measure (Liu [1]). Then, the triplet $(\Gamma, \Lambda, \mathrm{M})$ is called an uncertainty space.

Definition 2.3. An uncertain variable ξ is a measurable function from an uncertainty space $(\Gamma, \Lambda, \mathrm{M})$ to the set of real numbers, that is, for any Borel set B of $\Re$, we have (Liu [1]):

$$\{\gamma \in \Gamma \mid \xi(\gamma) \in B\} \in \Lambda. \tag{1}$$

Example 2.1. Suppose that Γ is a nonempty set containing three points γ_1, γ_2, and γ_3. Let Λ be the power set of Γ. Define $\mathrm{M}\{\emptyset\}=0$, $\mathrm{M}\{\gamma_1\}=0.2$, $\mathrm{M}\{\gamma_2\}=0.4$, $\mathrm{M}\{\gamma_3\}=0.6$, $\mathrm{M}\{\gamma_1,\gamma_2\}=0.4$, $\mathrm{M}\{\gamma_2,\gamma_3\}=0.8$, $\mathrm{M}\{\gamma_1,\gamma_3\}=0.6$ and $\mathrm{M}\{\Gamma\}=1$. It is easy to prove that M satisfies the normality, duality, and subadditivity axioms. Then, M is an uncertain measure, and the triplet $(\Gamma, \Lambda, \mathrm{M})$ is an uncertainty space. Define a real function $\xi(\gamma_1)=x_1$, $\xi(\gamma_2)=x_2$ and $\xi(\gamma_3)=x_3$. Since Λ is the largest σ-algebra over Γ, for any Borel set B of $\Re$, it is clear that $\{\xi \in B\} \in \Lambda$. Thus, ξ is a measurable function, that is, an uncertain variable, on uncertainty space $(\Gamma, \Lambda, \mathrm{M})$.

Confidence function

Let ξ be an uncertain variable. For theoretical studies, we always assume that we can define an uncertainty space $(\Gamma, \Lambda, \mathrm{M})$ with known uncertain measure such that ξ is a measurable function from this space to the set of real numbers. However, it is difficult to construct an uncertainty space in practice since there is no efficient construction approach for uncertain measure.

If Λ is finite and has a small number of elements, we may define an uncertain measure by enumerating the degree of beliefs for all events according to the expert knowledge.

However, if the cardinality of Λ is infinite, it is impossible to do in this way since the number of events is unmanageable. Therefore, an efficient construction approach for uncertain measure should be necessary. This paper defines a maximum-mean uncertain measure and proposes a construction approach based on the maximum and mean operations on confidence function. First, we introduce the concept of confidence function.

Let $\Gamma = \{x_1, x_2, \cdots\}$ be the universal set, i.e., the collection of candidates. Suppose that there are m experts with confidences $\{\nu_1, \nu_2, \cdots, \nu_m\}$, which are given by an expert of experts. Without loss of generality, we assume that each candidate associated with one and only one confidence under the following rules: (i) if some candidate receives no support from any expert, we assign it with confidence 0; (ii) if some candidate receives support from one and only one expert, for example, expert i, we assign ν_i to it; (iii) if some candidate receives supports from more experts, for example, experts $i_1, i_2, \cdots, i_{m_i}$, we assign it with the confidence $\max\left\{\nu_{i_1}, \nu_{i_2}, \cdots, \nu_{i_{m_i}}\right\}$.

A confidence function μ is a mapping from Γ to $[0, 1]$ such that $\mu(x_i)$ denotes the confidence on the ith candidate x_i, which has the following form:

$$\mu(x) = \begin{cases} \mu_1, & \text{if } x = x_1 \\ \mu_2, & \text{if } x = x_2 \\ \vdots & \vdots \\ \mu_n, & \text{if } x = x_n. \end{cases} \tag{2}$$

The confidence function indicates a relative preference among different candidates for the uncertain quantity. Generally speaking, we prefer to take the candidate that has the largest confidence value.

Maximum-mean uncertain measure

This section defines a maximum-mean uncertain measure based on the confidence function, which expresses the degree of belief that an uncertain event will occur.

First, let us consider the confidence on a set of candidates. As we know, each uncertain quantity has one and only one correct value. In other words, there is only one expert providing the correct estimation, and the others all provide the wrong estimation. The uncertainty arises from the difficulty that we do not know who is the correct expert. Therefore, if we believe that the uncertain quantity takes a value belonging to set A, the confidence should be the maximum value among all confidences on candidates in set A.

Example 4.1. Suppose that ξ is an uncertain quantity with confidence function:

$$\mu(x) = \begin{cases} 0.3, & \text{if } x = 10 \\ 0.5, & \text{if } x = 15 \\ 0.7, & \text{if } x = 20 \\ 1.0, & \text{if } x = 30. \end{cases}$$

Since candidate 30 has the maximum confidence, we prefer to believe that ξ will take value 30. On the other hand, for the uncertain event $\{\xi \neq 30\}$, the confidence should be the maximum confidence on candidates 10, 15, and 20, that is, $0.3 \vee 0.5 \vee 0.7 = 0.7$.

Let ξ be an uncertain quantity with confidence function μ. Based on the above analysis, for each Borel set A, we calculate the confidence of the uncertain event $\{\xi \in A\}$ by:

$$\mu_A = \max_{x_i \in A} \mu_i, \tag{3}$$

and calculate the confidence of its complement $\{\xi \in A^c\}$ by:

$$\mu_{A^c} = \max_{x_i \in A^c} \mu_i. \tag{4}$$

If $\mu_A > 0$, we believe that it is possible that ξ takes a value in set A, and the degree of belief should have a strong positive correlation with respect to the confidence such that a larger confidence leads to a larger degree of belief.

Theorem 4.1. Suppose that Γ is a nonempty set $\{x_1, x_2, \cdots\}$, and μ is a nonnegative real function satisfying:

$$\max_{x_i \in \Gamma} \mu_i > 0. \tag{5}$$

Let Λ be a σ-algebra over Γ. For each $A \in \Lambda$, we define:

$$\mathrm{M}\{A\} = \frac{\max\limits_{x_i \in A} \mu_i}{\max\limits_{x_i \in A} \mu_i + \max\limits_{x_i \in A^c} \mu_i}. \tag{6}$$

Then, the set function M is an uncertain measure (called the maximum-mean uncertain measure), and the triplet $(\Gamma, \Lambda, \mathrm{M})$ is an uncertainty space.

Proof: We prove that the set function M satisfies the normality axiom, duality axiom, and subadditivity axiom. First, according to condition (5), the normality can be obtained immediately from formula (6) by showing:

$$\max_{x_i \in \Gamma^c} \mu_i = \max_{x_i \in \emptyset} \mu_i = 0.$$

Furthermore, let us prove the duality. For any $A \in \Lambda$, it follows from (6) that:

$$\mathrm{M}\{A\} = \frac{\max\limits_{x_i \in A} \mu_i}{\max\limits_{x_i \in A} \mu_i + \max\limits_{x_i \in A^c} \mu_i}, \quad \mathrm{M}\{A^c\} = \frac{\max\limits_{x_i \in A^c} \mu_i}{\max\limits_{x_i \in A^c} \mu_i + \max\limits_{x_i \in A} \mu_i},$$

which implies that $\mathrm{M}\{A\} + \mathrm{M}\{A^c\} = 1$. Finally, we prove the subadditivity. For any events $A, B \in \Lambda$, without loss of generality, we assume that:

$$\max_{x_i \in A} \mu_i \le \max_{x_i \in B} \mu_i.$$

Then, it can be proved that:

$$\max_{x_i \in A \cup B} \mu_i = \max_{x_i \in A} \mu_i \vee \max_{x_i \in B} \mu_i = \max_{x_i \in B} \mu_i. \tag{7}$$

The argument breaks down into two cases. If:

$$\max_{x_i \in A \cap B^c} \mu_i < \max_{x_i \in A^c \cap B^c} \mu_i,$$

it follows from $B^c = (A \cap B^c) \cup (A^c \cap B^c)$ that:

$$\max_{x_i \in B^c} \mu_i = \max_{x_i \in A^c \cap B^c} \mu_i. \tag{8}$$

Then, Equations (7) and (8) prove that:

$$\frac{\max\limits_{x_i\in A\cup B}\mu_i}{\max\limits_{x_i\in A\cup B}\mu_i+\max\limits_{x_i\in A^c\cap B^c}\mu_i}=\frac{\max\limits_{x_i\in B}\mu_i}{\max\limits_{x_i\in B}\mu_i+\max\limits_{x_i\in B^c}\mu_i},$$

which implies that $\mathrm{M}\{A\cup B\}=\mathrm{M}\{B\}\le\mathrm{M}\{A\}+\mathrm{M}\{B\}$. Otherwise, we have:

$$\max_{x_i\in A\cap B^c}\mu_i\ge\max_{x_i\in A^c\cap B^c}\mu_i.$$

In this case, it can be proved that:

$$\max_{x_i\in B^c}\mu_i=\max_{x_i\in A\cap B^c}\mu_i\le\max_{x_i\in A}\mu_i. \tag{9}$$

It follows from $A^c=(A^c\cap B)\cup(A^c\cap B^c)$ that:

$$\begin{aligned}\max_{x_i\in A^c}\mu_i &= \max_{x_i\in A^c\cap B^c}\mu_i\vee\max_{x_i\in A^c\cap B}\mu_i\\ &\le \max_{x_i\in A\cap B^c}\mu_i\vee\max_{x_i\in A^c\cap B}\mu_i\\ &\le \max_{x_i\in A}\mu_i\vee\max_{x_i\in B}\mu_i\\ &= \max_{x_i\in B}\mu_i.\end{aligned} \tag{10}$$

According to inequalities (9) and (10), we have:

$$\mathrm{M}\{A\}\ge\frac{\max\limits_{x\in A}\mu(x)}{\max\limits_{x\in A}\mu(x)+\max\limits_{x\in B}\mu(x)},\quad \mathrm{M}\{B\}\ge\frac{\max\limits_{x\in B}\mu(x)}{\max\limits_{x\in A}\mu(x)+\max\limits_{x\in B}\mu(x)}$$

which implies that $\mathrm{M}\{A\}+\mathrm{M}\{B\}\ge 1\ge\mathrm{M}\{A\cup B\}$. The subadditivity is proved. The proof is complete.

Remark 4.1. As a special uncertain measure, the maximum-mean uncertain measure also satisfies the monotonicity theorem, null-additivity theorem, and asymptotic theorem.

Let ξ be an uncertain quantity with confidence function μ. Define an uncertainty space $(\Gamma,\Lambda,\mathrm{M})$ according to Theorem 4.1. Then, ξ can be well described as $\xi(x_i)=x_i$, $\forall x_i\in\Gamma$. A general procedure on generating an uncertainty space with maximum-mean uncertain measure for ξ is summarized as follows:

Step 1. Define the set of candidates $\Gamma=\{x_1,x_2,\cdots\}$.
Step 2. Define a confidence function $\mu:\Gamma\to[0,\infty)$.
Step 3. Take a σ-algebra Λ over Γ.
Step 4. Define a maximum-mean uncertain measure M according to formula (6).
Step 5. Define $\xi(x_i)=x_i$ for all $x_i\in\Gamma$.

Remark 4.2. If there is a candidate with confidence zero, for example, $\mu_1=0$, then, we have $\mathrm{M}\{\xi=x_1\}=0$. If all candidates have the same confidence, then, the degree of belief that ξ takes value x_i is $\mathrm{M}\{\xi=x_i\}=0.5$ for each $1\le i\le n$.

An illustrative example

This practical example illustrates how to construct an uncertainty space for an uncertain quantity. Yufu Ning is a middle-aged college professor. He is my friend, but I have no idea on the exact value of his age, except that it should take value in $\Gamma = \{41, 42, 43, 44, 45, 46, 47, 48, 49, 50\}$. In order to obtain a better understanding on it, I send a questionnaire to ten of our common friends. The first question is "How old is Prof. Ning," and the second question is "What are the degrees of belief that Prof. Ning is less than or equal to 45 and Prof. Ning is larger than 45." The received answers are summarized by Table 1, where the third column shows the candidates selected by these experts, the fourth column shows the confidences of experts, the fifth column shows the degrees of belief that Prof. Ning is less than or equal to 45, and the last column shows the degrees of belief that Prof. Ning is larger than 45. Here, the confidences of experts are given by Prof. Ning.

It follows from Table 1 that the confidence function is $\mu(41) = 0.4$, $\mu(44) = 0.8$, $\mu(45) = 0.8$, $\mu(46) = 0.9$, and $\mu(47) = 0.9$. Take Λ to be the power set of Γ. According to formula (6), we can define a maximum-mean uncertain measure M. Then, Ning's age may be treated as the identity uncertain variable ξ on uncertainty space $(\Gamma, \Lambda, \mathrm{M})$.

Now, we consider the maximum-mean uncertain measure of events $\{\xi \leq 45\}$ and $\{\xi > 45\}$. According to formula (6), it is easy to calculate that $\mathrm{M}\{\xi \leq 45\} = 0.8/(0.8 + 0.9) = 0.471$ and $\mathrm{M}\{\xi > 45\} = 0.9/(0.9 + 0.8) = 0.529$. On the other hand, according to Table 1, we aggregate the experts' knowledge as follows:

- Prof. Ning is less than or equal to 45 with:

$$\frac{\sum_{i=1}^{10} \nu_i \alpha_i}{\sum_{i=1}^{10} \nu_i \alpha_i + \sum_{i=1}^{10} \nu_i \beta_i} = 0.472.$$

- Prof. Ning is larger than 45 with:

$$\frac{\sum_{i=1}^{10} \nu_i \beta_i}{\sum_{i=1}^{10} \nu_i \alpha_i + \sum_{i=1}^{10} \nu_i \beta_i} = 0.528.$$

Therefore, the maximum-mean uncertain measure satisfies the experts' knowledge.

Conclusion

The main contribution of this paper is the constructive approach of an uncertainty space with maximum-mean uncertain measure for an uncertain variable. It is the first

Table 1 Answers on Prof. Ning's age

	Expert	Candidate	Confidence (ν_i)	$\leq 45 (\alpha_i)$	$> 45 (\beta_i)$
1	Dr. Chen	47	0.7	0.4	0.6
2	Dr. Gao	41	0.4	0.7	0.2
3	Prof. Ke	45	0.6	0.8	0.6
4	Dr. Li	45	0.8	0.5	0.4
5	Prof. Qin	47	0.9	0.5	0.5
6	Dr. Wen	45	0.4	0.3	0.7
7	Prof. Zhou	45	0.3	0.9	0.1
8	Prof. Zhang	46	0.9	0.0	1.0
9	Prof. Li	44	0.8	0.6	0.4
10	Prof. Zhu	45	0.7	0.8	0.8

such study on how to characterize a practical uncertain quantity mathematically based on expert knowledge, which makes it possible to apply uncertainty theory to real-life problems.

Acknowledgements
This work was supported by the Specialized Research Fund for the Doctoral Program of Higher Education of China (No. 20110009120036).

References
1. Liu, B: Uncertainty theory. 2nd ed. Springer-Verlag, Berlin, Germany (2007)
2. Liu, B: Theory and practice of uncertain programming. 2nd ed. Springer-Verlag, Berlin, Germany (2009)
3. Wang, XS, Gao, ZC, Guo, HY: Uncertain hypothesis testing for two experts' empirical data. Math. Comput. Model. **55**, 1478–1482 (2012)
4. Peng, J: Risk metrics of loss function for uncertain system. Fuzzy Optim. Decis. Making. **12**(1), 53–64 (2013)
5. Liu, B: Membership functions and operational law of uncertain sets. Fuzzy Optim. Decis. Making. **11**(4), 387–410 (2012)
6. Li, X, Liu, B: Hybrid logic and uncertain logic. J. Uncertain Syst. **3**(2), 83–94 (2009)
7. Gao, X, Gao, Y, Ralescu, DA: On Liu's inference rule for uncertain systems. Int. J. Uncertain. Fuzz. **18**(1), 1–11 (2010)
8. Liu, B: Uncertain set theory and uncertain inference rule with application to uncertain control. J. Uncertain Syst. **4**(2), 83–98 (2010)
9. Liu, B: Fuzzy process, hybrid process and uncertain process. J. Uncertain Syst. **2**(1), 3–16 (2008)
10. Yao, K, Li, X: Uncertain alternating renewal process and its application. IEEE Trans. Fuzzy Syst. **20**(6), 1154–1160 (2012)
11. Yao, K: Uncertain calculus with renewal process. Fuzzy Optim. Decis. Making. **11**(3), 285–297 (2012)
12. Barbacioru, IC: Uncertainty functional differential equations for finance. Surveys Math. Appl. **5**, 275–284 (2010)
13. Liu, B: Toward uncertain finance theory. J. Uncertain. Anal. Appl. **1**(1) (2013)
14. Liu, B: Uncertainty theory. 4th ed. [Online] Available: http://orsc.edu.cn/liu/ut.pdf
15. Liu, B: Uncertainty theory: a branch of mathematics for modeling human uncertainty. Springer-Verlag, Berlin, Germany (2010)
16. Gao, X: Some properties of continuous uncertain measure. Int. J. Uncertain. Fuzz. **17**(3), 419–426 (2009)
17. Liu, B: Some research problems in uncertainty theory. J. Uncertain Syst. **3**(1), 3–10 (2009)
18. Peng, ZX, Iwamura, K: Some properties of product uncertain measure. J. Uncertain Syst. **6**(4), 263–269 (2012)
19. Zadeh, LA: Fuzzy sets as a basis for a theory of possibility. Fuzzy Set. Syst. **1**(1), 3–28 (1978)
20. Zadeh, LA: Mathematical Frontiers of the Social and Policy Sciences. In: Hayes et al. (ed.), pp. 69–129. Westview Press, Boulder, Cororado, (1979)
21. Liu, B, Liu, Y: Expected value of fuzzy variable and fuzzy expected value models. IEEE Trans. Fuzzy Syst. **10**(4), 445–450 (2002)
22. Li, X, Liu, B: A sufficient and necessary condition for credibility measures. Int. J. Uncertain. Fuzz. **14**(5), 527–535 (2006)

Monty Hall game: a host with limited budget

Agustín Alvarez[1] and Juan P Pinasco[2*]

*Correspondence:
jpinasco@dm.uba.ar
[2]IMAS (UBA-CONICET) and Departamento de Matematica, FCEN, University of Buenos Aires, Intendente Guiraldes 2160, Ciudad Universitaria Pab I (1428), Buenos Aires, Argentina
Full list of author information is available at the end of the article

Abstract

In this paper we introduce a new version of the classical Monty Hall problem, where the host is trying to maximize the audience while restricted in its budget. This problem is related to the design of games with a predetermined outcome and decision-making process under uncertainty when the agent does not know if the received advice is favorable or not.

Keywords: Monty Hall; Probability; Game theory

Introduction

The Monty Hall problem appeared first in a letter to the *American Statistician* of S. Selvin [1], and it is a nice and controversial problem for introductory courses in probability, statistics, and game theory. The problem can be posed as follows:

> *You are playing a game on a TV show. There are three doors. One of them has a car in the backside, and the other ones have goats. You select one of the doors, say door No. 1. And before opening it, the host who knows where the car is, opens another door, say No. 3, which has a goat. And then he gives you the possibility to change, allowing you to pick door No. 2. What do you do?*

The problem posed in this way may lead to a lot of controversy, mainly because we do not know whether the behavior of the host had anything to do with your first choice or not. As Gill stated in [2], this is a problem of mathematical modeling, and the answer is not a probability but a decision, and the decision must be chosen in a setting of uncertainty.

Perhaps the host would open a door with a goat only when your first choice was right. In this case, it was not a good choice to change doors. But if the host always shows you a door with a goat after your first choice, then by changing the first choice you increase your probability of winning a car from 1/3 to 2/3. One of the easiest way of seeing this is the following: imagine that you play this game repeatedly, around 1/3 of the times your first choice is right, and then you do not win the car because you change doors; in the other 2/3 of the times your first choice is wrong, and changing to the door the host did not open will make you win the car. So, 2/3 of the times you will be winning the car.

There are many variants on this game, see the recent book of Rosenhouse [3]. Some of the analysis are based on game theory although there is only one player, since the host's behavior is completely determined. In all the variants we know, the game is analyzed from the point of view of the player, and the host acts in almost a deterministic way (of course, sometimes the host takes a decision at random). In several variants, the host's behavior correspond to one of two main cases: the *malicious host* who opens a door only when the player chooses the right door, and the *benevolent host* who opens a door only when the player chooses one of the wrong doors, see [4,5] and Chapter 5 in [3]. A combination of both cases can be found in [6], and it is assumed that the participant has no information on the proportion of times that the host behaves as a malicious or benevolent host, and he cannot change doors when the host does not open a door.

However, as Fernandez and Piron remark in [7], the host is another player in the game, with his own motivation, and he shows a door or not according to his own objectives. Hence, we wish to analyze here a different combination of both cases, based on the following assumptions on TV shows. First, we assume that the host tries to maximize the audience while minimizing the costs. Now, suppose the TV show wants to last long playing this game because of its success, and its budget is not compatible with giving a car in 2/3 of the programs. Let us say that in the long run (by using the law of large numbers) he can give a car with probability α each show, where

$$\frac{1}{3} \leq \alpha \leq \frac{2}{3}.$$

We will restrict our analysis to this case, since lower values of α can be easily allowed by adding more doors, and higher values can be obtained by adding more doors and then opening more than one with goats. This general case follows in almost the same way.

Moreover, we assume that the host believes that the success of the game is based on the tense moment in which the host has opened a door with a goat after the first choice of the player and gives the player the possibility of changing his choice. So the host plays with a strategy which fixes the probability α of the player of winning the car and maximizes the proportion of shows in which the host opens a door with a goat.

We can think of this problem as an inverse problem in game theory, since α is the minimax solution of the zero-sum game between the player and the host: it is the minimum expected prize that the player can win and the maximum expected prize that the host can pay.

Another difference with other versions of the Monty Hall problem is the following: the player will have the option of changing his first choice both when the host shows a door with a goat and also when he does not show anything. This is a variant that seems to be overlooked in previous works; although in many real game shows, the host offers the player the option to change his decision, asking repeatedly things like 'Are you sure?', 'Do you want to change your answer?', and 'Is that your final answer?' Usually, there exists a key question which finishes the option to change. We believe that it is more realistic to include both option - the player can change his choice even if no door is open and the host is constrained in the number of times he can offer to open a door - in this game.

Finally, let us mention that this model is related to a more complex multi-agent problem, where each agent is a driver choosing among few roads. A real-time device - like a radio station, GPS, and intelligent transport systems - can inform or not the state of the roads, reducing the travel time uncertainty. However, several little accidents in the same road (cars stopped due to flat tires or lack of gasoline) are not interesting enough to catch the attention of the media compared with a major accident or the effect on the traffic cannot be measured quickly, although the road starts to be congested. Here, the system acts as the host, and in the worst-case scenario, it gives the minimum information about traffic or the information reaches the agent when a change of roads is not possible due to communications delays. We refer the interested reader to [8-10] for models of route choice with real-time information, and a discussion about the behavioral mechanism of drivers in this kind of decision-making process.

The work is organized as follows: in Section 'The Monty Hall problem and our model', we introduce the parameters of the model and the rules of the game. In Section 'Optimal strategies', we formulate and solve an equivalent problem, and we find the optimal strategies of the host and the player. The equivalence of the problems is proved in Section 'The dual problem' where we solve the inverse problem giving the optimal strategies as functions of the probability parameter α. We compare the payoffs when the players is not allowed to change doors if the host does not open an extra door in Section 'A related model', and we conclude in the Section 'Conclusions'.

The Monty Hall problem and our model

A variant of the problem

Let us state the conditions of the game:

1. There are three doors, one of them has a car in the backside, and the other ones have goats.
2. The host keeps the winning probability fixed at some $\alpha \in [1/3, 2/3]$.
3. The host knows where the car is, and his strategy will be based in two numbers, say:

    - $m =$ Probability of the host showing a door with a goat given the first choice of the player was right.
    - $b =$ Probability of the host showing a door with a goat given the first choice of the player was wrong.

    We can associate as in [6] the letter m to *malevolence*, since the host is tempting the player to change his choice when the player has chosen the right door; and the letter b to *benevolence*, since the host is tempting the player to change his mind after he has chosen the wrong door.
    He will always use those values.
4. The host wants to maximize the number of times he opens a door.
5. The host understands that in the long run both numbers m and b will be well estimated by the show's followers since in each tv show, when the game is over, all three doors are shown to the viewers in order to demonstrate the transparency of the game. So, the host expects that the player will know m and b and that the player will act in order to maximize its probability of winning.
6. The player's strategy is based on two probabilities, say:

    - $c =$ Probability for the player changing given the host has open a door with a goat.
    - $n =$ Probability for the player changing given the host has not open a door with a goat.
    - The players will use the same values of m, b, n and c in each game.

The problem, now, is as follows: How the probabilities m, b, c and n must be chosen by the host and the player? Recall that the host is trying to maximize the expected number of times that he will open a door, keeping the probability α fixed, and the player is trying to win the car.

We have the following sequential game in each tv show between the host and the player:

- A car is hidden behind one of three doors and remains there until the game is finished.
- The player chooses a door.
- The host - knowing if this choice was right or wrong - decides to open a door or not by using the probabilities m or b, respectively. If he opens a door, he shows a door with a goat.
- The player can change his initial choice, knowing the host's decision, and the new information if the host shows a goat or not, according to the probabilities c or n.
- The game finishes and the host open all the doors.
- The player wins if and only if the car is behind the door he finally choose.

Observe that the host must pay the full value of the car when the player wins. The law of large numbers enables him to estimate the prizes as the number of shows times the winning probability, although the variance introduces a serious risk when expensive prizes are involved. A classical way to deal with this uncertainty is to obtain a specialized coverage from an insurance company and the cost will depend on the player's winning probability.

Optimal strategies

Let us call $\mathbb{P}(m, b, c, n)$ the player's probability to finally win the car when the host's strategy is (m, b) and the player's strategy is (c, n).

We define now

$$PFW(m, b) = \max_{(c,n)} \mathbb{P}(m, b, c, n),$$

which gives the best probability for the player to finally win the car if the host uses strategy (m, b).

Let us now calculate the probability for the host showing a door with a goat in terms of m and b. First of all we name some events as follows:

- O = 'The host opens a door with a goat'.
- F_R = 'Player's first choice was right'.
- F_W = 'Player's first choice was wrong'.
- PW = 'The player finally wins using his best possible strategy'.
- C = 'The player changes his first choice'.
- NC = 'The player does not change his first choice'.

We will call O^c the complementary event of O.

According to the law of total probability, we have

$$\mathbb{P}(O) = m \cdot \mathbb{P}(F_R) + b \cdot \mathbb{P}(F_W) = \frac{m}{3} + \frac{2b}{3}. \tag{1}$$

Also, if $(m, b) \neq (0, 0)$

$$\mathbb{P}(F_R|O) = \frac{\mathbb{P}(F_R \cap O)}{\mathbb{P}(O)} = \frac{\mathbb{P}(F_R) \cdot \mathbb{P}(O|F_R)}{\frac{m}{3} + \frac{2b}{3}} = \frac{\frac{m}{3}}{\frac{m}{3} + \frac{2b}{3}}. \tag{2}$$

and if $(m, b) \neq (1, 1)$

$$\mathbb{P}(F_{\mathrm{R}}|O^{\mathrm{c}}) = \frac{\mathbb{P}(F_{\mathrm{R}} \cap O^{\mathrm{c}})}{\mathbb{P}(O^{\mathrm{c}})} = \frac{\mathbb{P}(F_{\mathrm{R}}) \cdot \mathbb{P}(O^{\mathrm{c}}|F_{\mathrm{R}})}{\frac{1}{3}(1-m) + \frac{2}{3}(1-b)} = \frac{\frac{1}{3}(1-m)}{\frac{1}{3}(1-m) + \frac{2}{3}(1-b)}. \tag{3}$$

Now that we have made all these definitions, the host is willing to find

$$(m^{\star}(\alpha), b^{\star}(\alpha)) = \underset{PFW(m,b)\leq\alpha}{\arg\max}\ \frac{m}{3} + \frac{2b}{3}, \tag{4}$$

that is, the host is maximizing the number of times that he opens a door, keeping the player's winning probability bounded by α.

A dual problem

Let us consider the problem,

$$(\tilde{m}(\beta), \tilde{b}(\beta)) = \underset{\frac{m}{3}+\frac{2b}{3}\geq\beta}{\arg\min}\ PFW(m, b), \tag{5}$$

namely, the host chooses his probabilities in order to minimize the winning probability of the player, constrained to open the door in at least $100 \cdot \beta\%$ of the shows.

As we will see in the next section (see Lemma 4.1), it is equivalent to solve any of the problems (4) and (5). However, the minimization problem (5) is conditioned on a very simple restriction, and we can give explicitly one of the variables in terms of the other; problem (4) on the other hand is conditioned on PFW which is given by a piecewise function.

Now, if the host deviates from the optimal strategy, there are two possibilities: when he opens a door more times, he is playing with higher values of m and/or b, and the dual problem shows that the player will win more games than the ones allowed by the host's budget (recall that the player can detect the values of m and b). On the other hand, by reducing the number of time he opens a door, the player's winning probability decreases, and the host spent only a part of the budget.

So we will concentrate now in solving problem (5), and in Section 'The dual problem' we show that if this problem has a solution, this must be a solution of problem (4).

An expression for PFW

First of all we compute PFW. Notice that when the host opens a door with a goat, the player has only two options: either he keeps his first choice with a probability of winning $\mathbb{P}(F_{\mathrm{R}}|O)$ or he can change his first option to the other unknown door with a probability of winning of $1 - \mathbb{P}(F_{\mathrm{R}}|O)$. Whereas when the host does not open a door with a goat, he can stay with a probability of winning of $\mathbb{P}(F_{\mathrm{R}}|O^{\mathrm{c}})$ or he can choose any of the other two doors with probability of winning $\frac{1-\mathbb{P}(F_{\mathrm{R}}|O^{\mathrm{c}})}{2}$

Hence, we can calculate the probability for the player to finally win using the best possible strategy given the host's strategy is (m, b) as follows:

$$\begin{aligned}\mathbb{P}(PW) &= \mathbb{P}(PW \cap O) + \mathbb{P}(PW \cap O^{\mathrm{c}}) \\ &= \mathbb{P}(O)\mathbb{P}(PW|O) + \mathbb{P}(O^{\mathrm{c}})\mathbb{P}(PW|O^{\mathrm{c}}).\end{aligned}$$

Using the independence of C with F_R and F_W given O respectively, we have

$$\mathbb{P}(PW|O) = \max_{c,n}\{\mathbb{P}(F_R \cap NC|O) + \mathbb{P}(F_W \cap C|O)\}.$$

However, since n appears only in the case that the host does not open a door, we get

$$\begin{aligned}\mathbb{P}(PW|O) &= \max_{0\le c\le 1}\{(1-c)\mathbb{P}(F_R|O) + c\mathbb{P}(F_W|O)\} \\ &= \max\{\mathbb{P}(F_R|O), \mathbb{P}(F_W|O)\},\end{aligned} \tag{6}$$

since we will take $c = 0$ (respectively, $c = 1$) when $\mathbb{P}(F_R|O) > \mathbb{P}(F_W|O)$ (resp., when $\mathbb{P}(F_R|O) < \mathbb{P}(F_W|O)$). Clearly, the player is indifferent when $\mathbb{P}(F_R|O) = \mathbb{P}(F_W|O)$.

Analogously, we get

$$\mathbb{P}(PW|O^c) = \max\{\mathbb{P}(F_R|O^c), \frac{1}{2}\mathbb{P}(F_W|O^c)\}. \tag{7}$$

When $(m, b) = (0, 0)$,we have

$$PFW(0,0) = \mathbb{P}(PW) = \mathbb{P}(W|O^c) = \max\{\mathbb{P}(F_R|O^c), \frac{1}{2}\mathbb{P}(F_W|O^c)\} = \frac{1}{3},$$

and when $(m, b) = (1, 1)$

$$PFW(1,1) = \mathbb{P}(PW) = \mathbb{P}(PW|O) = \max\{\mathbb{P}(F_R|O), \mathbb{P}(F_W|O)\} = \frac{2}{3}.$$

For other values of (m, b), using Equations 1, 2, 3, 6, and 7 we get

$$\begin{aligned}\mathbb{P}(O) &= \frac{1}{3}m + \frac{2}{3}b, \\ \mathbb{P}(O^c) &= \frac{1}{3}(1-m) + \frac{2}{3}(1-b), \\ \mathbb{P}(PW|O) &= \max\left(\frac{\frac{1}{3}m}{\frac{1}{3}m + \frac{2}{3}b}, 1 - \frac{\frac{1}{3}m}{\frac{1}{3}m + \frac{2}{3}b}\right), \\ \mathbb{P}(PW|O^c) &= \max\left(\frac{\frac{1}{3}(1-m)}{\frac{1}{3}(1-m) + \frac{2}{3}(1-b)}, \frac{1 - \frac{\frac{1}{3}(1-m)}{\frac{1}{3}(1-m)+\frac{2}{3}(1-b)}}{2}\right).\end{aligned}$$

Then, $\mathbb{P}(PW) = PFW(m, b)$ is given by

$$\begin{aligned}PFW(m,b) = &\left(\tfrac{1}{3}m + \tfrac{2}{3}b\right)\max\left(\frac{\frac{1}{3}m}{\frac{1}{3}m+\frac{2}{3}b}, 1 - \frac{\frac{1}{3}m}{\frac{1}{3}m+\frac{2}{3}b}\right) \\ &+ \left(\tfrac{1}{3}(1-m) + \tfrac{2}{3}(1-b)\right)\max\left(\frac{\frac{1}{3}(1-m)}{\frac{1}{3}(1-m)+\frac{2}{3}(1-b)}, \frac{1-\frac{\frac{1}{3}(1-m)}{\frac{1}{3}(1-m)+\frac{2}{3}(1-b)}}{2}\right).\end{aligned}$$

The function *PFW*

We now analyze the function $PFW(m, b)$. In the previous formula for PFW there are two maxima, and we can think of it as a piecewise-defined function. The changes occur in the lines $b = m$ and $b = \frac{1}{2}m$.

Let us analyze the function in the segment $\frac{1}{3}m + \frac{2}{3}b = \beta$ inside the square $[0,1] \times [0,1]$. In this segment we have $b = (\beta - \frac{1}{3}m)\frac{3}{2}$. Then, for each β we can introduce the function $h_{1,\beta}$,

$$h_{1,\beta}(m) = PFW\left(m, (\beta - \frac{1}{3}m)\frac{3}{2}\right).$$

Replacing in the formula for PFW, a simple computation gives the following piecewise expression for $h_{1,\beta}(m)$:

$$h_{1,\beta}(m) = \begin{cases} \beta\left(1 - \frac{\frac{1}{3}m}{\beta}\right) + (1-\beta)\frac{\frac{1}{3}(1-m)}{1-\beta} & \text{if } m \le \beta \\ \beta\left(1 - \frac{\frac{1}{3}m}{\beta}\right) + (1-\beta)\frac{1-\frac{\frac{1}{3}(1-m)}{1-\beta}}{2} & \text{if } \beta < m \le \frac{3}{2}\beta \\ \beta\left(\frac{\frac{1}{3}m}{\beta}\right) \qquad + (1-\beta)\frac{1-\frac{\frac{1}{3}(1-m)}{1-\beta}}{2} & \text{if } \frac{3}{2}\beta \le m \le 1. \end{cases}$$

We will replace now the minimization problem (5) by a simpler one, which can be solved explicitly in terms of $h_{1,\beta}$.

The minimization problem

Let us consider first the auxiliary problem

$$(\hat{m}(\beta), \hat{b}(\beta)) = \underset{\frac{1}{3}m+\frac{2}{3}b=\beta}{\arg\min}\ PFW(m,b),$$

and let us define the function h_2 as

$$h_2(\beta) = PFW(\hat{m}(\beta), \hat{b}(\beta)). \tag{8}$$

Observe that we are minimizing now on the boundary of the restriction of problem (5). By inspecting $h_{1,\beta}$ we get

$$h_{1,\beta}(m) = \begin{cases} \beta - \frac{2m}{3} + \frac{1}{3} & \text{if } m \le \beta \\ \frac{\beta}{2} - \frac{m}{6} + \frac{1}{3} & \text{if } \beta < m \le \frac{3\beta}{2} \\ -\frac{\beta}{2} + \frac{m}{2} + \frac{1}{3} & \text{if } \frac{3\beta}{2} \le m \le 1. \end{cases}$$

Observe that $h_{1,\beta}(m)$ decreases when $0 \le m \le \frac{3\beta}{2}$, and increases when $\frac{3\beta}{2} < m \le 1$. We then find the following:

- If $\frac{3\beta}{2} \le 1$, then $h_{1,\beta}$ has a unique minimum at $m = \frac{3\beta}{2}$.
- If $\frac{3\beta}{2} \ge 1$, the function $h_{1,\beta}$ decreases in the whole interval $[0,1]$ and has its minimum at $m = 1$.

Hence, in order to compute the function h_2, let us note that

- If $\frac{3\beta}{2} \le 1$, then $\hat{m}(\beta) = \frac{3}{2}\beta$ which implies that $\hat{b}(\beta) = \frac{3\beta}{4}$.
- If $\frac{3\beta}{2} \ge 1$, $\hat{m}(\beta) = 1$ which implies that $\hat{b}(\beta) = (\beta - \frac{1}{3})\frac{3}{2}$.

Then h_2 is given by

$$h_2(\beta) = \begin{cases} \frac{\beta}{4} + \frac{1}{3} \text{ if } \beta \leq \frac{2}{3}, \\ \\ \frac{\beta}{2} + \frac{1}{6} \text{ if } \frac{2}{3} \leq \beta \leq 1, \end{cases} \tag{9}$$

so h_2 is increasing in β, which implies that $(\tilde{m}(\beta), \tilde{b}(\beta)) = (\hat{m}(\beta), \hat{b}(\beta))$ is the unique solution of (5).

Hence, we have

$$\begin{cases} \tilde{m}(\beta) = \frac{3\beta}{2},\ \tilde{b}(\beta) = \frac{3\beta}{4} & \text{if } \beta \leq \frac{2}{3}, \\ \\ \tilde{m}(\beta) = 1,\quad \tilde{b}(\beta) = (\beta - \frac{1}{3}) \cdot \frac{3}{2} & \text{if } \frac{2}{3} \leq \beta \leq 1. \end{cases} \tag{10}$$

Host's strategies

It's worth noticing that for $0 \leq \beta \leq \frac{2}{3}$ (or equivalently $\frac{1}{3} \leq \alpha \leq \frac{1}{2}$ as we show in Section 'The dual problem'), the optimum strategy of the host consist in having double malevolence, that is $\tilde{m}(\beta) = 2\tilde{b}(\beta)$. Whereas if $\frac{2}{3} < \beta \leq 1$ or equivalently $\frac{1}{2} < \alpha \leq \frac{2}{3}$, this is not possible since $\tilde{m}(\beta) = 1$ and $\tilde{b}(\beta) > \frac{1}{2}$.

The function h_2 defined in (8) represents the winning probability of a player following its best strategy given that the host is using its best strategy. It can be seen in (9) that this probability increases with a higher speed ($\frac{1}{2}$ instead of $\frac{1}{4}$) when the host is not able of having malevolence anymore while doubling the benevolence.

Player's strategies

We are assuming that in the long term, the player will estimate well the parameters m and b of the host. If the host opens an extra door, the player must compare $\mathbb{P}(F_R|O)$ with $\frac{1}{2}$ and

- If $\mathbb{P}(F_R|O) < \frac{1}{2}$, then the player must change door.
- If $\mathbb{P}(F_R|O) > \frac{1}{2}$, then the player must keep his first choice.
- If $\mathbb{P}(F_R|O) = \frac{1}{2}$, then is indistinct.

If the host does not open an extra door, the player must compare $\mathbb{P}(F_R|O^c)$ with $\frac{1}{3}$ and now:

- If $\mathbb{P}(F_R|O^c) < \frac{1}{3}$, then the player must change door.
- If $\mathbb{P}(F_R|O^c) > \frac{1}{3}$, then the player must keep his first choice.
- If $\mathbb{P}(F_R|O^c) = \frac{1}{3}$, then is indistinct.

If the host is using an optimal strategy, then there are two possibilities for these parameters:

- If $m = 2b$ then if the host shows an extra door, $\mathbb{P}(F_R|O) = \frac{1}{2}$, and changing is indistinct. In addition, if the host does not show an extra door, $\mathbb{P}(F_R|O^c) < \frac{1}{3}$, and the player must change door.
- If $m = 1$ and $b > \frac{1}{2}$ then if the host opens an extra door, $\mathbb{P}(F_R|O) < \frac{1}{2}$, and the player must change. In addition, if the host does not open an extra door, then $\mathbb{P}(F_R|O^c) = 0 < \frac{1}{3}$, and so the player must change door.

In summary, the player's strategy could be always changing and would be doing as best as possible if the host is using an optimal strategy.

The dual problem

Our goal is to show now the equivalence of solving (4) or (5). This gives the values of the desired optimal probabilities of the host $(m^\star, b^\star)$ as functions of α.

Lemma 4.1. *For any* $\alpha \in [\frac{1}{3}, \frac{2}{3}]$, *let*

$$(m^\star(\alpha), b^\star(\alpha)) = \underset{f(m,b)\leq\alpha}{\arg\max}\, g(m,b), \tag{11}$$

$$(\tilde{m}(\beta), \tilde{b}(\beta)) = \underset{g(m,b)\geq\beta}{\arg\min}\, f(m,b), \tag{12}$$

where $\beta = g(m^\star(\alpha), b^\star(\alpha))$, $g(m,b) = \frac{m}{3} + \frac{2b}{3}$ *and* $f(m,b) = PFW(m,b)$. *Then*

$$(m^\star(\alpha), b^\star(\alpha)) = (\tilde{m}(\beta), \tilde{b}(\beta)).$$

Proof. Let $\beta = g(m^\star(\alpha), b^\star(\alpha))$, this β will appear in the constraint in problem (12), for which we will find $(\tilde{m}(\beta), \tilde{b}(\beta))$. Observe that by the constraint of problem (11), we have

$$f(m^\star(\alpha), b^\star(\alpha)) \leq \alpha. \tag{13}$$

If $\beta = 1$, then the only possibility is that $(m^\star(\alpha), b^\star(\alpha)) = (1,1)$. Moreover, $(m,b) = (1,1)$ is the only point which satisfies the constraint in problem (12). Hence,

$$(m^\star(\alpha), b^\star(\alpha)) = (\tilde{m}(\beta), \tilde{b}(\beta)) = (1,1).$$

Indeed, we get the classical Monty Hall problem and in this case $f(m,b) = 2/3$.

Let us consider now the case $\beta < 1$.

Take (m,b) such that $g(m,b) > \beta$, then $f(m,b) > \alpha$. Otherwise, $(m^\star(\alpha), b^\star(\alpha))$ did not maximize (11).

If $g(m,b) = \beta < 1$, then $(m,b) \neq (1,1)$. Since g is continuous and (m,b) is not a local maxima, there exists a sequence $\{(m_n, b_n)\}_{n\geq 1}$ which converges to (m,b) with $g(m_n,b_n) > \beta$. In particular, $f(m_n,b_n) > \alpha$.

So, by continuity of f we have $f(m,b) \geq \alpha$. So we have seen that in any case that $g(m,b) \geq \beta$ then $f(m,b) \geq \alpha$.

Moreover, we have that $(m^\star(\alpha), b^\star(\alpha))$ satisfies

$$g(m^\star(\alpha), b^\star(\alpha)) = \beta,$$

and therefore

$$f(m^\star(\alpha), b^\star(\alpha)) \geq \alpha.$$

Then, by inequality (13), we get $f(m^\star(\alpha), b^\star(\alpha)) = \alpha$ so is a minimizer for (12) which is unique and so $(m^\star(\alpha), b^\star(\alpha)) = (\tilde{m}(\beta), \tilde{b}(\beta))$. □

To finish solving the original problem (4), we only need to find an expression for

$$\beta(\alpha) := g(m^\star(\alpha), b^\star(\alpha)) = \max_{f(m,b)\leq\alpha} g(m,b)$$

since we know from Lemma 4.1 that $(m^\star(\alpha), b^\star(\alpha)) = (\tilde{m}(\beta(\alpha)), \tilde{b}(\beta(\alpha)))$. The next lemma will show that $\beta(\alpha)$ is the inverse function of h_2 defined in (8).

Lemma 4.2. *Let K be any set, and $f : K \to \mathbb{R}$ and $g : K \to \mathbb{R}$ are such that*

$$\begin{aligned}\Phi(\alpha) &= \max_{f(k)\leq\alpha} g(k)\\ \Psi(\beta) &= \min_{g(k)\geq\beta} f(k)\end{aligned} \tag{14}$$

are well defined and that one of them (say Ψ) is strictly increasing in the interval I, then $\Psi : I \to Im(\Psi)$ is invertible and its inverse is $\Phi : Im(\Psi) \to I$.

Proof. Let $\alpha = \Psi(\beta)$, then there exists an element $k_0 \in K$ such that $g(k_0) \geq \beta$ and $f(k_0) = \alpha$. Actually $g(k_0) = \beta$, because if $g(k_0) = \beta' > \beta$ then $\Psi(x) \equiv \alpha$ in the interval $[\beta, \beta']$ which contradicts the fact of Ψ being strictly increasing.

Now, to prove the lemma is enough to see that $\Phi(\alpha) = \beta$. Since $f(k_0) \leq \alpha$, then $\Phi(\alpha) \geq g(k_0) = \beta$. Suppose $\Phi(\alpha) = \beta' > \beta$. Then, there exists k_1 such that $f(k_1) \leq \alpha$ and $g(k_1) = \beta' > \beta$. Therefore $\Psi(\beta') \leq \alpha = \Psi(\beta)$, being $\beta < \beta'$ which is absurd because Ψ is strictly increasing. □

Taking f and g as in Lemma 4.1 and $K = [0, 1] \times [0, 1] \subseteq \mathbb{R}^2$. Φ and Ψ of Lemma 4.2 are well defined because of the continuity of f and g and the compactness of K. Then h_2 defined in (8) can be seen as

$$h_2(\beta) = \min_{g(m,b)=\beta} f(m, b),$$

but since h_2 is strictly increasing, then

$$h_2(\beta) = \min_{g(m,b)\geq\beta} f(m, b).$$

Then $h_2 = \Psi$ and seeing that the formula (9) $\Psi : [0, 1] \to [\frac{1}{3}, \frac{2}{3}]$ is strictly increasing. Due to Lemma 4.2, $\beta(\alpha) : [\frac{1}{3}, \frac{2}{3}] \to [0, 1]$ is its inverse.

By inverting h_2, using $(m^\star(\alpha), b^\star(\alpha)) = (\tilde{m}(\beta(\alpha)), \tilde{b}(\beta(\alpha)))$ and (10) we get

$$(m^\star(\alpha), b^\star(\alpha)) = \begin{cases} (6\alpha - 2, 3\alpha - 1) & \text{if } \frac{1}{3} \leq \alpha \leq \frac{1}{2}, \\ (1, 3\alpha - 1) & \text{if } \frac{1}{2} \leq \alpha \leq \frac{2}{3}. \end{cases} \tag{15}$$

A related model

The same approach we have done can be applied for the Monty hall game described in [6]. In this version, the player is not allowed to change doors if the host does not open an extra door. If we call $PFW_c(m, b)$ the best probability for the player to finally win if the host uses strategy (m, b), then we find that

$$\begin{aligned} PFW_c(m, b) &= \mathbb{P}(O) \max\left(\mathbb{P}(F_R|O), 1 - \mathbb{P}(F_R|O)\right) + \mathbb{P}(O^c)\mathbb{P}(F_R|O^c) \\ &= \left(\frac{m + 2b}{3}\right) \max\left(\frac{m}{m + 2b}, 1 - \frac{m}{m + 2b}\right) \\ &\quad + \left(1 - \frac{m + 2b}{3}\right) \frac{1 - m}{3 - (m + 2b)}. \end{aligned}$$

As we did before, we analyzed the function in the segment $\frac{m+2b}{3} = \beta$ and defined

$$h^c_{1,b}(m) = PFW_c\left(m, (\frac{3\beta - m}{2})\right).$$

After some calculations, we get

$$h^c_{1,\beta}(m) = \begin{cases} \beta - \frac{2m}{3} + \frac{1}{3} & \text{if } m \leq \frac{3\beta}{2}, \\ \frac{1}{3} & \text{if } m \geq \frac{3\beta}{2}. \end{cases}$$

So we have the following:

- If $\frac{3\beta}{2} \leq 1$, then any $m \geq \frac{2\beta}{3}$ is a global minimum of $h^c_{1,\beta}(m)$ and its minimum value is $\frac{1}{3}$.
- If $\frac{3\beta}{2} \geq 1$, then $h^c_{1,\beta}(m)$ has its minima in $m = 1$ and its minimum value is $\beta - \frac{1}{3}$.

Repeating the previous argument, we define

$$(\hat{m}_c(\beta), \hat{b}_c(\beta)) = \underset{\frac{m+2b}{3}=\beta}{\arg\min}\, PFW_c(m, b),$$

and let us introduce the function h^c_2 defined as

$$h^c_2(\beta) = PFW_c(\hat{m}_c(\beta), \hat{b}_c(\beta)). \tag{16}$$

We get

$$h^c_2(\beta) = \begin{cases} \frac{1}{3} & \text{if } \beta \leq \frac{2}{3}, \\ \beta - \frac{1}{3} & \text{if } \beta \geq \frac{2}{3}. \end{cases} \tag{17}$$

Now, $h^c_2(\beta)$ is the percentage of times the player will win the car if both the player and the host use their optimum strategies (let us recall that β is the proportion of times an extra door is showed to the player). The comparison with $h_2(\beta)$ is shown in Figure 1.

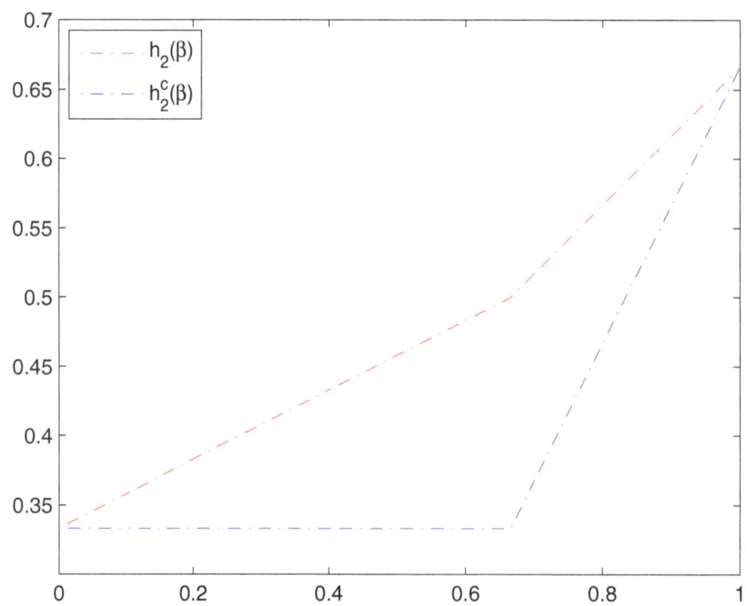

Figure 1 Comparison of $h_2(\beta)$. Percentage of times the player will win the car if both the player and the host use their optimum strategies as a function of β.

The argument at the end of Section 'The dual problem' can be repeated since $h_2^c(\beta)$ is strictly increasing in $[\frac{2}{3}, 1]$ to find $\beta_c(\alpha_c) : [\frac{1}{3}, \frac{2}{3}] \to [\frac{2}{3}, 1]$, the inverse function of $h_2^c(\beta)$. Now, by using

$$(m_c^\star(\alpha), b_c^\star(\alpha)) = (\tilde{m}_c(\beta_c(\alpha)), \tilde{b}_c(\beta_c(\alpha))),$$

and $\tilde{m}_c(\beta) = 1$, with $\tilde{b}_c(\beta) = \frac{3\beta-1}{2}$ if $\beta \geq \frac{2}{3}$, we get

$$(m_c^\star(\alpha), b_c^\star(\alpha)) = \left(1, \frac{3\alpha}{2}\right).$$

This formula can be compared with (15).

We see that the player's chances are worse than in the previous version of the game. From the host's perspective, in this version he can pay the same prizes as before, by opening an extra door more times. However, this is achieved by reducing the number of times the host opens a door when the player misses with his first choice and does not have the option to change doors, which is a very unfriendly policy.

Conclusions

We have analyzed a different version of the Monty Hall problem, where the player faces a host which has its own objectives (maximize the audience) and limited budget. From the host's perspective, he must solve an inverse problem in zero-sum game theory: to determine the payoffs of the game with a given minimax equilibria. From the player's point of view, this is a toy model of a decision process where the agent does not know if the received information is beneficial or not (i.e., the host is benevolent or malevolent), although the player knows the probability of each behavior.

We show that we can formulate the problem both in terms of the expected prize that the host will pay (α) and in terms of the proportion of times he opens a door (β). The later one is better for computations.

Competing interests

The authors declare that they have no competing interests.

Acknowledgements

AA is a fellow of University of Buenos Aires, and JPP is a member of CONICET. This research was partially supported by grants W276 and 20020100100400 from the University of Buenos Aires and by CONICET (Argentina) PIP 5478/1438.

Author details

[1]Instituto de Calculo, FCEN, University of Buenos Aires, Intendente Guiraldes 2160, Ciudad Universitaria Pab II (1428), Buenos Aires, Argentina. [2]IMAS (UBA-CONICET) and Departamento de Matematica, FCEN, University of Buenos Aires, Intendente Guiraldes 2160, Ciudad Universitaria Pab I (1428), Buenos Aires, Argentina.

References

1. Selvin, S: A problem in probability. Am. Statistician. **29**(1), 67 (1975)
2. Gill, R: The Monty Hall problem is not a probability puzzle (it's a challenge in mathematical modelling). Statistica Neerlandica **65**, 58–71 (2011)
3. Rosenhouse, J: The Monty Hall Problem. Oxford University Press, New York (2009)
4. Granberg, D: To switch or not to switch. In: vos Savant, M (ed.) The Power of Logical Thinking. St. Martin's Press, New York (1996)
5. Tierney, J: Behind Monty Hall's doors: puzzle, debate and answer? The New York Times (July 21 1991)
6. Schuller, JC: The malicious host: a minimax solution of the Monty Hall problem. J. Appl. Stat. **39**, 215–221 (2012)

7. Fernandez, L, Piron, R: Should she switch? A game-theoretic analysis of the Monty Hall problem. Math. Mag. **72**, 214–217 (1999)
8. Abdel-Aty, MA, Abdalla, MF: Examination of multiple mode/routechoice paradigms under ATIS. IEEE Trans. Intell. Transportation Syst. **7**, 332–348 (2006)
9. Ben-Elia, E, Shiftan, Y: Which road do I take? A learning-based model of route choice with real-time information. Transportation Res. Part A Policy Pract. **44**, 249–264 (2010)
10. Gao, S, Frejinger, E, Ben-Akiva, E: Adaptive route choices in risky traffic networks: a prospect theory approach. Transportation Res. Part C: Emerg. Technol. **18**, 727–740 (2010)

17

A bi-objective model for uncertain multi-modal shortest path problems

Yan Zhang, Pei Liu, Lixing Yang* and Yuan Gao

*Correspondence:
lxyang@bjtu.edu.cn
State Key Laboratory of Rail Traffic Control and Safety, Beijing Jiaotong University, Beijing 100044, China

Abstract

This paper employs uncertain programming to investigate the uncertain multi-modal shortest path problem, in which the arc weights (arc travel time, arc travel costs) associated with different transport modes are characterized by uncertain variables. By using the chance-constrained programming approach, we firstly formulate a bi-objective optimization model to minimize the total travel time and travel costs simultaneously with the given confidence levels. Moreover, using the basic concepts and properties in the uncertainty theory, we transform the proposed model into its deterministic crisp equivalent with an explicit proof. Finally, some numerical experiments are implemented to show the performance of the proposed approaches on a multi-modal transportation network with three specific modes.

Keywords: Multi-modal shortest path problem; Uncertainty theory; Chance-constrained programming; Bi-objective model

Introduction

The shortest path problem is widely applied to network optimization and has been studied by a lot of researchers. Classical shortest path problems focus on networks with deterministic arc weights (lengths), and Dijkstra [1], Bellman [2], and Dreyfus [3] have proposed some efficient algorithms which are still referenced widely now. However, due to the failure, maintenance, and other uncertain factors, arc weights are usually non-deterministic in a busy transportation network. In view of this fact, some researchers introduced probability theory into the shortest path problem and used probability distributions to describe the existing indeterminacy, such as Frank [4], Loui [5], Mirchandani [6], Yang et al. [7], and Yang and Zhou [8]. However, this method is typically imprecise when we are lack of *a priori* data information with respect to networks. With this concern, Dubois and Prade [9] first introduced a fuzzy shortest path problem. In the fuzzy set theory, the decision can be estimated by experts based on their experiences and professional judgments. Some other routing optimization with fuzzy information can be referred to Ji et al. [10], Hernandes et al. [11], and Yang et al. [12].

However, both probability theory and fuzzy set theory may lead to counterintuitive results [13]. In this case, uncertainty theory was founded to rationally deal with belief degrees, which enlightened a new approach to describe non-deterministic phenomena. Up to now, the uncertainty theory has been applied to many classical optimization problems, such as solid transportation problem [14], project scheduling problem [15], maximum flow problem [16], optimal assignment problem [17], uncertain graph [18], etc. As for the shortest path problem, Liu [19] introduced three concepts of uncertain path according to different decision criteria, including the expected shortest path, α-shortest path, and the most shortest path. He formulated three types of uncertain programming models and converted them into deterministic optimization ones. Gao [20] studied the uncertainty distribution of the shortest path length and proposed an effective method to find the α-shortest path and the most shortest path in an uncertain network. He pointed out that there existed an equivalence relation between the α-shortest path in an uncertain network and the shortest path in a corresponding deterministic network. Moreover, an effective algorithm was proposed to find the α-shortest path and the most shortest path. Zhou et al. [21] discussed the inverse shortest path problem on the graph with uncertain edge weights. This problem was formulated as an uncertain programming and was reformulated into a deterministic programming model.

Multi-modal transportation refers to a trip consisting of two or more means of transport to guide the passengers reach their destinations. The multi-modal shortest path problem is a significant generalization of the traditional shortest path problem, and it has been extensively investigated by a lot of researchers. Lozano and Storchi [22] found the shortest viable path in a multi-modal network using label correcting techniques. They defined the viable path as a path whose sequence of modes is feasible with respect to a set of constraints. Ma [23] presented an A^* label setting algorithm to solve a constrained shortest path problem in a multi-modal network. In this work, each link is characterized with a vector of resource consumption besides the travel time. Ambrosino and Sciomachen [24] proposed an approach for computing shortest routes in multi-modal networks with objectives of minimizing the overall time, cost, and users' discomfort. Liu et al. [25] designed an improved exact algorithm for a multi-criteria multi-modal shortest path problem with both arriving time window and transfer delaying. Galvez-Fernandez et al. [26] introduced a transfer graph approach, which was believed to better abstract the distributed nature of real transport information sources, to calculate the best paths in multi-modal networks. Yamani et al. [27] presented a fuzzy shortest path algorithm in multi-modal transportation networks, which concerned about not only the path cost but also the path time which consisted of travel time and delays.

Indeed, due to the complexity of the real-world situations, it is in general difficult to explicitly determine travel time or travel costs in a large-scale multi-modal network. What is more, it is also complicated when we consider the transfer time and costs from one mode to another. Thus, investigating the multi-modal shortest path problem in the indeterministic environment becomes a significant and challenging issue for the practical applications. We here note that finding the shortest path in a multi-modal network has been studied in various forms such as static [22], dynamic [26,28], stochastic [29,30], fuzzy [27], constrained [23], and multi-criteria [24,25,31,32]. However, to the best of our knowledge, few studies have been considered in the uncertain environment within the framework of uncertain programming.

This paper aims to investigate the multi-modal shortest path problem, in which both travel time and travel costs are regarded as uncertain variables. Specifically, the travel time includes travel time on the travel arcs and transfer time spent on the mode change. The travel costs consist of cost on each arc and the fixed cost. We intend to formulate a chance-constrained programming model with two objectives minimizing the travel time and travel cost simultaneously. The results of this research will provide a fundamental framework for investigating the shortest path problems with both multi-modes and uncertainty characteristics.

The remainder of this paper is organized as follows. Section 'Preliminaries' introduces some basic concepts and properties of uncertainty theory used throughout this paper. Section 'Model formulations' makes a description of the problem and formulates a chance-constrained programming model with two objectives. In Section 'Crisp equivalent of the model', we demonstrate how to convert the model into its crisp equivalent. In Section 'Numerical experiments', some experiments are given to illustrate the performance of the proposed model. Finally, some conclusions are made in Section 'Conclusions'.

Preliminaries

As this research will investigate the problem of interest within the framework of uncertainty theory, we next first introduce some basic knowledge in this field for the completeness of this paper. Uncertainty theory was proposed by Liu [13], and it provided an axiomatic system to handle the imprecise information. Some foundational definitions and results in uncertainty theory are introduced below.

Definition 2.1. [13] Let Γ be a nonempty set, and L be a σ-algebra over Γ. Each element Λ in L is called a measurable set. Then, the triplet (Γ, L, M) is called an uncertainty space if the uncertain measure M satisfies the following axioms:

Axiom 1. (*Normality Axiom*) $M\{\Gamma\} = 1$ for the universal set Γ.
Axiom 2. (*Duality Axiom*) $M\{\Lambda\} + M\{\Lambda^c\} = 1$ for any event Λ.
Axiom 3. (*Subadditivity Axiom*) For every countable sequence of events $\Lambda_1, \Lambda_2, \cdots$, we have

$$M\left\{\bigcup_{i=1}^{\infty} \Lambda_i\right\} \leq \sum_{i=1}^{\infty} M\{\Lambda_i\}.$$

Based on the uncertainty space, a formal definition of the uncertain variable will be introduced for better describing the uncertainties mathematically.

Definition 2.2. [13] An uncertain variable is a measurable function ξ from an uncertainty space (Γ, L, M) to the set of real numbers such that $\{\xi \in B\}$ is an event for any Borel set B.

Definition 2.3. [13] The uncertainty distribution Φ of an uncertain variable ξ is defined by:

$$\Phi(x) = M\{\xi \leq x\}$$

for any real number x.

Theorem 2.1. *([33], Measure Inversion Theorem) Let ξ be an uncertain variable with continuous uncertainty distribution Φ. Then for any real number x, we have:*

$$M\{\xi \leq x\} = \Phi(x), M\{\xi > x\} = 1 - \Phi(x). \tag{1}$$

Definition 2.4. [33] Let ξ be an uncertain variable with regular uncertainty distribution $\Phi(x)$. Then, the inverse function $\Phi^{-1}(\alpha)$ is called the inverse uncertainty distribution of ξ.

Definition 2.5. [13] The uncertain variables $\xi_1, \xi_2, \cdots, \xi_n$ are said to be independent if:

$$M\left\{\bigcap_{i=1}^{n}\{\xi_i \in B_i\}\right\} = \bigwedge_{i=1}^{n} M\{\xi_i \in B_i\}$$

for any Borel set $B_1, B_2, \cdots, B_n$ of real numbers.

There are some kinds of special uncertain variables namely the linear uncertain variable, the zigzag uncertain variable, the normal uncertain variable and the lognormal uncertain variable. As an example, the following will describe corresponding properties of the zigzag uncertain variable.

Definition 2.6. [13] An uncertain variable ξ is called zigzag if it has a zigzag uncertainty distribution

$$\Phi(x) = \begin{cases} 0, & \text{if } x \le a, \\ (x-a)/2(b-a), & \text{if } a \le x \le b, \\ (x+c-2b)/2(c-b), & \text{if } b \le x \le c, \\ 1, & \text{if } x \ge c. \end{cases} \tag{2}$$

denoted by $Z(a, b, c)$, where a, b, and c are real numbers with $a<b<c$.

Theorem 2.2. *[13] Assume that ξ_1 and ξ_2 are independent zigzag uncertain variables $Z(a_1, b_1, c_1)$ and $Z(a_2, b_2, c_2)$, respectively. Then, the sum $\xi_1 + \xi_2$ is also a zigzag uncertain variable $Z(a_1 + a_2, b_1 + b_2, c_1 + c_2)$, i.e.,*

$$Z(a_1, b_1, c_1) + Z(a_2, b_2, c_2) = Z(a_1 + a_2, b_1 + b_2, c_1 + c_2). \tag{3}$$

The product of a zigzag uncertain variable $Z(a, b, c)$ and a scalar number $k > 0$ is also a zigzag uncertain variable $Z(ka, kb, kc)$, i.e.,

$$k \cdot Z(a, b, c) = Z(ka, kb, kc). \tag{4}$$

To characterize the structural features of an uncertain variable, the concepts of critical values, including optimistic value and pessimistic value, are formally defined as follows.

Definition 2.7. [13] Let ξ be an uncertain variable.

(1) The optimistic value of ξ is defined by:

$$\xi_{\sup}(\alpha) = \sup\{r | M\{\xi \ge r\} \ge \alpha\}, \alpha \in (0, 1], \tag{5}$$

(2) The pessimistic value of ξ is defined by:

$$\xi_{\inf}(\alpha) = \inf\{r | M\{\xi \le r\} \ge \alpha\}, \alpha \in (0, 1]. \tag{6}$$

Model formulations

Consider a multi-modal network $G = (V, E, M)$, in which V is the set of nodes and E is the set of arcs. M represents the transport modes set that consists of three different modes, namely, private, metro, and bus. Each arc is denoted by an ordered node pair (i, j, k) to indicate connections between the adjacent nodes of mode k. The whole network thus can be divided into three subnetworks as the private network, metro network, and bus network correspondingly. It should be noted that multi-modal network is characterized by weights on both nodes and arcs. Weights on arcs are generally used to represent the travel time or travel cost spent on some kind of vehicle. Besides, weights should be added on the transfer nodes if a mode change happens. The weights on the transfer nodes may indicate the transfer time or transfer cost when a passenger transfers from one mode to another. For the simplicity of our model, we shall divide a transfer node into different nodes according to the number of modes at this node. Additionally, dummy arcs will be added between these newly created nodes to indicate a modal transfer. Take a small network shown in Figure 1 as an example. Obviously, there exist three modes at node a. Therefore, we divide a into three different nodes as a, b, and c with modes private, metro, and bus, respectively, and three dummy arcs should be added to the network in order to represent the mode change between different subnetworks. In this case, the three subnetworks can be connected by the given transfer arcs. Then, we can give weights to the transfer arcs to indicate the transfer time or costs. For convenience of modeling formulation, the transfer will be treated as a special kind of mode, namely, the transfer mode. Thus, four modes are included in the mode set M, which is denoted as $M = \{p, b, m, t\}$.

There are various measures to evaluate a multi-modal path, while in this paper, we are concerned with the total travel time and travel costs. In a multi-modal network, the total travel time mainly includes in-vehicle time within a certain subnetwork and transfer time when there is a transfer. Generally, a transfer between different modes needs the traveler to walk to the stop (or station) of the following mode, then he/she may have to wait for the bus or train coming. After getting on the vehicle, more time may be spent for waiting for the vehicle to depart the stop. In summary, the transfer time consists of the passenger's walking time, waiting time, and the vehicle's stopping time. To make the problem easy to be dealt with, we consider the transfer time as a whole. Let ξ_{ijk} represent time spent on arc (i, j, k) with a specific mode k. Therefore, it is the in-vehicle travel time when k denotes the private mode, the bus mode, or the metro mode. While when k means a transfer mode, ξ_{ijk} is actually the transfer time. Define a decision variable x_{ijk}: $x_{ijk} = 1$ denotes arc (i, j, k)

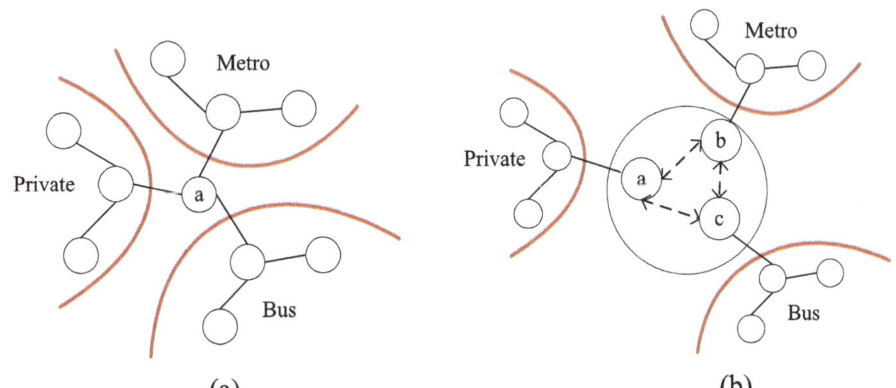

Figure 1 Schematic diagram of multi-modal network. (a) Physical network. **(b)** Transfer network.

is selected by the path, and $x_{ijk} = 0$ otherwise. Then, the total travel time of a multi-modal path can be expressed as follows:

$$f_1 = \sum_{(i,j,k)\in E} x_{ijk}\xi_{ijk}. \tag{7}$$

In addition, we need a flow balance constraint to generate the feasible path, which is given as below. (In this constraint, O and D represent the origin node and destination node of the network, respectively).

$$\sum_{(i,j,k)\in E} x_{ijk} - \sum_{(j,i,k)\in E} x_{jik} = \begin{cases} 1, & i{=}O, \\ -1, & i{=}D, \\ 0, & \text{otherwise.} \end{cases} \tag{8}$$

It is no doubt that minimizing the total travel time cannot meet all the demands of the travelers. Each mode has its own monetary cost, and sometimes, it may be contrary to the travel time. For instance, the private mode is faster than the bus mode, but it has much higher monetary cost as well. With this concern, the travel costs also need to be considered in the model. In this paper, we take two types of costs into consideration: costs on travel arcs and the fixed cost. Denote c_{ijk} to represent the travel cost on arc (i, j, k), where $k \in M\backslash\{t\}$. The fixed cost represents the ticket price of the public modes. It always happens along with the occurrence of the transfer. For example, when a passenger transfers from the bus subnetwork to the metro subnetwork, he/she then needs to pay for the fixed cost to the metro subnetwork. Therefore, if any arc (i, j, k) of a certain mode k is selected by a path, the fixed cost will occur. To depict this relationship, we introduce another decision variable y_k as:

$$y_k = \begin{cases} 1, \text{ if } x_{ijk} = 1, \text{ for a certain } (i,j,k) \in E, \text{ and } k \in M\backslash\{t\}, \\ 0, \text{ otherwise.} \end{cases} \tag{9}$$

Obviously, $y_k = 1$ indicates the fixed cost happens, and 0 otherwise. Let d_k denote the fixed cost when a certain arc of mode $k(k \in M\backslash\{t\})$ is taken by a path. Then, we can easily get the total travel costs as:

$$f_2 = \sum_{(i,j,k)\in E} x_{ijk}c_{ijk} + \sum_{k\in M\backslash\{t\}} y_k d_k. \tag{10}$$

Indeed, travelers would pay more attention to the times of transfer, and most travelers would not like to transfer too many times. So, it is rational to limit the number of the modes adopted by a path. Then, we have the following constraint to ensure that the number of modal transfers is not greater than N.

$$\sum_{k\in M\backslash\{t\}} y_k \le N. \tag{11}$$

For the convenience of formulating the model, we would like to summarize the above notations in Table 1.

In this paper, we intend to formulate a bi-objective integer programming model to minimize both the travel time and travel costs. Different decision criteria have been used to evaluate the shortest path problem under uncertain environment. One of the widely used criteria is the critical value. Uncertain programming using the critical value

Table 1 Notations used in formulations

Notations	Definition
V	= set of nodes in the multi-modal network.
E	= set of arcs in the multi-modal network.
M	= set of the transport modes, $M = \{p, b, m, t\}$.
p	= the private mode.
b	= the bus mode.
m	= the metro mode.
t	= the transfer mode.
i, j	= index of nodes, $i, j \in N$.
(i, j, k)	= connections between the adjacent nodes of mode k.
ξ_{ijk}	= travel time/transfer time on arc (i, j, k).
c_{ijk}	= travel cost on arc (i, j, k), $k \in M\backslash\{t\}$.
d_k	= fixed cost when entering mode k.
x_{ijk}	= 1, if arc (i, j, k) is on the path, 0 otherwise, $(i, j, k) \in E$.
y_k	= 1, if $x_{ijk} = 1$ for a certain $(i, j, k) \in E$, 0 otherwise, $k \in M\backslash\{t\}$.

is called the chance-constrained programming. Enlightened by this, we shall formulate a chance-constrained programming model, which is shown as follows:

$$
\begin{cases}
\min \overline{f_1}, \overline{f_2} \\
s.t. \\
\qquad M\{\sum_{(i,j,k)\in E} x_{ijk}\xi_{ijk} \le \overline{f_1}\} \ge \alpha & (12) \\
\qquad M\{\sum_{(i,j,k)\in E} x_{ijk}c_{ijk} + \sum_{k\in M\backslash\{t\}} y_k d_k \le \overline{f_2}\} \ge \beta & (13) \\
\qquad \sum_{(i,j,k)\in E} x_{ijk} - \sum_{(j,i,k)\in E} x_{jik} = 1, \ i = O, & (14) \\
\qquad \sum_{(i,j,k)\in E} x_{ijk} - \sum_{(j,i,k)\in E} x_{jik} = 0, \ i \ne O, D, & (15) \\
\qquad \sum_{(i,j,k)\in E} x_{ijk} - \sum_{(j,i,k)\in E} x_{jik} = -1, \ i = D, & (16) \\
\qquad \sum_{k\in M\backslash\{t\}} y_k \le N & (17) \\
\qquad x_{ijk}, y_k \in \{0, 1\}, \forall (i, j, k) \in E, k \in M\backslash\{t\} & (18)
\end{cases}
$$

In the above model, α and β are predetermined confidence levels. The objective function implies optimizing the total travel time and total travel cost. Constraints (12) and (13) mean that the total travel time and travel cost will be less than $\overline{f_1}$ and $\overline{f_2}$ with confidence levels α and β, respectively. Constraints (14), (15), and (16) ensure that a feasible path can be generated in the network. Constraint (17) implies the total transfer times will not exceed the given limit N. It is noted that the travel time and travel cost on arcs are imprecise due to the complexity of the actual network. Hence, the corresponding variables like ξ_{ijk}, c_{ijk}, and d_k are treated as uncertain variables.

Crisp equivalent of the model

It is easy to know that if the involved uncertain variables are complex, the model may be difficult to be dealt with. So it is necessary for us to transform the model into its crisp equivalent. In this section, we shall introduce how the model can be transformed into a deterministic form.

Theorem 4.1. *[33] Let ξ and η be two independent uncertain variables. The corresponding optimistic values and pessimistic values are $\xi_{\sup}(\alpha)$ and $\xi_{\inf}(\alpha)$, $\eta_{\sup}(\alpha)$ and $\eta_{\inf}(\alpha)$, respectively. Then, we have:*

(I) $(\xi+\eta)_{\sup}(\alpha) = \xi_{\sup}(\alpha) + \eta_{\sup}(\alpha)$,
(II) $(\xi+\eta)_{\inf}(\alpha) = \xi_{\inf}(\alpha) + \eta_{\inf}(\alpha)$.

Theorem 4.2. *[33] Let ξ be an independent uncertain variable, k be a scalar number with $k > 0$. Then, we have:*

(I) $(k\xi)_{\sup}(\alpha) = k \cdot \xi_{\sup}(\alpha)$,
(II) $(k\xi)_{\inf}(\alpha) = k \cdot \xi_{\inf}(\alpha)$.

Theorem 4.3. *Let ξ and η be two independent uncertain variables, k_1 and k_2 are scalar numbers with $k_1 > 0, k_2 > 0$. Then we have:*

(I) $(k_1\xi + k_2\eta)_{\sup}(\alpha) = k_1 \cdot \xi_{\sup}(\alpha) + k_2 \cdot \eta_{\sup}(\alpha)$,
(II) $(k_1\xi + k_2\eta)_{\inf}(\alpha) = k_1 \cdot \xi_{\inf}(\alpha) + k_2 \cdot \eta_{\inf}(\alpha)$.

It is worth noting that the critical value of an uncertain variable is associated with the uncertainty distribution according to the Measure Inversion Theorem. Just for completeness, two theorems will be cited to illustrate this point.

Theorem 4.4. *[34] Suppose that ξ is an uncertain variable with continuous uncertainty distribution $\Phi(x)$ when $0 < \Phi(x) < 1$, and $g(x,\xi) = h(x) - \xi$, α is a confidence level in $(0,1)$. Then, we have $M\{g(x,\xi) \le 0\} \ge \alpha$ if and only if $h(x) \le f_\xi(\alpha)$, where $f_\xi(\alpha) = \Phi^{-1}(1-\alpha)$.*

Theorem 4.5. *[34] Suppose that ξ is a continuous uncertain variable with increasing uncertainty distribution $\Phi(x)$ when $0 < \Phi(x) < 1$, $g(x,\xi) = h(x) - \xi$ and α is a confidence level in (0,1). Then, $M\{g(x,\xi) \ge 0\} \ge \alpha$ if and only if $h(x) \ge f_\xi(\alpha)$, where $f_\xi(\alpha) = \Phi^{-1}(\alpha)$.*

Based on the aforementioned theorems, we can easily deduce the following conclusions:

Theorem 4.6. *Let ξ be a continuous uncertain variable. The proposed critical value can also be expressed as:*

$$\xi_{\sup}(\alpha) = \sup\left\{r | r \le \Phi_\xi^{-1}(1-\alpha)\right\} = \Phi_\xi^{-1}(1-\alpha), \alpha \in (0,1], \tag{19}$$

$$\xi_{\inf}(\alpha) = \inf\left\{r | r \ge \Phi_\xi^{-1}(\alpha)\right\} = \Phi_\xi^{-1}(\alpha), \alpha \in (0,1]. \tag{20}$$

With the aid of relevant theorems given above, the following theorem intends to show the equivalent model.

Theorem 4.7. *In the established chance-constrained programming model, we assume that ξ_{ijk}, c_{ijk}, and d_k are all independent continuous uncertain variables. The equivalent model is as follows:*

$$\begin{cases} \min \left\{ \sum_{(i,j,k)\in E} x_{ijk} \cdot \Phi_{\xi_{ijk}}^{-1}(\alpha), \sum_{(i,j,k)\in E} x_{ijk} \cdot \Phi_{c_{ijk}}^{-1}(\beta) + \sum_{k\in M\backslash\{t\}} y_k \cdot \Phi_{d_k}^{-1}(\beta) \right\} \\ s.t.\ (14) \sim (18). \end{cases} \tag{21}$$

Proof. In the model, the two objective functions are to minimize $\overline{f_1}, \overline{f_2}$ with the constraint conditions:

$$M\left\{ \sum_{(i,j,k)\in E} x_{ijk}\xi_{ijk} \le \overline{f_1} \right\} \ge \alpha, \quad M\left\{ \sum_{(i,j,k)\in E} x_{ijk}c_{ijk} + \sum_{k\in M\backslash\{t\}} y_k d_k \le \overline{f_2} \right\} \ge \beta.$$

That is to say we need obtain the minimum values of $\overline{f_1}, \overline{f_2}$ which meet the above conditions, i.e.,

$$\inf\left\{ \overline{f_1} \,|M\left\{ \sum_{(i,j,k)\in E} x_{ijk}\xi_{ijk} \le \overline{f_1} \right\} \ge \alpha \right\}, \ \inf\left\{ \overline{f_2} \,|M\left\{ \sum_{(i,j,k)\in E} x_{ijk}c_{ijk} + \sum_{k\in M\backslash\{t\}} y_k d_k \le \overline{f_2} \right\} \ge \beta \right\}.$$

According to the definition of pessimistic value, it means that $\overline{f_1}, \overline{f_2}$ are the pessimistic values. By utilizing the properties of addition and multiplication, the constraints can be rewritten as the form of pessimistic values in the objective functions, expressed as follows:

$$\sum_{(i,j,k)\in E} x_{ijk} \cdot (\xi_{ijk})_{\inf}(\alpha), \quad \sum_{(i,j,k)\in E} x_{ijk} \cdot (c_{ijk})_{\inf}(\beta) + \sum_{k\in M\backslash\{t\}} y_k \cdot (d_k)_{\inf}(\beta).$$

Based on Equation 20, the pessimistic value of an uncertain variable is the inverse function with respect to α. Hence, we can easily obtain the equivalent model. The proof is thus completed. □

Remark. In the shortest path problem, the optimal objectives are related to the confidence levels α, β, i.e., if the confidence levels satisfy $\alpha_1 < \alpha_2$ and $\beta_1 < \beta_2$, then the corresponding optimal objectives satisfy: $\min \overline{f}_1 < \min \overline{f}'_1$ and $\min \overline{f}_2 < \min \overline{f}'_2$, because the uncertainty distributions of uncertain variables are monotone nondecreasing functions. Here, $\overline{f}_1, \overline{f}'_1, \overline{f}_2$, and $\overline{f}'_2$ indicate the optimal objective values with $\alpha = \alpha_1$, $\alpha = \alpha_2$, $\beta = \beta_1$, and $\beta = \beta_2$. □

Corollary 4.1. *Especially, suppose that ξ_{ijk}, c_{ijk}, and d_k are all independent zigzag uncertain variables with $\xi_{ijk} \sim \left(\xi^1_{ijk}, \xi^2_{ijk}, \xi^3_{ijk}\right)$, $c_{ijk} \sim \left(c^1_{ijk}, c^2_{ijk}, c^3_{ijk}\right)$, and $d_k \sim \left(d^1_k, d^2_k, d^3_k\right)$, respectively. Then, according to Theorem 4.7, we have the following crisp equivalent of the above model.*

$$\begin{cases} \min\ g_1(x),\ g_2(x,y) \\ s.t.\ (14) \sim (18). \end{cases} \tag{22}$$

where

$$g_1(x,y) = \begin{cases} (1-2\alpha) \sum_{(i,j,k)\in E} \xi^1_{ijk} x_{ijk} + 2\alpha \sum_{(i,j,k)\in E} \xi^2_{ijk} x_{ijk}, & \text{if } \alpha \le 0.5 \\ (2-2\alpha) \sum_{(i,j,k)\in E} \xi^2_{ijk} x_{ijk} + (2\alpha - 1) \sum_{(i,j,k)\in E} \xi^3_{ijk} x_{ijk}, & \text{if } \alpha > 0.5 \end{cases} \tag{23}$$

and

$$g_2(x,y) = \begin{cases} (1-2\beta) r_1(x,y) + 2\beta r_2(x,y), & \text{if } \beta \le 0.5 \\ (2-2\beta) r_2(x,y) + (2\beta - 1) r_3(x,y), & \text{if } \beta > 0.5 \end{cases} \tag{24}$$

$$r_1(x,y) = \sum_{(i,j,k)\in E} c^1_{ijk} x_{ijk} + \sum_{k\in M\setminus\{t\}} d^1_k y_k$$

$$r_2(x,y) = \sum_{(i,j,k)\in E} c^2_{ijk} x_{ijk} + \sum_{k\in M\setminus\{t\}} d^2_k y_k \tag{25}$$

$$r_3(x,y) = \sum_{(i,j,k)\in E} c^3_{ijk} x_{ijk} + \sum_{k\in M\setminus\{t\}} d^3_k y_k$$

Numerical experiments

In this section, we give a multi-modal network which consists of 47 nodes and 87 arcs, as shown in Figure 2, to illustrate the proposed model. There exist four modes in the network, namely, private (mode 1), bus (mode 2), metro (mode 3), and transfer (mode 4) as we have mentioned in Section 'Model formulations'. In a real-world situation, the private mode will be adopted only at the origin node. That is to say, one would not enter the private mode from other modes. So, it just generates fixed cost when entering into the bus subnetwork and the metro subnetwork from other modes, which follow $d_2 \sim Z(2,3,5)$ and $d_3 \sim Z(3,5,9)$, respectively. For the convenience of calculations, we assume that all weights on arcs are zigzag uncertain variables; the time and costs on travel arcs and time on transfer arcs are illustrated in Tables 2 and 3, respectively.

Before starting the experiments, we first need to transform our bi-objective model into a single-objective one. In literature, there were a variety of methods to handle the multi-objective programming problems, such as the linear weighting method, ideal point method, layered method, min-max method, etc. In this paper, we prefer to adopt a constraint-based method, which chooses one objective as the main objective function while the other objectives are transformed into constraints within the given thresholds.

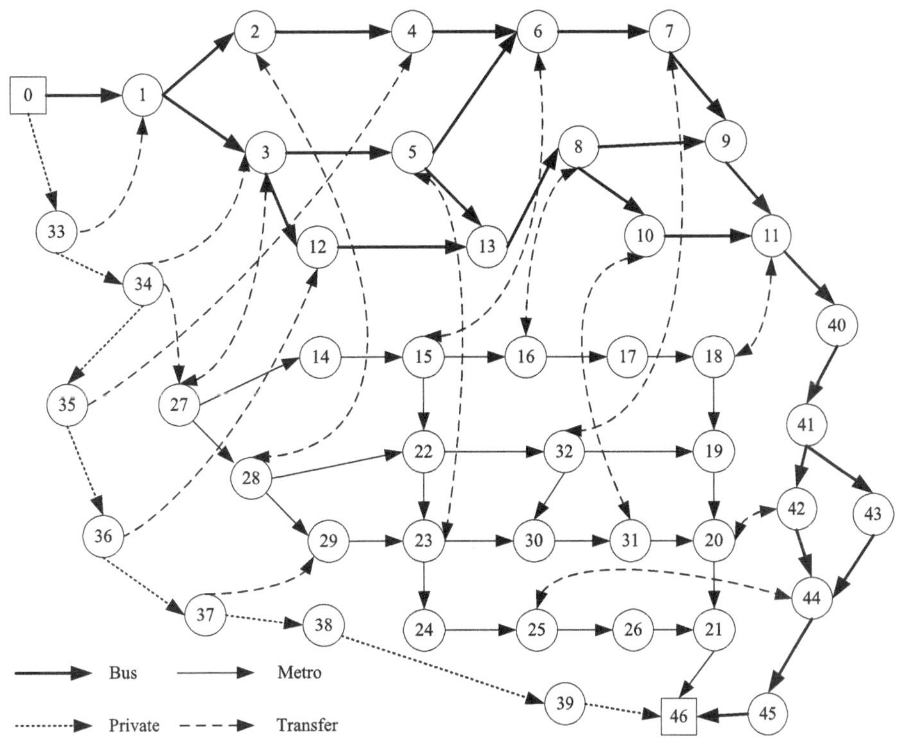

Figure 2 Multi-modal transportation network.

As for the problem of interest, the total cost will be treated as the main goal that needs to be optimized. To produce a single-objective model, we firstly solve the model with single-objective function $g_1(x)$, given as follows.

$$\begin{cases} \min g_1(x) \\ s.t. \\ \qquad \sum\limits_{(i,j,k)\in E} x_{ijk} - \sum\limits_{(j,i,k)\in E} x_{jik} = 1, \; i = O, \\ \qquad \sum\limits_{(i,j,k)\in E} x_{ijk} - \sum\limits_{(j,i,k)\in E} x_{jik} = 0, \; i \neq O, D, \\ \qquad \sum\limits_{(i,j,k)\in E} x_{ijk} - \sum\limits_{(j,i,k)\in E} x_{jik} = -1, \; i = D, \\ \qquad \sum\limits_{k\in M\backslash\{t\}} y_k \leq N \\ \qquad x_{ijk}, y_k \in \{0,1\}, \forall (i,j,k) \in E, k \in M\backslash\{t\} \end{cases} \tag{26}$$

Assume that we obtain the optimal objective value $\overline{g_1}(x')$. Then, $\overline{g_1}(x')$ with a threshold value δ will be regarded as upper limit to $g_1(x)$, which forms a new constraint to the model. In this case, the initial model can be transformed into the following single objective model.

Table 2 Time and costs on travel arcs of the multi-modal network

Arcs	Arc time (ξ_{ijk})	Arc costs (c_{ijk})	Arcs	Arc time (ξ_{ijk})	Arc costs (c_{ijk})
(0,33,1)	Z(4, 5, 8)	Z(8, 9,12)	(42,44,2)	Z(1,2,5)	Z(1,2,4)
(33,34,1)	Z(3, 4, 9)	Z(7, 9,14)	(43,44,2)	Z(2,5,6)	Z(2,4,5)
(34,35,1)	Z(15,17,20)	Z(10,12,13)	(44,45,2)	Z(3,5,8)	Z(2,3,5)
(35,36,1)	Z(12,14,18)	Z(9,10,13)	(45,46,2)	Z(2,3,5)	Z(2,3,6)
(36,37,1)	Z(14,17,18)	Z(12,14,17)	(27,14,3)	Z(1,3,4)	Z(1,2,4)
(37,38,1)	Z(9,12,14)	Z(11,14,16)	(27,28,3)	Z(1,2,4)	Z(1,2,4)
(38,39,1)	Z(22,25,27)	Z(18,24,26)	(14,15,3)	Z(1,2,4)	Z(1,3,4)
(39,46,1)	Z(11,15,17)	Z(11,14,16)	(15,16,3)	Z(2,3,5)	Z(1,3,4)
(0, 1,2)	Z(4, 7, 9)	Z(1, 3, 4)	(16,17,3)	Z(1,2,4)	Z(1,2,5)
(1, 2,2)	Z(3, 7, 8)	Z(2, 3, 6)	(17,18,3)	Z(2,3,5)	Z(1,2,4)
(1, 3,2)	Z(4, 5, 7)	Z(2, 5, 6)	(28,22,3)	Z(3,4,6)	Z(3,4,7)
(2, 4,2)	Z(5, 7, 8)	Z(2, 4, 5)	(22,32,3)	Z(4,5,7)	Z(2,3,6)
(3, 5,2)	Z(4, 7, 8)	Z(3, 6, 7)	(32,19,3)	Z(5,7,8)	Z(4,6,7)
(4, 6,2)	Z(2, 5, 7)	Z(1, 3, 4)	(28,29,3)	Z(1,2,4)	Z(2,4,7)
(5, 6,2)	Z(6, 9,10)	Z(2, 5, 6)	(29,23,3)	Z(1,3,4)	Z(2,4,7)
(5,13,2)	Z(3, 5, 6)	Z(2, 3, 5)	(23,30,3)	Z(1,2,4)	Z(2,5,7)
(3,12,2)	Z(1, 3, 4)	Z(1, 3, 4)	(30,31,3)	Z(2,3,5)	Z(2,4,5)
(12,13,2)	Z(3, 5, 6)	Z(1, 3, 4)	(31,20,3)	Z(2,3,5)	Z(1,2,4)
(13, 8,2)	Z(2, 5, 6)	Z(1, 4, 5)	(18,19,3)	Z(2,4,5)	Z(1,2,4)
(6, 7,2)	Z(2, 3, 5)	Z(1, 2, 4)	(19,20,3)	Z(2,3,5)	Z(2,4,5)
(8, 9,2)	Z(3, 5, 6)	Z(2, 4, 5)	(20,21,3)	Z(3,4,6)	Z(2,3,5)
(7, 9,2)	Z(2, 3, 6)	Z(1, 3, 4)	(15,22,3)	Z(2,3,5)	Z(1,2,4)
(8,10,2)	Z(1, 2, 4)	Z(1, 2, 5)	(22,23,3)	Z(1,2,5)	Z(2,3,5)
(9,11,2)	Z(2, 3, 5)	Z(1, 2, 4)	(23,24,3)	Z(1,2,4)	Z(2,5,6)
(10,11,2)	Z(1, 2, 4)	Z(1, 2, 4)	(24,25,3)	Z(3,4,6)	Z(2,5,7)
(11,40,2)	Z(4, 7, 9)	Z(1, 3, 4)	(25,26,3)	Z(1,2,4)	Z(2,3,5)
(40,41,2)	Z(1, 3, 5)	Z(1, 2, 4)	(26,21,3)	Z(1,2,5)	Z(1,2,4)
(41,42,2)	Z(1, 2, 4)	Z(1, 2, 4)	(21,46,3)	Z(1,2,4)	Z(1,2,5)
(41,43,2)	Z(2, 4, 5)	Z(1, 3, 4)			

Table 3 Transfer time on transfer arcs of the multi-modal network

Arcs	Arc time (ξ_{ijk})	Arcs	Arc time (ξ_{ijk})
(33, 1,4)	Z(3, 4, 6)	(6,15,4)	Z(5, 6, 8)
(34, 3,4)	Z(4, 5, 7)	(8,16,4)	Z(3, 4, 6)
(34,27,4)	Z(3, 4, 6)	(16, 8,4)	Z(4, 6, 7)
(27, 3,4)	Z(4, 6, 7)	(10,31,4)	Z(5, 8, 9)
(3,27,4)	Z(2, 3, 5)	(31,10,4)	Z(6, 7, 9)
(36,12,4)	Z(4, 5, 7)	(7,32,4)	Z(8,10,11)
(35, 4,4)	Z(2, 3, 5)	(32, 7,4)	Z(7, 8,10)
(28, 2,4)	Z(2, 3, 5)	(11,18,4)	Z(2, 3, 5)
(2,28,4)	Z(3, 4, 6)	(18,11,4)	Z(2, 5, 6)
(37,29,4)	Z(4, 6, 9)	(20,42,4)	Z(1, 2, 4)
(5,23,4)	Z(8, 9,11)	(42,20,4)	Z(2, 5, 6)
(23, 5,4)	Z(7, 9,10)	(25,44,4)	Z(2, 3, 6)
(15, 6,4)	Z(4, 5, 7)	(44,25,4)	Z(1, 2, 4)

Table 4 Optimal solutions with different α and β ($\delta = 5$)

	Parameters	Optimal objective values	Shortest path
Single Objective	$\alpha = 0.6$ $\beta = 0.6$	$g_1(x) = 37.2$ $g_2(x,y) = 35.8$	$0 \to 33 \to 34 \to 27 \to 28 \to 29 \to 23 \to 24 \to 25 \to 26 \to 21 \to 46$ $0 \to 1 \to 3 \to 12 \to 13 \to 8 \to 10 \to 31 \to 20 \to 21 \to 46$
Double Objectives	$\delta = 5$	$g_1(x) = 42.0$ $g_2(x,y) = 38.6$	$0 \to 1 \to 2 \to 28 \to 22 \to 23 \to 30 \to 31 \to 20 \to 21 \to 46$
Single Objective	$\alpha = 0.7$ $\beta = 0.7$	$g_1(x) = 42.4$ $g_2(x,y) = 39.6$	$0 \to 33 \to 34 \to 27 \to 28 \to 29 \to 23 \to 24 \to 25 \to 26 \to 21 \to 46$ $0 \to 1 \to 3 \to 12 \to 13 \to 8 \to 10 \to 31 \to 20 \to 21 \to 46$
Double Objectives	$\delta = 5$	$g_1(x) = 47.2$ $g_2(x,y) = 42.8$	$0 \to 1 \to 3 \to 5 \to 23 \to 24 \to 25 \to 26 \to 21 \to 46$
Single Objective	$\alpha = 0.8$ $\beta = 0.8$	$g_1(x) = 46.4$ $g_2(x,y) = 43.4$	$0 \to 1 \to 2 \to 28 \to 29 \to 23 \to 24 \to 25 \to 26 \to 21 \to 46$ $0 \to 1 \to 3 \to 12 \to 13 \to 8 \to 10 \to 31 \to 20 \to 21 \to 46$
Double Objectives	$\delta = 5$	$g_1(x) = 50.8$ $g_2(x,y) = 46.2$	$0 \to 1 \to 3 \to 5 \to 23 \to 24 \to 25 \to 26 \to 21 \to 46$
Single Objective	$\alpha = 0.9$ $\beta = 0.9$	$g_1(x) = 50.2$ $g_2(x,y) = 47.2$	$0 \to 1 \to 2 \to 28 \to 29 \to 23 \to 24 \to 25 \to 26 \to 21 \to 46$ $0 \to 1 \to 3 \to 12 \to 13 \to 8 \to 10 \to 31 \to 20 \to 21 \to 46$
Double Objectives	$\delta = 5$	$g_1(x) = 54.4$ $g_2(x,y) = 49.6$	$0 \to 1 \to 3 \to 5 \to 23 \to 24 \to 25 \to 26 \to 21 \to 46$
Single Objective	$\alpha = 1.0$ $\beta = 1.0$	$g_1(x) = 54.0$ $g_2(x,y) = 51.0$	$0 \to 1 \to 2 \to 28 \to 29 \to 23 \to 24 \to 25 \to 26 \to 21 \to 46$ $0 \to 1 \to 3 \to 12 \to 13 \to 8 \to 10 \to 31 \to 20 \to 21 \to 46$
Double Objectives	$\delta = 5$	$g_1(x) = 59.0$ $g_2(x,y) = 52.0$	$0 \to 1 \to 3 \to 5 \to 23 \to 30 \to 31 \to 20 \to 21 \to 46$

$$\begin{cases} \min g_2(x,y) \\ s.t. \\ \quad \sum_{(i,j,k)\in E} x_{ijk} - \sum_{(j,i,k)\in E} x_{jik} = 1,\ i = O, \\ \quad \sum_{(i,j,k)\in E} x_{ijk} - \sum_{(j,i,k)\in E} x_{jik} = 0,\ i \neq O, D, \\ \quad \sum_{(i,j,k)\in E} x_{ijk} - \sum_{(j,i,k)\in E} x_{jik} = -1,\ i = D, \\ \quad \sum_{k\in M\backslash\{t\}} y_k \leq N \\ \quad \boldsymbol{x}_{ijk}, y_k \in \{0,1\}, \forall (i,j,k) \in E, k \in M\backslash\{t\} \\ \quad g_1(x) \leq \overline{g_1}(x') + \delta \end{cases} \tag{27}$$

With this method, the LINGO optimization software is employed to generate the shortest path from origin node 0 to destination node 46 in the multi-modal network. It is noted that α, β and δ are three parameters which will influence the optimal objective values. To capture the corresponding relations between the parameters and the objective functions, we shall implement two sets of experiments as follows:

(1) Optimal solutions with different α and β: As the optimal solution is closely related to parameters α and β, we firstly use different critical values to produce the shortest paths, with the fixed parameter $\delta = 5$. The results are listed in Table 4, in which the confidence levels in critical values are set as 0.6, 0.7, 0.8, 0.9, and 1.0, respectively. Figure 3 shows the variation tendency of the optimal objectives with respect to different α and β values. Obviously, both the objective values of $g_1(x)$ and $g_2(x,y)$ are increasing with respect to the relevant parameters when δ is fixed. This conclusion just coincides with the Remark.

(2) Optimal solutions with different δ: The second set of experiments focus on the other index δ. Table 5 illustrates the relationship between the optimal objectives and different δ values, in which α and β are all set as 0.9. When we depict the results in Figure 4, we can find that the optimal objectives will not change any more when δ is set to be not less than 7; moreover, the optimal objective value of $g_2(x,y)$ is decreasing with respect to the increase of δ, while $g_1(x)$ varies with an opposite tendency. When the objective value of

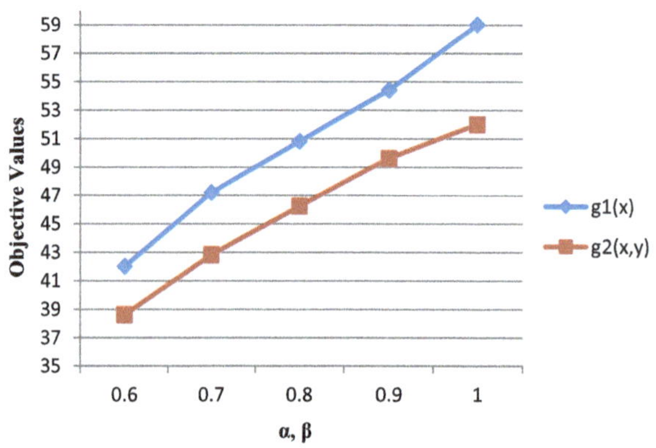

Figure 3 Objective values with different α and β ($\delta = 5$).

Table 5 Optimal solutions with different δ ($\alpha = 0.9, \beta = 0.9$)

δ	Objective values
1	$g_1(x) = 50.2$
	$g_2(x,y) = 55.2$
2	$g_1(x) = 51.4$
	$g_2(x,y) = 54.2$
3	$g_1(x) = 51.4$
	$g_2(x,y) = 54.2$
4	$g_1(x) = 54.0$
	$g_2(x,y) = 52.4$
5	$g_1(x) = 54.4$
	$g_2(x,y) = 49.6$
6	$g_1(x) = 55.6$
	$g_2(x,y) = 48.6$
7	$g_1(x) = 56.8$
	$g_2(x,y) = 47.2$
8	$g_1(x) = 56.8$
	$g_2(x,y) = 47.2$
9	$g_1(x) = 56.8$
	$g_2(x,y) = 47.2$
10	$g_1(x) = 56.8$
	$g_2(x,y) = 47.2$
11	$g_1(x) = 56.8$
	$g_2(x,y) = 47.2$
12	$g_1(x) = 56.8$
	$g_2(x,y) = 47.2$

$g_2(x, y)$ reaches 47.2, which is the optimal objective when $g_2(x, y)$ is treated as the single objective, the figure will not change any more.

Conclusions

Uncertainty theory is a branch of mathematics to study the characteristics of nondeterministic phenomenon. In this paper, we applied uncertainty theory to investigating the multi-modal shortest path problem in which the arc time and arc costs are represented

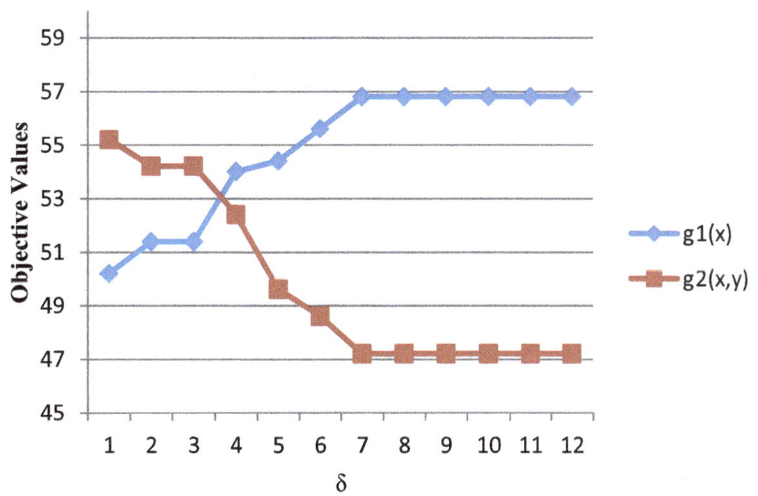

Figure 4 Objective values with different δ ($\alpha = 0.9, \beta = 0.9$).

by uncertain variables. Considering two objectives, which are to minimize the total travel time and travel costs, we formulated a chance-constrained programming model for the problem of interest. For handling convenience, the model was then transformed into its deterministic crisp equivalent. Additionally, the bi-objective model was simplified into a single objective model by converting one objective function into a new constraint with the given threshold. Finally, the results of the numerical experiments deriving from the LINGO software demonstrated the performance of the proposed approaches. Actually, multi-modal shortest path problems have been widely solved by the label-setting algorithm [35], label-correcting algorithm [25], and genetic algorithm [36,37], which can also be considered in our further studies.

References

1. Dijkstra, EW: A note on two problems in connexion with graphs. Numerische Mathematik. **1**(1), 269–271 (1959)
2. Bellman, R: On a routing problem. Q. Appl. Math. **16**, 87–90 (1958)
3. Dreyfus, SE: An appraisal of some shortest-path algorithms. Oper. Res. **17**(3), 395–412 (1969)
4. Frank, H: Shortest paths in probabilistic graphs. Oper. Res. **17**(4), 583–599 (1969)
5. Loui, RP: Optimal paths in graphs with stochastic or multidimensional weights. Commun. ACM. **26**(9), 670–676 (1983)
6. Mirchandani, PB: Shortest distance and reliability of probabilistic networks. Comput. Oper. Res. **3**(4), 347–355 (1976)
7. Yang, L, Yang, X, You, C: Stochastic scenario-based time-stage optimization model for the least expected time shortest path problem. Int. J. Uncertainty, Fuzziness Knowledge-Based Syst. **21**(Supp01), 17–33 (2013)
8. Yang, L, Zhou, X: Constraint reformulation and Lagrangian relaxation-based solution algorithm for a least expected time path problem. Transp. Res. Part B. **59**, 22–44 (2014)
9. Dubois, D, Prade, H: Fuzzy Sets and Systems: Theory and Applications. Academic Press, New York (1980)
10. Ji, X, Iwamura, K, Shao, Z: New models for shortest path problem with fuzzy arc lengths. Appl. Math. Model. **31**, 259–269 (2007)
11. Hernandes, F, Lamata, MT, Verdegay, JL, Yamakami, A: The shortest path problem on networks with fuzzy parameters. Fuzzy Sets Syst. **158**(14), 1561–1570 (2007)
12. Yang, L, Zhou, X, Gao, Z: Credibility-based rescheduling model in a double-track railway network: a fuzzy reliable optimization approach. Omega. **48**, 75–93 (2014)
13. Liu, B: Uncertainty Theory. Springer-Verlag, Berlin (2007)
14. Baidya, A, Bera, UK, Maiti, M: A solid transportation problem with safety factor under different uncertainty environments. J. Uncertain. Anal. Appl. **1**, Article 18 (2013)
15. Ke, H: Uncertain random time-cost trade-off problem. J. Uncertain. Anal. Appl. **2**, Article 23 (2014)
16. Sheng, Y, Gao, J: Chance distribution of the maximum flow of uncertain random network. J. Uncertain. Anal. Appl. **2**, Article 15 (2014)
17. Zhang, B, Peng, J: Uncertain programming model for uncertain optimal assignment problem. Appl. Math. Model. **37**(9), 6458–6468 (2013)
18. Gao, Y, Yang, L, Li, S: On distribution function of the diameter in uncertain graph. Inform. Sci. **296**, 61–74 (2015)
19. Liu, W: Uncertain programming models for shortest path problem with uncertain arc lengths. In: Proceedings of the first international conference on uncertainty theory, pp. 148–153, Urumchi, China, (2010)
20. Gao, Y: Shortest path problem with uncertain arc length. Comput. Math. Appl. **62**, 2591–2600 (2011)
21. Zhou, J, Yang, F, Wang, K: An inverse shortest path problem on an uncertain graph. J. Netw. **9**(9), 2353–2359 (2014)
22. Lozano, A, Storchi, G: Shortest viable path algorithm in multimodal networks. Transportation Res. Part A. **35**, 225–241 (2001)
23. Ma, T: An A^* label-setting algorithm for multimodal resource constrained shortest path problem. Procedia - Soc. Behav. Sci. **111**, 330–339 (2014)
24. Ambrosino, D, Sciomachen, A: An algorithmic framework for computing shortest routes in urban multimodal networks with different criteria. Procedia - Soc. Behav. Sci. **108**, 139–152 (2014)
25. Liu, L, Yang, J, Mu, H, Li, X, Wu, F: Exact algorithm for multi-criteria multi-modal shortest path with transfer delaying and arriving time-window in urban transit network. Appl. Math. Model. **38**(9–10), 2613–2629 (2014)
26. Galvez-Fernandez, C, Khadraoui, D, Ayed, H, Habbas, Z, Alba, E: Distributed approach for solving time-dependent problems in multimodal transport networks. Adv. Oper. Res. **2009** (2009)
27. Yamani, OE, Mouncif, H, Rida, M: A fuzzy TOPSIS approach for finding shortest path in multimodal transportation networks. Int. J. Comput. Optimization. **1**(2), 95–111 (2014)
28. Zografos, KG, Androutsopoulos, KN: Algorithms for itinerary planning in multimodal transportation networks. IEEE Trans. Intel. Transportation Syst. **9**, 175–184 (2008)
29. Niksirat, M, Ghatee, M, Hashemi, SM: Multimodal K-shortest viable path problem in Tehran public transportation network and its solution applying ant colony and simulated annealing algorithms. Appl. Math. Model. **36**, 5709–5726 (2012)
30. Spiess, H, Florian, M: Optimal strategies: a new assignment model for transit networks. Transportation Res. Part B. **23**, 83–102 (1989)

31. Artigues, C, Huguet, M, Gueye, F, Schettini, F, Dezou, L: State-based accelerations and bidirectional search for bi-objective multi-modal shortest paths. Transportation Res. Part C. **27**, 233–259 (2013)
32. Disser, Y, Müller-Hannemann, M, Schnee, M: Multi-criteria shortest paths in time-dependent train networks. Exp. Algorithms Lecture Notes Comput. Sci. **5038**, 347–361 (2008)
33. Liu, B: Uncertainty theory: a branch of mathematics for modeling human uncertainty. Springer-Verlag, Berlin (2010)
34. Yang, L, Liu, P, Li, S, Gao, Y, Ralescu, DA: Reduction methods of type-2 uncertain variables and their applications to solid transportation problem. Inform. Sci. **291**, 204–237 (2015)
35. Horn, MET: An extended model and procedural framework for planning multi-modal passenger journeys. Transportation Res. Part B. **37**, 641–660 (2003)
36. Yu, H, Lu, F: A multi-modal route planning approach with an improved genetic algorithm. Adv. Geo-Spatial Inform. Sci. **38**, 193–202 (2012)
37. Abbaspour, RA, Samadzadegan, F: An evolutionary solution for multimodal shortest path problem in metropolises. Comput. Sci. Inform. Syst. **7**(4), 789–811 (2010)

18

Extraction methods for uncertain inference rules by ant colony optimization

Ling Chen, Yun Sun* and Yuanguo Zhu

*Correspondence:
chinalsy_881220@163.com
School of Science, Nanjing
University of Science and
Technology, Nanjing 210094, China

Abstract

In recent years, the research on data mining methods has received increasing attention. In this paper, we design an uncertain system with the extracted uncertain inference rules to solve the classification problems in data mining. And then, two extraction methods integrated with ant colony optimization are proposed for the generation of the uncertain inference rules. Finally, two applications are given to verify the effectiveness and superiority of the proposed methods.

Keywords: Uncertain inference rule; Uncertain system; Ant colony optimization algorithm; Rules extraction; Data classification

Introduction

Nowadays, databases and computer networks, coupled with the use of advanced automated data generation and collection tools, are widely used in many different fields such as finance, E-commerce, logistics, etc. As a result, the amount of data that people have to deal with is dramatically increasing. People hope to carry out scientific research, business decision, or business management on the basis of the analysis of the existing data. However, the current data analysis tools have difficulty in processing the data in depth. To compensate for this deficiency, there come the data mining techniques. Data mining is the computational process of discovering some interesting, potentially useful patterns in large data sets. Those patterns can be concepts, rules, laws, and modes. The overall goal of data mining is to extract information from a data set and transform it into an understandable structure for further use. Data mining helps us to discover valuable information and knowledge. Data mining is applied to many fields in reality. There are many successful examples [1] of data mining in business and science research. For instance, data mining is widely used in financial data analysis, telecommunication, retail, and biomedical research. Therefore, the study of data mining technology has an important practical significance.

The main jobs of data mining are data description, data classification, data dependency, data compartment analysis, data regression, data aggregate, and data prediction. What data classification does is to find a couple of models or functions that can accurately describe the characteristics of the data sets. Then, we can identify the categories of the previously unknown data. After obtaining the models or functions from the set of training data with data mining algorithms, we use many methods to describe the output such as classification rules (if-then), decision trees, mathematical formula, and neutral network.

There are a variety of approaches in data mining. For mining objects in different fields, many different specified methods are invented. The approaches we usually used are statistical methods, machine learning methods, and modern intelligent optimization methods. The statistical methods are very effective methods from the start. In addition, many other data mining methods are invented based on the statistical methods. When dealing with classification problems, Bayesian classification and Bayesian belief network are important classification methods that based on the statistical principle. Machine learning methods are mainly used to solve the conceptual learning, pattern classification, and pattern clustering problems. The core content of machine learning is inductive learning. And there already exist a number of mature technology methods, such as decision tree method for classification problems. Decision trees method is one of the most popular classification methods. The early decision trees algorithm is ID3 method. Later, based on ID3, many algorithms such as C4.5 method [2] are proposed. Besides, there are some variants of the decision trees algorithm including incremental tree structure ID4, ID5, and expandable tree structure SLIQ for massive data set.

In recent years, intelligent optimization algorithms are widely applied into data mining. Neutral network is a simulation model for complex system with nonlinear relations. It is very suitable to deal with complex nonlinear relations in spatial data. Researchers have already proposed different network models to realize the clustering, classification, regression, and pattern recognition of the data. Furthermore, many evolution algorithms such as simulated annealing algorithm are introduced into neutral network algorithm as the optimization strategies. Genetic algorithm is a global search algorithm that simulates the biological evolution and genetic mechanism. It plays an important role in optimization and classification machine learning. Mixed algorithms of genetic algorithm and other algorithms, such as decision trees, neutral network, have been applied to the data mining technology. Ant colony optimization algorithm is a bionic optimization algorithm that simulates the behavior of the ants. Based on that, a data mining technique ant-miner [3] was invented. And Herrera [4] applied it to fuzzy rules learning. However, ant colony optimization algorithm has some weakness such as slow convergence, random initial solutions. For this reason, some improved ant colony optimization algorithms are proposed. Zhu proposed an improved ant colony optimization algorithm (ACOA) [5] and a mutation ant colony optimization algorithm (MACO) [6] to speed up the algorithms and avoid the solutions getting stuck in local optimums. Hybrid genetic ant colony optimization [7] and hybrid particle swarm ant colony optimization algorithm [8] significantly improve the performance of the original ant colony optimization algorithm.

The real world is so complex that human being may face different types of indeterminacy everyday. To get a better understanding of the real world, many mathematical tools are created. One of them is probability theory which is used to model indeterminacy from samples. However, in many cases, no samples are available to estimate a probability distribution. In this situation, we have no choice but to invite some domain experts to evaluate the belief degree that each event may occur. We cannot use probability theory to deal with belief degree since human beings usually overweight unlikely events which makes the belief degrees deviate far from the frequency. In view of this, Liu [9] founded uncertainty theory based on normality axiom, duality axiom, subadditivity axiom, and product measure axiom. It has become a powerful mathematical tool dealing with indeterminacy.

Many researchers have done a lot of theoretical work related to uncertainty theory. In 2008, Liu [10] presented the uncertain differential equation. Later, the existence and uniqueness theorem was given [11]. And the stability of uncertain differential equation was discussed [12,13]. Also, some analysis and numerical methods for solving uncertain differential equation were proposed. With uncertain differential equation describing the evolution of the system, we may solve some practical problems. Peng and Yao [14] studied an option pricing models for stocks. Zhu [15] proposed an uncertain optimal control model in 2010.

In [16,17], Liu proposed and studied the uncertain systems based on the concepts of uncertain sets, membership functions, and uncertain inference rules. An uncertain system is a function from its inputs to outputs based on the uncertain inference rule. Usually, an uncertain system consists of five parts: inputs, rule-base, uncertain inference rules, expected value operator, and outputs. Following that, Gao et al. [18] generalized uncertain inference rules and described uncertain systems with them. Peng and Chen [19] proved that uncertain systems are universal approximator and then demonstrated that the uncertain controller is a reasonable tool. Gao [20] designed an uncertain inference controller that successfully balanced an inverted pendulum with 5×5 if-then rules. What is more important is that this uncertain inference controller has a good ability of robustness.

On the basis of uncertainty theory, we consider two extraction methods for uncertain inference rules by ant colony optimization algorithm. In the next section, we review the ant colony optimization algorithm and give some basic concepts about uncertain sets. Then, we formulate a model to extract inference rules based on data set. And then, we propose an extraction method for uncertain inference rules by ant colony optimization algorithm with a mutation operation. Finally, we combine the ant colony optimization algorithm with simulated annealing algorithm to speed up the extraction method. In the last section, we discuss two typical classification problems in data mining with our results.

Preliminary

In this section, we review the ant colony optimization algorithm. And then, we give some basic concepts on uncertainty sets.

Ant colony optimization algorithm

Ant colony optimization algorithm, initiated by Dorigo, is a heuristic optimization approach. It simulates the behavior of real ants when they forage for food which relies on the pheromone communication. In ant colony optimization algorithm, each path of artificial ants walking from the food sources to the nest is a candidate solution to the problem. When walking on the path, the ants will release pheromone which evaporates over time. And the artificial ants will lay down more pheromone on the path corresponding to the better solution. While one ant has many paths to go, it will make a choice according to the amount of the pheromone on the paths. The more pheromone there is on the path, the better the solution is. As a result, bad paths will disappear since the pheromone evaporates over time. And good paths will be reserved since ants walking on it increases the pheromone levels. Finally, one path which is used by most of the ants is left. Then, the optimal solution to the problem is obtained.

Consider the following optimization problem:

$$\begin{cases} \min f(x) \\ s.t. \\ \quad g(x) \geq 0 \\ \quad x \in D \end{cases} \tag{1}$$

where x is the decision variable in the domain D. And $f(x)$ is the objective function while $g(x)$ is the constraint function.

We can use ant colony optimization algorithm to obtain the optimal solution to the problem (1). The parameters in the algorithm are initial pheromone τ_0, ant transfer probability p, number of ants M, pheromone evaporation rate ρ, and number of iterations T. The procedures are as follows.

Step 1 Randomly generate a feasible solution x_0 and set optimal solution $s = x_0$. Initialize all pheromone trails with the same pheromone level τ_0. Set $k \leftarrow 0$.

Step 2 The artificial ant generates a walking path x in some probability p according to the pheromone trails. If $x \in D$, then go to Step 3; otherwise, repeat Step 2 until $x \in D$.

Step 3 Repeat Step 2 until for each ant and generate M feasible solutions. Let s_k be the best solution in this iteration.

Step 4 If $f(s_k) < f(s)$, then $s \leftarrow s_k$ and update the pheromone trails according to the optimal solution in the current iteration.

Step 5 If $k < T$, then $k \leftarrow k + 1$ and go to Step 2; otherwise, terminate.

Step 6 Report the optimal solution.

Uncertain set

Let Γ be a nonempty set and $\mathcal{L}$ be σ-algebra over Γ. Each $\Lambda \in \mathcal{L}$ is called an event. For any Λ, $\mathcal{M}\{\Lambda\} \in [0, 1]$. The set function $\mathcal{M}$ defined on $\mathcal{L}$ is called an uncertain measure if it satisfies the following three axiom: $\mathcal{M}\{\Gamma\} = 1$; $\mathcal{M}\{\Lambda\} + \mathcal{M}\{\Lambda^c\} = 1$ for any $\Lambda \in \mathcal{L}$; $\mathcal{M}\left\{\bigcup_{i=1}^{\infty} \Lambda_i\right\} \leq \sum_{i=1}^{\infty} \mathcal{M}\{\Lambda_i\}$ for all $\Lambda_1, \Lambda_2, \cdots \in \mathcal{L}$. Then, the triplet $(\Gamma, \mathcal{L}, \mathcal{M})$ is called an uncertainty space [9]. The product uncertain measure $\mathcal{M}$ is an uncertain measure satisfying $\mathcal{M}\left\{\prod_{i=1}^{\infty} \Lambda_k\right\} = \bigwedge_{i=1}^{\infty} \mathcal{M}_k\{\Lambda_k\}$, where Λ_k are arbitrarily chosen events from $\mathcal{L}_k$ for $k = 1, 2, \cdots$, respectively.

Definition 1. [16] *An uncertain set is a function ξ from an uncertainty space $(\Gamma, \mathcal{L}, \mathcal{M})$ to a collection of sets of real numbers such that both $\{B \subset \xi\}$ and $\{\xi \subset B\}$ are events for any Borel set B.*

Example 1. Take $(\Gamma, \mathcal{L}, \mathcal{M})$ to be $\{\gamma_1, \gamma_2, \gamma_3\}$ with power set $\mathcal{L}$. Then, the set-valued function

$$\xi(\gamma) = \begin{cases} [1, 3], & \text{if } \gamma = \gamma_1 \\ [2, 4], & \text{if } \gamma = \gamma_2 \\ [3, 5], & \text{if } \gamma = \gamma_3 \end{cases}$$

is an uncertain set on $(\Gamma, \mathcal{L}, \mathcal{M})$.

Definition 2. [16] *The uncertain sets $\xi_1, \xi_2, \xi_3, \cdots, \xi_n$ are said to be independent if for any Borel sets $B_1, B_2, B_3, \cdots, B_n$, we have*

$$\mathcal{M}\left\{\bigcap_{i=1}^{n}(\xi_i^* \subset B_i)\right\} = \bigwedge_{i=1}^{n}\mathcal{M}\left\{\xi_i^* \subset B_i\right\}$$

and

$$\mathcal{M}\left\{\bigcup_{i=1}^{n}(\xi_i^* \subset B_i)\right\} = \bigvee_{i=1}^{n}\mathcal{M}\left\{\xi_i^* \subset B_i\right\}$$

where ξ_i^* are arbitrarily chosen from $\left\{\xi_i, \xi_i^c\right\}$, $i = 1, 2, \cdots, n$, respectively.

Definition 3. [21] *An uncertain set ξ is said to have a membership function μ if for any Borel set B of real numbers, we have*

$$\mathcal{M}\{B \subset \xi\} = \inf_{x \in B}\mu(x), \mathcal{M}\{\xi \subset B\} = 1 - \sup_{x \in B^c}\mu(x).$$

The above equations will be called measures inversion formulas.

Remark 1. When an uncertain set ξ does have a membership function μ, it follows from the first measure inversion formula that

$$\mu(x) = \mathcal{M}\{x \in \xi\}.$$

Example 2. An uncertain set ξ is called triangular if it has a membership function

$$\mu(x) = \begin{cases} \frac{x-a}{b-a}, & a \le x \le b \\ \frac{x-c}{b-c}, & b \le x \le c \end{cases} \tag{2}$$

denoted by (a, b, c) where a, b, c are real numbers with $a < b < c$.

Definition 4. [21] *A membership function μ is said to be regular if there exists a point x_0 such that $\mu(x_0) = 1$, and $\mu(x)$ is unimodal about the mode x_0. That is, $\mu(x)$ is increasing on $(-\infty, x_0]$ and decreasing on $[x_0, +\infty)$.*

Definition 5. [16] *Let ξ be an uncertain set. Then, the expected value of ξ is defined by*

$$E[\xi] = \int_0^{+\infty}\mathcal{M}\{\xi \succeq r\}\mathrm{d}r - \int_{-\infty}^{0}\mathcal{M}\{\xi \preceq r\}\mathrm{d}r$$

provided that at least one of the two integrals is finite and

$$\mathcal{M}\{\xi \succeq r\} = \frac{1}{2}(\mathcal{M}\{\xi \geq r\} + 1 - \mathcal{M}\{\xi < r\}),$$

$$\mathcal{M}\{\xi \preceq r\} = \frac{1}{2}(\mathcal{M}\{\xi \leq r\} + 1 - \mathcal{M}\{\xi > r\}).$$

Theorem 1. [13] *Let ξ be an uncertain set with regular membership function μ. Then*

$$E[\xi] = x_0 + \frac{1}{2}\int_{x_0}^{+\infty}\mu(x)\mathrm{d}x - \frac{1}{2}\int_{-\infty}^{x_0}\mu(x)\mathrm{d}x, \tag{3}$$

where x_0 is a point such that $\mu(x_0) = 1$.

Example 3. Let ξ be a triangular uncertain set denoted by (a, b, c). Then, according to Theorem 1, we have

$$E[\xi] = \frac{a + 2b + c}{4}.$$

In fact, it follows from Equations 2 and 3 that

$$\begin{aligned} E[\xi] &= b + \frac{1}{2}\int_b^c \frac{x - c}{b - c}\mathrm{d}x - \frac{1}{2}\int_a^b \frac{x - a}{b - a}\mathrm{d}x \\ &= b - \frac{1}{4}(b - c) - \frac{1}{4}(b - a) \\ &= \frac{a + 2b + c}{4}. \end{aligned}$$

Uncertain inference rule

Here, we introduce concepts of the uncertain inference and uncertain system. Inference rules are the key points of the inference systems. In fuzzy systems, CRI approach [22], Mamdani inference rules [23] and Takagi-Sugeno inference rules [24] are the most common used inference rules. Fuzzy if-then inference rules use fuzzy sets to describe the antecedents and the consequents. Unlike fuzzy inference, both antecedents and consequents in uncertain inference are characterized by uncertain sets. Uncertain inference [16] is a process of deriving consequences from human knowledge via uncertain set theory. First, we introduce the following inference rule.

Inference Rule 1. [16] *Let $\mathbb{X}$ and $\mathbb{Y}$ be two concepts. Assume a rule 'if $\mathbb{X}$ is an uncertain set ξ, then $\mathbb{Y}$ is an uncertain set η'. From $\mathbb{X}$ is a constant a, we infer that $\mathbb{Y}$ is an uncertain set*

$$\eta^* = \eta|_{a \in \xi}$$

which is the conditional uncertain set of η given $a \in \xi$. The inference rule is represented by

Rule: If $\mathbb{X}$ is ξ, then $\mathbb{Y}$ is η

From: $\mathbb{X}$ is a constant a

Infer: $\mathbb{Y}$ is $\eta^* = \eta|_{a \in \xi}$

Theorem 2. [16] *Let ξ and η be independent uncertain sets with membership functions μ and ν, respectively. If ξ^* is a constant a, then the Inference Rule 1 yields that η^* has a membership function*

$$\nu^*(y) = \begin{cases} \frac{\nu(y)}{\mu(a)}, & \text{if } \nu(y) < \frac{\mu(a)}{2} \\ \frac{\nu(y)+\mu(a)-1}{\mu(a)}, & \text{if } \nu(y) > 1 - \frac{\mu(a)}{2} \\ 0.5, & \text{otherwise.} \end{cases}$$

Based on Inference Rule 1, Gao et al. [18] proposed the multi-input, multi-if-then-rule inference rules.

Inference Rule 2. [13] *Let* $\mathbb{X}_1, \mathbb{X}_2, \cdots, \mathbb{X}_m, \mathbb{Y}$ *be concepts. Assume rules 'if* $\mathbb{X}_1$ *is* ξ_{i1} *and* $\cdots$ *and* $\mathbb{X}_m$ *is* ξ_{im}*, then* $\mathbb{Y}$ *is* η_i*' for* $i = 1, 2, \cdots, k$*. From* $\mathbb{X}_1$ *is a constant* a_1 *and* $\cdots$ *and* $\mathbb{X}_m$ *is a constant* a_m*, we infer that*

$$\eta^* = \sum_{i=1}^{k} \frac{c_i \cdot \eta_i|_{(a_1 \in \xi_{i1}) \cap (a_2 \in \xi_{i2}) \cap \cdots \cap (a_m \in \xi_{im})}}{c_1 + c_2 + \cdots + c_k}, \tag{4}$$

where the coefficients are determined by

$$c_i = \mathcal{M}\{(a_1 \in \xi_{i1}) \cap (a_2 \in \xi_{i2}) \cap \cdots \cap (a_m \in \xi_{im})\}$$

for $i = 1, 2, \cdots, k$*. The inference rule is represented by*

Rule 1: If $\mathbb{X}_1$ is ξ_{11} and $\cdots$ and $\mathbb{X}_m$ is ξ_{1m}, then $\mathbb{Y}$ is η_1
Rule 2: If $\mathbb{X}_1$ is ξ_{21} and $\cdots$ and $\mathbb{X}_m$ is ξ_{2m}, then $\mathbb{Y}$ is η_2
$\cdots$
Rule k: If $\mathbb{X}_1$ is ξ_{k1} and $\cdots$ and $\mathbb{X}_m$ is ξ_{km}, then $\mathbb{Y}$ is η_k
From: $\mathbb{X}_1$ is a_1 and $\cdots$ and $\mathbb{X}_m$ is a_m
Infer: $\mathbb{Y}$ is determined by Eq. (4)

Theorem 3. [13] *Assume* $\xi_{i1}, \xi_{i2}, \cdots, \xi_{im}, \eta_i$ *are independent uncertain sets with membership functions* $\mu_{i1}, \mu_{i2}, \cdots, \mu_{im}, \nu_i$, $i = 1, 2, \cdots, k$*, respectively. If* $\xi_1^*, \xi_2^*, \cdots, \xi_m^*$ *are constants* $a_1, a_2, \cdots, a_m$*, respectively, then the Inference Rule 2 yields*

$$\eta^* = \sum_{i=1}^{k} \frac{c_i \cdot \eta_i^*}{c_1 + c_2 + \cdots + c_k}$$

where η_i^* *are uncertain sets whose membership functions are given by*

$$\nu_i^*(y) = \begin{cases} \frac{\nu_i(y)}{c_i}, & \text{if } \nu_i(y) < \frac{c_i}{2} \\ \frac{\nu_i(y)+c_i-1}{\mu(a)}, & \text{if } \nu_i(y) > 1 - \frac{c_i}{2} \\ 0.5, & \text{otherwise} \end{cases}$$

and $c_i = \min_{1 \le l \le m} \mu_{il}(a_l)$ *are constants.*

Uncertain system

Uncertain system, proposed by Liu [16], is a function from its inputs to outputs based on the uncertain inference rule. Usually, an uncertain system consists of five parts: inputs that are crisp data to be fed into the uncertain system; a rule-base that contains a set of if-then rules provided by the experts; an uncertain inference rule that infers uncertain consequents from the uncertain antecedents; an expected value operator that converts the uncertain consequents to crisp values; and outputs that are crisp data yielded from the expected value operator.

Now, we consider an uncertain system with m crisp inputs $\alpha_1, \alpha_2, \cdots, \alpha_m$, and n crisp outputs $\beta_1, \beta_2, \cdots, \beta_n$. We have the following if-then rules:

If $\mathbb{X}_1$ is ξ_{11} and $\cdots$ and $\mathbb{X}_m$ is ξ_{1m}, then $\mathbb{Y}_1$ is η_{11} and $\mathbb{Y}_2$ is η_{12} and $\cdots$ and $\mathbb{Y}_n$ is η_{1n}
If $\mathbb{X}_1$ is ξ_{21} and $\cdots$ and $\mathbb{X}_m$ is ξ_{2m}, then $\mathbb{Y}_1$ is η_{21} and $\mathbb{Y}_2$ is η_{22} and $\cdots$ and $\mathbb{Y}_n$ is η_{2n}
$\cdots$
If $\mathbb{X}_1$ is ξ_{k1} and $\cdots$ and $\mathbb{X}_m$ is ξ_{km}, then $\mathbb{Y}_1$ is η_{k1} and $\mathbb{Y}_2$ is η_{k2} and $\cdots$ and $\mathbb{Y}_n$ is η_{kn}

Thus, according to Inference Rule 1 and 2, we can infer that $\mathbb{Y}_j (j = 1, 2, \cdots, n)$ are

$$\eta_j^* = \sum_{i=1}^{k} \frac{c_i \cdot \eta_{ij}|_{(a_1 \in \xi_{i1}) \cap (a_2 \in \xi_{i2}) \cap \cdots \cap (a_m \in \xi_{im})}}{c_1 + c_2 + \cdots + c_k},$$

where $c_i = \mathcal{M}\{(a_1 \in \xi_{i1}) \cap (a_2 \in \xi_{i2}) \cap \cdots \cap (a_m \in \xi_{im})\}$ for $i = 1, 2, \cdots, k$. Then, by using the expected value operator, we obtain

$$\beta_j = E\left[\eta_j^*\right]$$

for $j = 1, 2, \cdots, n$. Now, we construct a function from crisp inputs $\alpha_1, \alpha_2, \cdots, \alpha_m$ to crisp outputs $\beta_1, \beta_2, \cdots, \beta_n$, i.e.,

$$(\beta_1, \beta_2, \cdots, \beta_n) = f(\alpha_1, \alpha_2, \cdots, \alpha_m).$$

Then, we get an uncertain system f. For the uncertain system we proposed, we have the following theorem.

Theorem 4. [13] *Assume that $\xi_{i1}, \xi_{i2}, \cdots, \xi_{im}$ and $\eta_{i1}, \eta_{i2}, \cdots, \eta_{in}$ are independent uncertain sets with membership functions $\mu_{i1}, \mu_{i2}, \cdots, \mu_{im}, \nu_{i1}, \nu_{i2}, \cdots, \nu_{in}$, $i = 1, 2, \cdots, k$, respectively. Then, the uncertain system from $\alpha_1, \alpha_2, \cdots, \alpha_m$ to $\beta_1, \beta_2, \cdots, \beta_n$ is*

$$b_j = \sum_{i=1}^{k} \frac{c_i \cdot E[\eta_{ij}^*]}{c_1 + c_2 + \cdots + c_k},$$

where $j = 1, 2, \cdots, n$ and η_{ij}^* are uncertain sets whose membership functions are given by

$$\nu_{ij}^*(y) = \begin{cases} \frac{\nu_{ij}(y)}{c_i}, & \text{if } \nu_{ij}(y) < \frac{c_i}{2} \\ \frac{\nu_{ij}(y)+c_i-1}{\mu(a)}, & \text{if } \nu_{ij}(y) > 1 - \frac{c_i}{2} \\ 0.5, & \text{otherwise} \end{cases}$$

and $c_i = \min_{1 \le l \le m} \mu_{il}(a_l)$ are constants.

Next, we discuss the expected value of a special triangular uncertain set. Without loss of generality, we assume $n = 1$. Then the uncertain system proposed in the above becomes:

$$b = \sum_{i=1}^{k} \frac{c_i \cdot E[\eta_i^*]}{c_1 + c_2 + \cdots + c_k}, \tag{5}$$

$$\nu_i^*(y) = \begin{cases} \frac{\nu_i(y)}{c_i}, & \text{if } \nu_i(y) < \frac{c_i}{2} \\ \frac{\nu_i(y)+c_i-1}{\mu(a)}, & \text{if } \nu_i(y) > 1 - \frac{c_i}{2} \\ 0.5, & \text{otherwise,} \end{cases} \tag{6}$$

$$c_i = \min_{1 \le l \le m} \mu_{il}(a_l). \tag{7}$$

Theorem 5. *Assume we have an uncertain system with m inputs and 1 output consisting of k inference rules. The antecedents of the rules are represented by the uncertain sets ξ_i with membership functions $\mu_{i1}, \mu_{i2}, \cdots, \mu_{im}, i = 1, 2, \cdots, k$. And the consequent is represented by an triangular uncertain set $\eta_i = (\alpha_i, \beta_i, \gamma_i)$ with a membership function ν_i, where* *the coefficients satisfy*

$$\alpha_i + \gamma_i = 2\beta_i, \quad i = 1, 2, \cdots, k. \tag{8}$$

We have

$$E\left[\eta_i^*\right] = \beta_i, \quad i = 1, 2, \cdots, k.$$

Proof. Given the m input data $a_1, a_2, \cdots, a_m$, we can calculate c_i from Equation 7. Then, we can get the membership functions ν_i^* of the consequence uncertain sets η_i^* according to Equation 6. Next, the computation of the expected value of uncertain consequence breaks into three cases.

Case 1: Assume $c_i/2 = 0.5$. We can immediately have $\nu_i^*(y) = \nu_i(y)$, thus

$$E[\eta_i^*] = \frac{\alpha_i + 2\beta_i + \gamma_i}{4} = \beta_i.$$

Case 2: Assume $c_i/2 < 0.5$. Let y_{11}^i and y_{12}^i $(y_{11}^i < y_{12}^i)$ be the two points that satisfy the equation $\nu_i(y) = c_i/2$. Similarly, y_{21}^i and y_{22}^i $(y_{21}^i < y_{22}^i)$ satisfy the equation $\nu_i(y) = 1 - c_i/2$. Since the membership function of a triangular uncertain set has a symmetry property, we have

$$y_{11}^i + y_{12}^i = 2\beta_i, \quad y_{21}^i + y_{22}^i = 2\beta_i. \tag{9}$$

Then, we can rewrite the membership function of η_i as follows:

$$\nu_i^* = \begin{cases} \frac{\nu_i(y)}{c_i}, & \text{if } \alpha_i \le y < y_{11}^i \\ \frac{\nu_i(y)+c_i-1}{c_i}, & \text{if } y_{21}^i \le y < y_{22}^i \\ \frac{\nu_i(y)}{c_i}, & \text{if } y_{12}^i \le y < \gamma_i \\ 0.5, & \text{otherwise.} \end{cases} \tag{10}$$

And $\nu_i^*(\beta_i) = 1$. Together with Equations 3, 8, and 9, we have

$$\begin{aligned} \mathrm{E}[\eta_i^*] &= \beta_i + \frac{1}{2}\left(\int_{\beta_i}^{y_{22}^i} \frac{\nu_i(y) + c_i - 1}{c_i}\mathrm{d}y + \int_{y_{22}^i}^{y_{12}^i} 0.5\mathrm{d}y + \int_{y_{12}^i}^{\gamma_i} \frac{\nu_i(y)}{c_i}\mathrm{d}y\right) \\ &\quad - \frac{1}{2}\left(\int_{y_{21}^i}^{\beta_i} \frac{\nu_i(y) + c_i - 1}{c_i}\mathrm{d}y + \int_{y_{11}^i}^{y_{21}^i} 0.5\mathrm{d}y + \int_{\alpha_i}^{y_{11}^i} \frac{\nu_i(y)}{c_i}\mathrm{d}y\right) \\ &= \beta_i + \frac{1}{2}\left(\int_{\beta_i}^{y_{22}^i} \frac{\nu_i(y) + c_i - 1}{c_i}\mathrm{d}y - \int_{y_{21}^i}^{\beta_i} \frac{\nu_i(y) + c_i - 1}{c_i}\mathrm{d}y\right) \\ &\quad + \frac{1}{2}\left(\int_{y_{12}^i}^{\gamma_i} \frac{\nu_i(y)}{c_i}\mathrm{d}y - \int_{\alpha_i}^{y_{11}^i} \frac{\nu_i(y)}{c_i}\mathrm{d}y\right) \\ &\quad + \frac{1}{2}\left(\int_{y_{22}^i}^{y_{12}^i} 0.5\mathrm{d}y - \int_{y_{11}^i}^{y_{21}^i} 0.5\mathrm{d}y\right) \\ &= \beta_i. \end{aligned}$$

Case 3: Assume $c_i > 0.5$. Similarly, we have $E[\eta_i^*] = \beta_i$. Thus, we have proved the theorem. □

Problem formulation

In this section, we propose an extraction model to obtain uncertain inference rules.

Let $X = (x_1, x_2, \cdots, x_n)$ be the decision vector, which represents a rule base consisting of n rules. Each rule has m antecedents which are described by Q uncertain sets and one consequent which is described by R uncertain sets. Each variable x_i represents a sequence $x_{i1}x_{i2}\cdots x_{im}x_{im+1}$, where $x_{ij} \in \{0, 1, 2, \cdots, Q\}(i = 1, 2, \cdots, n; j = 1, 2, \cdots, m)$ represent the antecedents of the inference rule. And $x_{im+1} \in \{0, 1, 2, \cdots, R\}(i = 1, 2, \cdots, n)$ represent the consequent. Thus, each variable of decision vector represents one inference rule. Some $x_{ij} = 0$ means this antecedent is not included. And some $x_{im+1} = 0$ means this inference rule will not be included in the rule base. For example, assume that we have one inference rule consists of 4 antecedents and 1 consequent. They are described by 5 uncertain sets which refer to five descriptions: very low, low, medium, high, and very high. We use $1, 2, 3, 4, 5$ to denote them. Thus, sequence "23045", for example, represents the rule: "if input 1 is low, input 2 is medium, and input 4 is high, then the output is very high".

Uncertain systems can be used for classification. But which uncertain system is better depends on the rule base. Here, we try to find best rule base by comparing the mean absolute errors of the origin output and the system output. That is,

$$\text{MAE} = \frac{1}{P}\sum_{i=1}^{P} |o_i - t_i|, \tag{11}$$

where P is the number of training data, $o_i, t_i(i = 1, 2, \cdots, P)$ are the system outputs and origin outputs, respectively. If we find the rule base with the least mean absolute error, we extract the uncertain inference rules successfully. We can obtain the system outputs by Equation 5. However, they may not be integers. To avoid this nonsense, for a classification problem with C classes, we can divide interval that covers all the system outputs into C subintervals. Then, if the output from Equation 5 is in the ith subinterval, we have $o_i = i$. Thus, we transfer the classification problem to the following optimization model:

$$\begin{cases} \min\limits_{X} F(X) = \text{MAE} \\ s.t. \\ \quad X = (x_1, x_2, \cdots, x_n) \\ \quad x_i = x_{i1}\cdots x_{im}x_{im+1} \\ \quad x_{ij} \in \{0, 1, \cdots, Q\} \\ \quad x_{im+1} \in \{0, 1, \cdots, R\} \\ \quad i = 1, 2, \cdots, n \\ \quad j = 1, 2, \cdots, m \end{cases}$$

Extraction method for uncertain inference rules with mutations

In this section, we propose the extraction method for uncertain inference rules with mutations by ant colony optimization algorithm.

As stated before, each x_i is a sequence of m values in $\{0, 1, 2, \cdots, Q\}$ and 1 value in $\{0, 1, 2, \cdots, R\}$. Without loss of generality, we set $Q = R$. Each number in $\{0, 1, 2, \cdots, Q\}$ is a node. Let ants walking across these nodes. Ants choose the next node in probability based on the pheromone levels in the $Q + 1$ choices at every step. Once ants move $m + 1$ steps, a candidate decision variable is generated. After repeat this process n times, we get a candidate solution. After all ants finish their walk, update the pheromone trails. Denote the pheromone trail by $\tau_{i;k,j}(t)$ associated to the node j at step k of x_i in iteration t. The procedures are described as follows.

(1) ***Initialization***: Randomly generate a feasible solution X_0, and set the optimal solution $\hat{X} = X_0$. Set $\tau_{i;k,j}(0) = \tau_0$, $i = 1, 2, \cdots, n$, $k = 1, 2, \cdots, m+1$, $j = 0, 1, 2, \cdots, Q$, where τ_0 is a fixed parameter.

(2) ***Ant movement***: At each step k after building the sequence $x_{i1}x_{i2}\cdots x_{ik}$, select the next node in probability following

$$p_{k;k+1} = \frac{\tau_{i;k+1,j}(t)}{\sum_{q=0}^{Q} \tau_{i;k+1,q}(t)}. \tag{12}$$

In this way, we could get a sequence $x_{i1}x_{i2}\cdots x_{im+1}$. To speed up the algorithm, we mutate this sequence to get a new candidate sequence. The mutation is made as follows: randomly add 1 or subtract 1 to each element x_{ij} in the sequence; if the element is 0, the mutated element is 1; if the element is Q, the mutated element is $Q-1$. Assume X' is the mutated solution, if $\Delta F = F(X') - F(X) \leq 0$, then $X \leftarrow X'$; otherwise, keep the current solution. If Q is very large, we could repeat this mutation until some termination condition is satisfied.

(3) ***Pheromone Update***: At each iteration t, let $\hat{X}$ be the optimal solution found so far and X_t be the best feasible solution in the current iteration. Assume $F(\hat{X})$ and $F(X_t)$ are the corresponding objective function values.

If $F(X_t) < F(\hat{X})$, then $\hat{X} \leftarrow X_t$.

Reinforce the pheromone trails on nodes of $\hat{X}$ and evaporate the pheromone trails on the left nodes:

$$\tau_{i;j,k}(t) = \begin{cases} (1-\rho)\tau_{i;j,k}(t-1) + \rho g(\hat{X}), & \text{if } (k,j) \in \hat{X} \\ (1-\rho)\tau_{i;j,k}(t-1), & \text{otherwise} \end{cases} \tag{13}$$

where ρ $(0 < \rho < 1)$ is the evaporation rate, $g(x)(0 < g(x) < +\infty)$ is a function with that $g(x) \geq g(y)$ if $F(x) < F(y)$, for example, $g(x) = L/(|F(x)|+1)$ is a function satisfying the condition where $L > 0$.

Let τ_0 be the initial value of pheromone trails, n be the number of decision variables, M be the number of ants, ρ be evaporation rate and T be the number of iterations. Now, we summarize the algorithm as follows.

Step 1 Initialize all pheromone trails with the same pheromone level τ_0. Randomly generate a feasible solution X_0, and set optimal solution $\hat{X} = X_0$. Set $l \leftarrow 0$.

Step 2 Ant movement in probability following Equation 12. Generate a decision variable x_i after $m+1$ steps.

Step 3 Repeat Step 2 until $X = (x_1, x_2, \cdots, x_n)$ is generated; mutate every x_i: thus, generate a new decision vector $X' = (x_1', x_2', \cdots, x_n')$; if $\Delta F = F(X') - F(X) \leq 0$, then $X \leftarrow X'$.

Step 4 Repeat Step 2 and Step 3 for all M ants.

Step 5 Calculate the system outputs by Equation 5. Then, calculate the objective function values for the M candidate solutions by Equation 11. Denote the best solution in this iteration by X_l.

Step 6 If $F(X_l) < F(\hat{X})$, then $\hat{X} \leftarrow X_l$; update the pheromone trails according to Equation 13.

Step 7 $l \leftarrow l+1$; if $l = T$, terminate; otherwise, go to Step 2.

Step 8 Report the optimal solution $\hat{X}$.

With this algorithm above, we obtain an uncertain rule base. Then, we successfully design an uncertain system and can use it for classification.

Extraction method for uncertain inference rules with SA

In the previous section, to speed up the algorithm, we introduce a mutation operation. Here, we introduce the simulated annealing algorithm as the local search operation.

Simulated annealing algorithm was initiated by Metropolis in 1953, applied to portfolio optimization by Kirkpatrick [25] in 1983. The name and inspiration come from annealing in metallurgy, a technique involving heating and controlled cooling of a material to increase the size of its crystals and reduce their defects. Simulated annealing algorithm is excellent at avoiding getting stuck in local optimums. It has a good robust property and is universal and easy to implement.

For optimization problem (1), we can use simulated annealing algorithm to search for the optimal solution. The algorithm is as follows.

Step 1 Randomly generate a initial solution x_0; $x \leftarrow x_0$; $k \leftarrow 0$; $t_0 \leftarrow t_{max}$(initial temperature);

Step 2 If the temperature satisfies the inner cycle termination criterion, go to Step 3; otherwise, randomly choose a point $x^{'}$ in the neighborhood $N(x)$, calculate $\Delta f = f(x^{'}) - f(x)$. If $\Delta f \leq 0$, then $x \leftarrow x^{'}$; otherwise, according to Metropolis acceptance criterion, if $\exp(-\Delta f/t_k) > random(0, 1)$, then $x \leftarrow x^{'}$. Repeat Step 2.

Step 3 $t_{k+1} = d(t_k)$ (temperature decrease); $k \leftarrow k + 1$; if the termination criterion is satisfied, stop and report the optimal solution; otherwise, go to Step 2.

In this section, we combine ant colony optimization algorithm and simulated annealing algorithm. In each iteration of ant colony optimization algorithm, we get a feasible solution. Then, we use it as the initial solution of the simulated annealing algorithm to get a neighbor solution. This neighbor solution will be accepted in probability. And for each decision vector $X = (x_1, x_2, \cdots, x_n)$, $x_i = x_{i1}x_{i2}\cdots x_{im+1}$, we build the neighbor solution as follows: for each x_i, for some randomly generated p and q $(1 \leq p < q \leq m)$, reverse the order of the sequence $x_{ip}\cdots x_{iq}$, i.e., $x_i^{'} = x_{i1}\cdots x_{ip-1}x_{iq}x_{iq-1}\cdots x_{ip+1}x_{ip}x_{iq+1}\cdots x_{im+1}$. For example, assume x_i is 0123456, $p = 2$, $q = 6$, and the neighbor solution $x_i^{'}$ is 0543216. In this way, we obtain a neighbor solution $X^{'}$. If $\Delta F = F(X^{'}) - F(X) \leq 0$, $X \leftarrow X^{'}$; otherwise, if $\exp(-\Delta F/t_k) > random(0, 1)$, then $X \leftarrow X^{'}$; otherwise, abandon this neighbor solution. Still denote the pheromone trail by $\tau_{i;k,j}(t)$. The procedure are described as follows.

(1) ***Initialization***: Generate a feasible solution X_0 randomly and set the optimal solution $\hat{X} = X_0$. Set $\tau_{i;k,j}(0) = \tau_0$, $i = 1, 2, \cdots, n$, $k = 1, 2, \cdots, m+1$, $j = 0, 1, 2, \cdots, Q$, where τ_0 is a fixed parameter.

(2) ***Ant movement***: At each step k after building the sequence $x_{i1}x_{i2}\cdots x_{ik}$, select the next node in probability following Equation 12. In this way, we could get a sequence $x_{i1}x_{i2}\cdots x_{im+1}$. In order to expand the search range, we use simulated annealing algorithm to search locally around the solution at this step. Assume the neighbor solution is $X^{'}$. If $\Delta F = F(X^{'}) - F(X) \leq 0$, $X \leftarrow X^{'}$; otherwise, if $\exp(-\Delta F/t_k) > random(0, 1)$ where t_k is the current temperature and $t_k \to 0$ when $k \to \infty$, then $X \leftarrow X^{'}$; otherwise, abandon this neighbor solution and still choose the original feasible solution.

(3) ***Pheromone Update***: Let $\hat{X}$ be the optimal solution found so far and X_t be the best feasible solution in the current iteration t. Assume $F(\hat{X})$ and $F(X_t)$ are the corresponding objective function values. To avoid the optimal solution $\hat{X}$ getting stuck in local optimums, we also use acceptance function here.

If $F(X_t) < F(\hat{X})$, then $\hat{X} \leftarrow X_t$.

Build a neighbor solution $\hat{X}'$.

If $F(\hat{X}') \leq F(\hat{X})$, then $\hat{X} \leftarrow \hat{X}'$;

If $F(\hat{X}') > F(\hat{X})$, check the Metropolis acceptance criterion, i.e., if $\exp(-\Delta\hat{F}/T_t) > random(0,1)$, $T_t \to 0, t \to \infty$, then $X^* \leftarrow \hat{X}'$.

Reinforce the pheromone trails on the nodes of $\hat{X}$ and X^* and evaporate the pheromone trails on the left nodes:

$$\tau_{i;j,k}(t) = \begin{cases} (1-\rho)\tau_{i;j,k}(t-1) + \rho g(\hat{X}), & \text{if } (k,j) \in \hat{X} \\ (1-\rho)\tau_{i;j,k}(t-1) + \frac{\rho}{2} g(\hat{X}), & \text{if } (k,j) \in X^* \\ (1-\rho)\tau_{i;j,k}(t-1), & \text{otherwise} \end{cases} \tag{14}$$

where, ρ $(0 < \rho < 1)$ is the evaporate rate, and $g(x)$ $(0 < g(x) < +\infty)$ is a function with that $g(x) \geq g(y)$ if $F(x) < F(y)$. For example, $g(x) = L/(|F(x)|+1)$ is an available function if $L > 0$.

Now, we summarize the algorithm as follows.

Step 1 Initialize all pheromone trails with the same pheromone level τ_0. Randomly generate a feasible solution X_0, and set optimal solution $\hat{X} = X_0$. Set $t \leftarrow 0$.

Step 2 Ant movement in probability following Equation 12. Generate a decision variable x_i after $m+1$ steps.

Step 3 Repeat Step 2 until decision vector $X = (x_1, x_2, \cdots, x_n)$ is generated. Build the neighbor solution X'. If $\Delta F = F(X') - F(X) \leq 0, X \leftarrow X'$; otherwise, if $\exp(-\Delta F/t_k) > random(0,1)$ where t_k is the current temperature and $t_k \to 0$ when $k \to \infty$, then $X \leftarrow X'$.

Step 4 Repeat Step 2 and Step 3 until all ants finish their walk, and generate M candidate solutions.

Step 5 Calculate the system outputs by Equation 5. Then, calculate the objective function values for the M candidate solutions by Equation 11. Denote the best solution in this iteration by X_t.

Step 6 If $F(X_t) < F(\hat{X})$, then $\hat{X} \leftarrow X_t$. Build the neighbor solution of $\hat{X}$, which is denoted by $\hat{X}'$. If $\Delta\hat{F} = F(\hat{X}') - F(\hat{X}) \leq 0$, then $\hat{X} \leftarrow \hat{X}'$; otherwise, if Metropolis acceptance criterion is satisfied, i.e., if $\exp(-\Delta\hat{F}/T_t) > random(0,1)$, $T_t \to 0, t \to \infty$, then $X^* \leftarrow \hat{X}'$.

Step 7 Update the pheromone trails according to Equation 14.

Step 8 $t \leftarrow t+1$; if $t = \mathrm{T}$, terminate; otherwise, go to Step 2.

Step 9 Report the optimal solution $\hat{X}$.

Table 1 Parameters

	a_p	b_p	c_p
$p = 1$	0.5	1.01	1.52
$p = 2$	1.7	2.74	4.48
$p = 3$	5	6.07	7.14

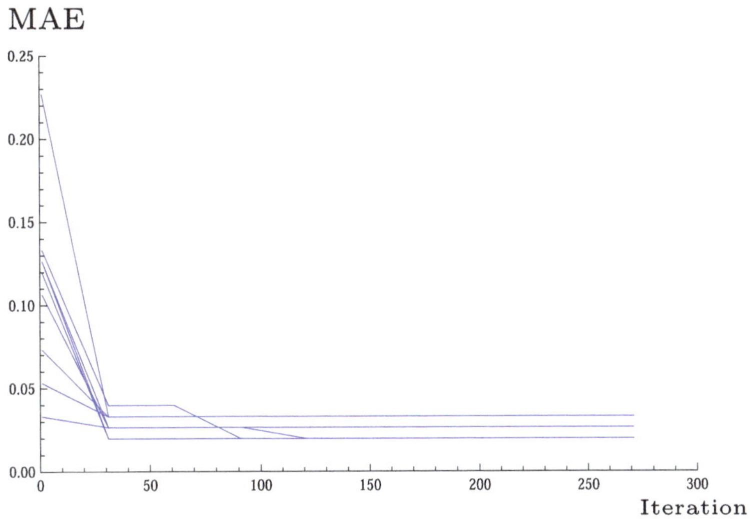

Figure 1 Results of method A.

Experiments

In this section, we use our two extraction methods to extract uncertain inference rules. And then use the uncertain systems to solve some classification problems. We applied our methods to the IRIS [26] classification problem and the Wisconsin Breast Cancer (WBC) [27] classification problem.

IRIS classification

IRIS data set is the typical date set in data classification. It contains 150 instances of 3 classes, which are Setosa, Versicolor, and Virginica. Each class has 50 instances. Each instance has 4 attributes which are sepal length (SL), sepal width (SW), petal length (PL),

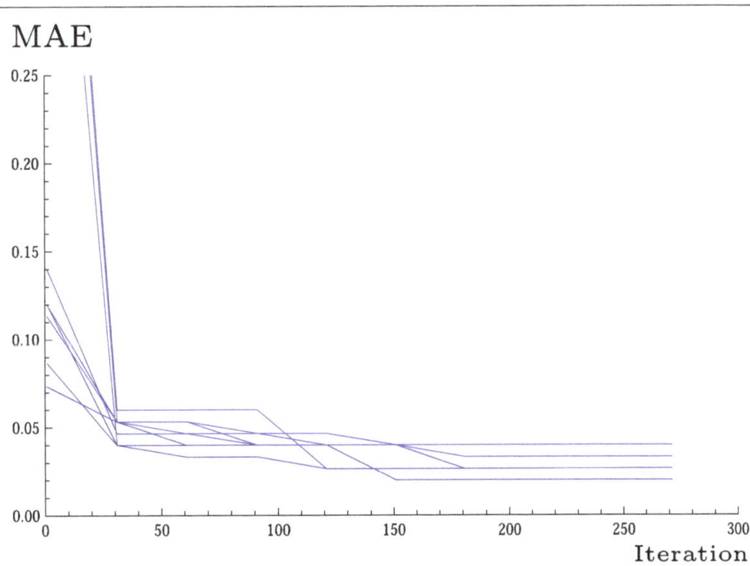

Figure 2 Results of method B.

Table 2 Accuracy comparison

Method	Paper	Accuracy rate (%)
Method A	This paper	97.33
Method B	This paper	97.5
C4.5	[2]	94.0
ACOA	[5]	96.6
MACO	[6]	95.53
HNFQ	[28]	98.67

Table 3 IRIS classification rules extracted by method A

IF				THEN
SL	SW	PL	PW	Class
1	3	1	3	1
1	0	1	1	1
1	2	3	2	1
1	1	2	1	2
2	1	0	3	2
3	2	0	2	3
1	1	3	3	3

Table 4 IRIS classification rules extracted by method B

IF				THEN
SL	SW	PL	PW	Class
3	2	3	1	2
1	1	0	0	2
0	2	1	1	3
0	1	1	3	1
1	1	3	3	2
1	1	3	1	1
2	1	1	2	1

Table 5 Parameters

	a_p	b_p	c_p
$p = 1$	0.3	1.01	1.72
$p = 2$	2	6.07	10.14

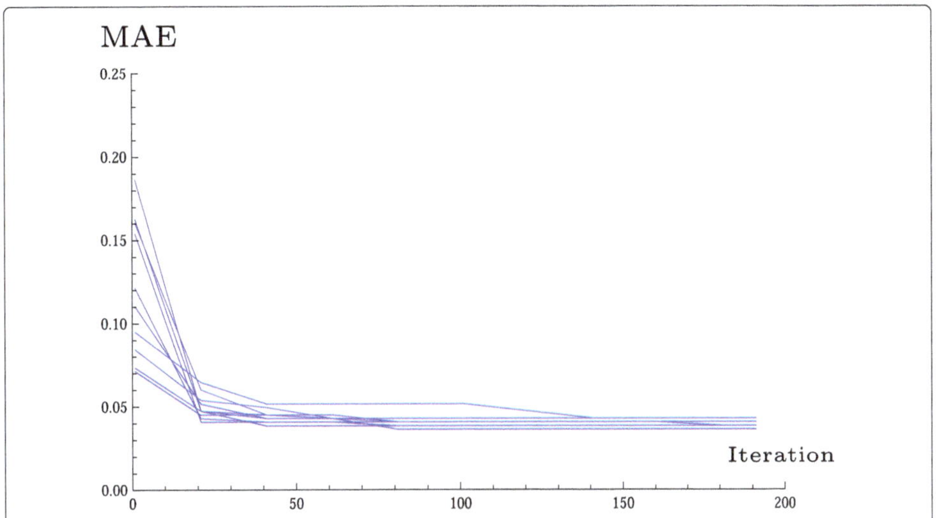

Figure 3 Results of method A.

and petal width (PW). They are described by 3 uncertain sets: low (1), medium (2), and high (3). The membership functions are

$$\mu_q(x) = \exp\left(-\frac{(x - V_q)^2}{2\beta^2}\right),$$

where x is the input, $\beta = 0.618$ and $V_q = \frac{q-1}{2}$, $q = 1, 2, 3$. Based on these 4 attributes, we try to infer which class does the instance belong to. We use 3 triangular uncertain sets $\eta_p = (a_p, b_p, c_p)$ $(p = 1, 2, 3)$ to describe the possible classes (class 1: Setosa; class 2: Versicolor; class 3: Virginica). And the parameters $a_p, b_p, c_p \in \mathcal{R}$ are listed in Table 1.

First, we normalize the data to $[0, 1]$ to simplify the computation. IRIS data set is our training set while it is also used for testing. Then, we set maximum number of rules $n = 10$, number of ants $M = 10$, evaporate rate $\rho = 0.3$, and number of iterations $T = 300$. Each algorithm runs ten times. The results are in Figures 1 and 2. Denote the extraction

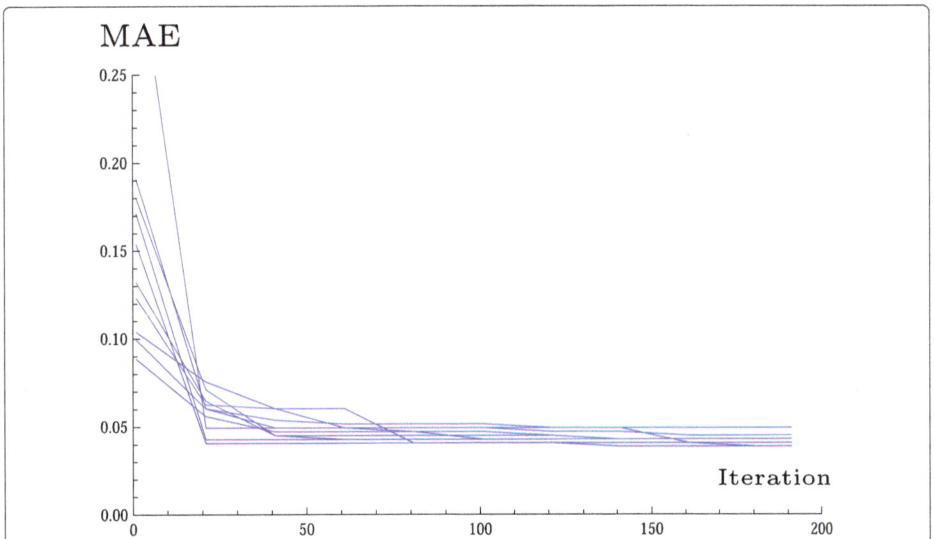

Figure 4 Results of method B.

Table 6 Accuracy rate comparison

Method	Paper	Accuracy rate (%)
Method A	This paper	98.3
Method B	This paper	98.33
C4.5	[2]	94.25
ACOA	[5]	97.91
MACO	[6]	97.07
FMM	[29]	97.86

method with mutation by A and the method with SA by B. It can be seen that the method A converges fast at about 120th iteration. And method B converges a little slower at about 150th iteration.

Then, we can classify the IRIS data with the uncertain systems we introduced. We find the average accuracy rates of the two methods are 97.33% and 97.5%, respectively. Comparison with other methods are listed in Table 2.

List the rule bases we get with the highest accuracy rates (98.0% and 98.67%, respectively) in Tables 3 and 4. Note that although the maximum number of rules is 10, the final rule bases we obtain has only 7 rules.

Wisconsin Breast Cancer classification

Wisconsin Breast Cancer data set is a common medical date set. It contains 699 instances of 2 classes, which are sick and healthy. Two hundred forty-one instances are sick and 458 instances are healthy. Each instances has 9 attributes, which are clump thickness (CT), uniformity of cell size (UCS), uniformity of cell shape (UCCS), marginal adhesion (MA), single epithelial cell size (SPCS), bare nuclei (BN), bland chromatin (BC), normal nucleoli (NN), and mitoses (MT). They are described by 5 uncertain sets: very low (1), low (2), medium (3), high (4), and very high (5). The membership functions are

$$\mu_q(x) = \exp\left(-\frac{(x - V_q)^2}{2\beta^2}\right),$$

where x is the input, $\beta = 0.4247$, and $V_q = \frac{q-1}{2}$, $q = 1, 2, 3, 4, 5$. Based on these attributes, we diagnose whether one instance is sick or not. We use 2 triangular uncertain sets $\eta_p = (a_p, b_p, c_p)$ $(p = 1, 2)$ to describe the possible classes (sick and healthy). And the parameters $a_p, b_p, c_p \in \mathcal{R}$ are listed in Table 5.

Table 7 WBC classification rules extracted by method A

IF									THEN
CT	UCS	UCCS	MA	SPCS	BN	BC	NN	MT	Class
1	5	0	3	2	0	4	2	1	2
1	2	1	1	4	4	2	1	0	1
1	3	5	2	3	2	1	1	4	2
1	3	4	2	3	1	2	2	1	1
2	4	4	1	1	2	4	5	1	2
3	3	4	5	4	3	2	4	4	2
5	2	4	0	3	0	0	2	1	1
2	4	4	1	1	2	4	5	1	1
2	4	2	3	5	3	2	5	5	2

Table 8 WBC classification rules extracted by method B

IF									THEN
CT	UCS	UCCS	MA	SPCS	BN	BC	NN	MT	Class
5	3	3	3	3	2	3	2	4	2
0	0	0	0	0	0	0	4	0	1
4	4	4	4	0	1	1	1	4	2
1	4	4	1	1	2	1	5	1	2
1	1	1	2	0	3	5	5	5	2
0	3	0	0	0	0	0	1	2	1

First, we normalize the data to [0, 1] to simplify the computation. The first 460 instances are used for training while the left 239 instances are used for testing. Then, we set maximum number of rules $n = 10$, number of ants $M = 20$, evaporate rate $\rho = 0.3$, and number of iterations $T = 200$. Each algorithm runs ten times. The results are in Figures 3 and 4. We still find that method A converges faster than method B. Method A stabilizes at about 50th iteration while method B stabilizes until about 80th iteration.

Then, we test the uncertain systems we get with the later 239 instances. We find the average accuracy rates of the two methods on the training set are 96.0% and 96.26%, respectively. Using the uncertain system with the highest accuracy rate of each method on the test set, we find the accuracy rates are 98.37% and 98.33%. Comparison with other methods are listed in Table 6.

The rule base with the highest accuracy rates (98.37% and 98.33%, respectively) on the test set are listed in Tables 7 and 8. Method A gives us a rule base of 9 rules, and method B provides a rule base of 6 rules.

We apply our two extraction methods to the classification problems of IRIS data set and WBC data set. Compare our results with other researchers' work, we can find that both methods have higher accuracy rate than ACOA and MACO in two classification problems. And for IRIS data set, accuracy rates of method A and B are lower than HNFQ but higher than C4.5. For WBC data set, their accuracy rates are higher than C4.5 and FMM.

Conclusions

In this paper, we designed an uncertain system for data classification. And we proposed two extraction methods for uncertain inference rules by using ant colony optimization algorithm. Then, we applied our methods to IRIS classification problem and WBC classification problem. Our methods are shown to be superior in accuracy to some existing methods.

Acknowledgements
This work is supported by the National Natural Science Foundation of China (No.61273009).

References

1. Kantardzic, M: Data Mining: Concepts, Models, Methods, and Algorithms. 2nd ed. Wiley, Hoboken (2011)
2. Quinlan, JR: Improved use of continuous attributes in C4.5. J. Artif. Intell. Res. **4**(1), 77–90 (1996)
3. Parpinelli, RS, Lopes, HS, Freitas, AA: Data mining with an ant colony optimization algorithm. IEEE Trans. Evolut. Comput. **6**(4), 321–332 (2002)
4. Casillas, J, Cordón, O, Herrera, F: Learning fuzzy rules using ant colony optimization algorithms. In: Proceedings of the 2nd International Workshop on Ant Algorithms: From Ant Colonies to Artificial Ants, pp. 13–21, Brussels, (2000)

5. Zhu, Y: Ant colony optimization-based hybrid intelligent algorithms. World J. Modell. Simul. **2**(5), 283–289 (2006)
6. Zhu, Y: An intelligent algorithm: MACO for continuous optimization models. J. Intell. Fuzzy Syst. **24**, 31–36 (2013)
7. Lee, Z, Su, S, Chuang, C, Liu, K: Genetic algorithm with ant colony optimization (GA-ACO) for multiple sequence alignment. Appl. Soft Comput. **8**(1), 55–78 (2008)
8. Shelokar, PS, Siarry, P, Jayaraman, VK, Kulkarni, BD: Particle swarm and ant colony algorithms hybridized for improved continuous optimization. Appl. Math. Comput. **188**(1), 129–142 (2007)
9. Liu, B: Uncertainty Theory. 2nd ed. Springer, Berlin (2007)
10. Liu, B: Fuzzy process, hybrid process and uncertain process. J. Uncertain Syst. **2**(1), 3–16 (2008)
11. Chen, X, Liu, B: Existence and uniqueness theorem for uncertain differential equations. Fuzzy Optimization Decis. Mak. **9**(1), 69–81 (2010)
12. Liu, B: Some research problems in uncertainty theory. J. Uncertain Syst. **3**(1), 3–10 (2009)
13. Liu, B: Uncertainty Theory: A Branch of Mathematics for Modeling Human Uncertainty. Springer, Berlin (2010)
14. Peng, J, Yao, K: A new option pricing model for stocks in uncertainty markets. Int. J. Oper. Res. **8**(2), 18–26 (2011)
15. Zhu, Y: Uncertain optimal control with application to a portfolio selection model. Cybern. Syst. **41**(7), 535–547 (2010)
16. Liu, B: Uncertain set theory and uncertain inference rule with application to uncertain control. J. Uncertain Syst. **4**(2), 83–98 (2010)
17. Liu, B: Uncertain logic for modeling human language. J. Uncertain Syst. **5**(1), 3–20 (2011)
18. Gao, X, Gao, Y, Ralescu, DA: On Liu's inference rule for uncertain systems. Int. J. Uncertain. Fuzz. Knowledged-Based Syst. **18**(1), 1–11 (2010)
19. Peng, Z, Chen, X: Uncertain systems are universal approximators. J. Uncertainty Anal. Appl. 2, Article, 13 (2014)
20. Gao, Y: Uncertain inference control for balancing inverted pendulum. Fuzzy Optimization Decis. Mak. **11**(4), 481–492 (2012)
21. Liu, B: Membership functions and operational law of uncertain sets. Fuzzy Optimization Decis. Mak. **11**(4), 387–410 (2012)
22. Zadeh, LA: Outline of a new approach to the analysis of complex systems and decision processes. IEEE Trans. Syst. Man Cybern. **3**(1), 28–44 (1973)
23. Mamdani, EH: Applications of fuzzy algorithms for control of a simple dynamic plant. Proc. Institution Electr. Eng. Control Sci. **121**(12), 1585–1588 (1974)
24. Takagi, K, Sugeno, M: Fuzzy identification of system and its applications to modeling and control. IEEE Trans. Syst. Man Cybern. **15**(1), 116–132 (1985)
25. Kirkpatrick, S, Gelatt, CD, Vecchi, MP: Optimization by simmulated annealing. Science. **220**(4598), 671–680 (1983)
26. Iris dataset (1936). https://archive.ics.uci.edu/ml/datasets/Iris
27. Wisconsin Breast Cancer Dataset (1992). https://archive.ics.uci.edu/ml/datasets/Breast+Cancer+Wisconsin+(Original)
28. de Souza, FJ, Vellasco, M, Pacheco MA: Hierarchical neuro-fuzzy quadtree models. Fuzzy Sets Syst. **130**(2), 189–205 (2002)
29. Gabrys, B, Bargiela, A: General fuzzy min-max neural network for clustering and classification. IEEE Trans. Neural Networ. **11**(3), 769–783 (2000)

19

Optimal consumer search with prospect utility in hybrid uncertain environment

Chi Zhou, Wansheng Tang* and Ruiqing Zhao

*Correspondence: tang@tju.edu.cn
Institute of Systems Engineering, Tianjin University, Tianjin 300072, China

Abstract

This paper studies a consumer search model with prospect utility in a hybrid uncertain environment. Hybrid uncertainty consists of the uncertainty of the consumer's valuation for each product and the randomness of stockout. An optimal search strategy is designed for the consumer to search or buy a product. The results show that the presence of prospect utility can result in low purchase threshold. By comparative statics, we demonstrate that the consumer's purchase threshold for low-value product increases with the risk coefficients for gains and losses, while the purchase threshold for high-value product decreases with the risk coefficients for gains and losses. In addition, we find that the consumer who is more sensitive to the loss will search for much less products. Finally, some numerical examples are given to illustrate the relationships between the model without and with prospect utility and further verify the effectiveness and credibility of the conclusions.

Keywords: Consumer search; Prospect theory; Loss aversion; Optimal stopping; Uncertain random variable

Introduction

Reference points and the different perceptions of gains and losses play an important role in actual consumer economics [1]. In a commodity market, consumers' purchase decisions depend not only on their valuations for the current product but also on their reference valuations for the past observed product. Kahneman and Tversky [2] and Tversky and Kahneman [3,4] suggest that the perception of potential losses may be much more important in determining behavior than the perceptions of potential gains. This is known as prospect utility which is fear of loss rather than hope of gain that limits consumer's purchase behavior. These characteristics are often ignored in dynamic consumer search which do not take into account prospect utility in searching differentiated but substitutable products.

Generally speaking, there is uncertain information in the existing patterns of consumer search models. The number of differentiated but substitutable products is assumed to be finite in the market. Probability distributions of the products' value are usually assumed to be known to consumers before searching, but consumers do not know which firm will offer which product value. In the real world, some information, such as quality, pattern, and size of one product, cannot be exactly predicted by consumers in advance, thus resulting in the unknown frequencies. Therefore, probability theory is no longer applicable to characterize such kind of uncertain information, because the lack of the frequency leads to the unknowable probability distribution. Since the valuation of the consumer to each product is subjective and the products are heterogeneous, the valuations can be characterized as independent but not identically distributed uncertain variables [5]. Moreover, products are randomly out of stock at a point in time. This is a realistic assumption. For example, a study of more than 71,000 consumers in 29 countries shows that they have little patience for stockouts.Once facing with a stockout, 21% to 43% of consumers will actually go to another store to buy a substitute product[a]. The valuation is an uncertain variable and the availability of previous product is random; thus, the maximum valuation between the current product and the best previous product can be characterized as an uncertain random variable [6]. Hence, the so-called twofold uncertainty is involved in this paper which consists of the uncertain product value and the random stockout.

This paper provides a simple explanation for how prospect utility affects consumers' purchase behaviors in a hybrid uncertain environment. We study a consumer search model with prospect utility in which the consumer searches sequentially and evaluates each product on the basis of multiple attributes. The multiple attributes of a product include the price, quality, color, pattern, material, and size of a product [7]. In the search process, the consumer exhibits loss aversion and decides when to stop searching for new substitute products by buying one of the products which have been encountered so far. An optimal stopping rule and an optimal search strategy are presented to maximize the benefit of the consumer. To discuss the effects of prospect utility on purchase threshold, we make a comparison between the model without and with prospect utility. By comparative statics, we further investigate the effects of risk coefficients for gains and losses, coefficient of loss aversion on the consumer's purchase threshold.

As a result of intercomparison between the model without and with prospect utility, we find that the purchase decisions of a consumer under prospect utility exhibit low purchase levels. Moreover, we show that the consumer's purchase threshold is decreasing with the search cost and is increasing with the probability of stockout. The product that has a good reputation will be preferred in the optimal search strategy. Our analysis also reveals some intriguing comparative statics. In our model, the consumer's purchase threshold for low-value product[b] monotonically increases as the risk coefficients for gains and losses become larger; the purchase threshold for high-value product decreases as the risk coefficients for gains and losses increase, holding other parameters of the model constant. Our findings also offer a theoretical explanation of the experimental evidence that consumers generally stop searching too early [8]. We demonstrate the consumer's purchase threshold is decreasing with the coefficient of loss aversion. This implies that if loss aversion is larger, the consumer will search for much less products because of low purchase threshold. Finally, our numerical analysis shows that the reservation value in the model with prospect utility will be lower than that in the model without prospect utility in a wider number range. These results further illustrate the feasibility of the comparative statics.

The remainder of this article is organized as follows: In the 'Literature review' section, we present our literature review. The 'Preliminaries' section recalls some fundamental concepts and formulas about uncertainty theory and chance theory. 'The model without prospect utility' section describes the consumer search model without prospect utility. 'The model without prospect utility' section studies the same model with prospect utility. We present an optimal search strategy and investigate the relationship between the reservation values without and with prospect utility. The 'Comparative statics' section discusses the effects of prospect utility's parameters on the reservation value. The numerical examples are provided in the 'Numerical examples' section, and we conclude in the 'Conclusion' section.

Literature review

The consumer search problem has attracted wide attention in the academic literature. Stahl [9] discussed a simple consumer search problem in which consumers search sequentially with perfect recall and a finite number of identical stores compete by choosing prices for a homogeneous good. Cachon et al. [10] studied a model in which consumers search among multiple competing firms for products. Branco et al. [7] established a continuous-time search model of gradual learning, in which consumers incur search costs to learn product information and update their expected utility of the product at each search occasion. Arbatskaya [11] presented an ordered search model that yields an equilibrium in which there is price dispersion, prices and profits decline in the order of search, and consumers with lower search costs search longer and obtain better deals. Based on the ordered search model, Wilson [12] studied a problem that consumers are allowed to choose the optimal order in which to search firms and firms are able to obfuscate this order by deliberately increasing search costs and prices. Zhou [13] improved the ordered search model in which consumers search both for price and product fitness. Chen and Zhang [14] considered a consumer search model in a homogeneous product market where uninformed consumers search for price information. Haan and Moraga-González [15] studied the model that when consumers search for products, they first visit the firm whose advertising is most salient. Chen and He [16] studied a model of product differentiation in which the auction of advertisement positions is embedded in a market game of consumer search.

Recent research trends in consumer search investigate the issue of boundedly rational consumer behavior. For example, Hardie et al. [17] developed a multinomial logit formulation of a reference-dependent choice model. They assumed that consumer choice is influenced by the position of brands relative to multiattribute reference points and loss aversion is exhibited. Weng [18] considered the sequential search problem where sellers search for the best price from a known distribution and exhibit asymmetric regretting and rejoicing. Zhou [19] considered a consumer search model and examined the implications of consumer reference dependence for market competition. He also finds that consumer loss aversion in the price dimension intensifies competition while that in the product dimension softens competition. Kuksov and Wang [20] also examined how the competitive strategy and profitability of firms are affected by the presence of consumer loss aversion in the price dimension. The above literature has neither studied any consumer's search strategy in conjunction with nonlinear prospect utility nor discussed the probabilistic stockout of products.

The prospect theory originates in the work of Kahneman and Tversky [2]. The most prominent feature of prospect theory is loss aversion, that is, people care much more about losses relative to their reference level than about gains. It (in particular, loss aversion) has been widely applied to management and operations fields. For example, Bowman et al. [21] studied a consumption-savings model based on loss aversion that implies a fundamental asymmetry in consumption behavior. Popescu and Wu [22] considered the dynamic pricing problem of a monopolist firm in a market with repeated interactions. They presented purchase decisions by assessing prices as discounts or surcharges relative to the reference price in the spirit of prospect theory. Foellmi et al. [23] studied the Cass-Koopmans-Ramsey model of optimal economic growth in the presence of loss aversion and habit formation. They showed that prospect utility can cause the economy to remain in a steady state with low consumption and low capital.

The existing literature depicted the uncertain information in consumer search problems by random variable. However, due to the lack of historical data and the impact of subjective factor, describing the uncertainty as randomness is not adequately reasonable. Uncertain measure is an alternative tool to characterize such uncertain information that has no any historical data and cannot be exactly predicted in advance. Liu [5] gave the basic concepts about uncertainty theory, such as uncertain variable, uncertainty distribution, expected value, and variance. From then on, uncertainty theory attracts considerable attention among researchers in many fields as an important role of mathematical model to deal with human uncertainty. For instance, Mu [24] studied the implementation of the employment relationship problem between the enterprise and the rural migrant worker, where enterprise's assessment on the rural migrant worker's own income at home is characterized as an uncertain variable. Lan and Du [25] considered a warranty contracting problem faced by a buyer who purchases products from a supplier and then sells them to consumers under uncertain demand. Zhou et al. [26] studied a dynamic recruitment problem with enterprise performance in the uncertain environment and presented an optimal search strategy for the firm's employment decisions. Wang et al. [27] considered a wage contract design problem faced by an employer who employs an employee to work in labor market and characterized the employee's ability as an uncertain variable. Wu et al. [28] studied the agency problem with multiple uncertain information. They introduced confidence level to quantify the degree of the agent's risk aversion and examined the effects of the agent's attitude toward risk on the optimal contract and the principal's income.

As an important extension of uncertainty theory, chance theory was pioneered by Liu [6] for modeling complex systems with not only uncertainty but also randomness. And then, many researchers proposed uncertain random programming [29], uncertain random process [30], uncertain random graph, and uncertain random network [31]. In addition, Sheng and Yao [32] provided some formulas to calculate the variance of uncertain random variables through chance distribution and inverse chance distribution. Zhou et al. [33] proposed uncertain random multi-objective programming for optimizing multiple, non-commensurable, and conflicting objectives.

Preliminaries

This section will introduce some basic results in uncertainty theory and chance theory. The former is a branch of axiomatic mathematics for modeling belief degrees, and the latter is a methodology for modeling complex systems with not only uncertainty but also randomness.

Uncertainty theory

Let Γ be a nonempty set and $\mathcal{L}$ a σ-algebra over Γ. Each element Λ in $\mathcal{L}$ is called an event. Liu [5] defined an uncertain measure by the following axioms:

Axiom 1. (*Normality axiom*) $\mathcal{M}\{\Gamma\} = 1$ *for the universal set* Γ.
Axiom 2. (*Duality axiom*) $\mathcal{M}\{\Lambda\} + \mathcal{M}\{\Lambda^c\} = 1$, $\forall \Lambda \in \mathcal{L}$.
Axiom 3. (*Subadditivity axiom*) $\mathcal{M}\left\{\bigcup_{i=1}^{\infty} \Lambda_i\right\} \leq \sum_{i=1}^{\infty} \mathcal{M}\{\Lambda_i\}$, $\forall \Lambda_i \in \mathcal{L}$, $i = 1, 2, \ldots$

The triplet $(\Gamma, \mathcal{L}, \mathcal{M})$ is called an uncertainty space. Furthermore, Liu [34] defined a product uncertain measure by the fourth axiom:

Axiom 4. (*Product axiom*) *Let* $(\Gamma_k, \mathcal{L}_k, \mathcal{M}_k)$ *be uncertainty spaces for* $k = 1, 2, \ldots$. *Then, the product uncertain measure* $\mathcal{M}$ *is an uncertain measure satisfying:*

$$\mathcal{M}\left\{\prod_{k=1}^{\infty} \Lambda_k\right\} = \bigwedge_{k=1}^{\infty} \mathcal{M}_k\{\Lambda_k\},$$

where Λ_k *are arbitrarily chosen events from* $\mathcal{L}_k$ *for* $k = 1, 2, \ldots$, *respectively.*

An uncertain variable was defined by Liu [5] as a function ξ from an uncertainty space $(\Gamma, \mathcal{L}, \mathcal{M})$ to the set of real numbers such that $\{\xi \in B\} = \{\nu \in \Gamma \,|\, \xi(\nu) \in B\}$ is an event for any Borel set B. The uncertainty distribution Φ of an uncertain variable ξ was defined by Liu [5] as $\Phi(x) = \mathcal{M}\{\xi \leq x\}$ for any real number x. Peng and Iwamura [35] verified that a function $\Phi : \mathcal{R} \to [0, 1]$ is an uncertainty distribution if and only if it is a monotone increasing function except $\Phi(x) \equiv 0$ and $\Phi(x) \equiv 1$. An uncertainty distribution $\Phi(x)$ is said to be regular if it is a continuous and strictly increasing function with respect to x at which $0 < \Phi(x) < 1$, and:

$$\lim_{x \to -\infty} \Phi(x) = 0, \quad \lim_{x \to +\infty} \Phi(x) = 1.$$

Let ξ be an uncertain variable with regular uncertainty distribution $\Phi(x)$. The inverse function $\Phi^{-1}(\alpha)$ is called the inverse uncertainty distribution of ξ [36]. Liu [37] also verified that a function $\Phi^{-1}(\alpha) : (0, 1) \to \mathcal{R}$ is an inverse uncertainty distribution if and only if it is a continuous and strictly increasing function with respect to α.

The expected value of an uncertain variable ξ was defined by Liu [5] as follows:

$$\mathrm{E}[\xi] = \int_0^{+\infty} \mathcal{M}\{\xi \geq r\}\mathrm{d}r - \int_{-\infty}^{0} \mathcal{M}\{\xi \leq r\}\mathrm{d}r$$

provided that at least one of the two integrals is finite. If ξ has an uncertainty distribution Φ, then the expected value may be calculated by:

$$\mathrm{E}[\xi] = \int_0^{+\infty} (1 - \Phi(x))\mathrm{d}x - \int_{-\infty}^{0} \Phi(x)\mathrm{d}x,$$

or equivalently, $\mathrm{E}[\xi] = \int_{-\infty}^{+\infty} x\mathrm{d}\Phi(x)$. If Φ is also regular, then:

$$\mathrm{E}[\xi] = \int_0^1 \Phi^{-1}(\alpha)\mathrm{d}\alpha.$$

By the literature [34], the uncertain variables $\xi_1, \xi_2, \ldots, \xi_n$ are said to be independent if:

$$\mathcal{M}\left\{\bigcap_{i=1}^{n} (\xi_i \in B_i)\right\} = \bigwedge_{i=1}^{n} \mathcal{M}\{\xi_i \in B_i\}$$

for any Borel sets $B_1, B_2, \ldots, B_n$ of real numbers. Moreover, if ξ and η are independent uncertain variables with finite expected values, then:

$$\mathrm{E}[d_1\xi + d_2\eta] = d_1\mathrm{E}[\xi] + d_2\mathrm{E}[\eta]$$

for any real numbers d_1 and d_2, which is referred to as the linearity of expected value operator.

Lemma 1. *(Liu [36]) Let $\xi_1, \xi_2, \ldots, \xi_n$ be independent uncertain variables with regular uncertainty distributions $\Phi_1, \Phi_2, \ldots, \Phi_n$, respectively. If the function $f(x_1, x_2, \ldots, x_n)$ is strictly increasing with respect to $x_1, x_2, \ldots, x_m$ and strictly decreasing with respect to $x_{m+1}, x_{m+2}, \ldots, x_n$, then $\xi = f(\xi_1, \xi_2, \ldots, \xi_n)$ is also an uncertain variable with the inverse uncertainty distribution:*

$$\Psi^{-1}(\alpha) = f\left(\Phi_1^{-1}(\alpha), \ldots, \Phi_m^{-1}(\alpha), \Phi_{m+1}^{-1}(1-\alpha), \ldots, \Phi_n^{-1}(1-\alpha)\right).$$

Liu and Ha [38] proved that the uncertain variable ξ has an expected value:

$$\mathrm{E}[\xi] = \int_0^1 f\left(\Phi_1^{-1}(\alpha), \ldots, \Phi_m^{-1}(\alpha), \Phi_{m+1}^{-1}(1-\alpha), \ldots, \Phi_n^{-1}(1-\alpha)\right) \mathrm{d}\alpha.$$

Lemma 2. *(Liu and Ha [38]) Let ξ be an uncertain variable on the uncertainty space $(\Theta, \mathcal{L}, \mathcal{M})$ with uncertain distribution $\Phi(x)$. If $g : \mathcal{R} \to \mathcal{R}$ is a monotone function, the expect value of $g(\xi)$ is:*

$$\mathrm{E}\left[g(\xi)\right] = \int_{-\infty}^{+\infty} g(x)\, d\Phi(x).$$

Lemma 3. *(Zhou et al. [26]) Let ξ be an uncertain variable with uncertainty distribution function $\Phi(\cdot)$. Assume that the support of ξ is $[a, b]$, we have:*

$$\mathrm{E}[\max\{\xi, z\}] = b - z\Phi(z) - \int_z^b \Phi(x)dx,$$

where $0 \le a \le z \le b$.

The definition of first-order uncertain dominance [39] is based on the uncertainty distribution functions of uncertain variables as follows:

Definition 1. *Let ζ_1 and ζ_2 be two uncertain variables with the uncertainty distribution functions $\Psi_1(x)$ and $\Psi_2(x)$, respectively. We say that ζ_1 uncertainly dominates ζ_2 in the first order, denoted by $\zeta_1 \succeq \zeta_2$, if and only if $\Psi_1(x) \le \Psi_2(x)$ for all $x \in \mathcal{R}$.*

Chance theory

Let $(\Gamma, \mathcal{L}, \mathcal{M})$ be an uncertainty space and let $(\Omega, \mathcal{A}, \Pr)$ be a probability space. Then, the product $(\Gamma, \mathcal{L}, \mathcal{M}) \times (\Omega, \mathcal{A}, \Pr)$ is called a chance space. Essentially, it is another triplet $(\Gamma \times \Omega, \mathcal{L} \times \mathcal{A}, \mathcal{M} \times \Pr)$, where $\Gamma \times \Omega$ is the universal set, $\mathcal{L} \times \mathcal{A}$ is the product σ-algebra, and $\mathcal{M} \times \Pr$ is the product measure.

The universal set $\Gamma \times \Omega$ is the set of all ordered pairs of the form (ν, ω), where $\nu \in \Gamma$ and $\omega \in \Omega$. That is, $\Gamma \times \Omega = \{(\nu, \omega) | \nu \in \Gamma, \omega \in \Omega\}$. The product σ-algebra $\mathcal{L} \times \mathcal{A}$ is the smallest σ-algebra containing measurable rectangles of the form $\Lambda \times A$, where $\Lambda \in \mathcal{L}$ and $A \in \mathcal{A}$. Any element Θ in $\mathcal{L} \times \mathcal{A}$ is called an event in the chance space. Then, the chance measure of Θ was defined by Liu [6] as:

$$\mathrm{Ch}\{\Theta\} = \int_0^1 \Pr\{\omega \in \Omega | \mathcal{M}\{\nu \in \Gamma | (\nu, \omega) \in \Theta\} \ge x\}\, dx.$$

Liu [6] proved that the chance measure $\mathrm{Ch}\{\Theta\}$ is a monotone increasing function of Θ and $\mathrm{Ch}\{\Lambda \times A\} = \mathcal{M}\{\Lambda\} \times \Pr\{A\}$ for any $\Lambda \in \mathcal{L}$ and any $A \in \mathcal{A}$. Especially, it holds that $\mathrm{Ch}\{\emptyset\} = 0$, $\mathrm{Ch}\{\Gamma \times \Omega\} = 1$. Liu [6] also proved that the chance measure is self-dual. That is, for any event Θ, we have $\mathrm{Ch}\{\Theta\} + \mathrm{Ch}\{\Theta^c\} = 1$. Furthermore, Hou [40] proved that the chance measure is subadditive.

An uncertain random variable was defined by Liu [6] as a function η from a chance space $(\Gamma, \mathcal{L}, \mathcal{M}) \times (\Omega, \mathcal{A}, \Pr)$ to the set of real numbers such that $\{\eta \in B\}$ is an event in $\mathcal{L} \times \mathcal{A}$ for any Borel set B. In addition, Liu [6] defined the chance distribution of an uncertain random variable η as $\Phi(x) = \text{Ch}\{\eta \leq x\}$ for any $x \in \mathcal{R}$.

Assume $\kappa_1, \kappa_2, \ldots, \kappa_m$ are independent random variables with probability distributions $K_1, K_2, \ldots, K_m$, and $\tau_1, \tau_2, \ldots, \tau_n$ are independent uncertain variables with uncertainty distributions $\Upsilon_1, \Upsilon_2, \ldots, \Upsilon_n$, respectively. Liu [29] proved that the uncertain random variable:

$$\eta = f(\kappa_1, \ldots, \kappa_m, \tau_1, \ldots, \tau_n)$$

has a chance distribution $\Phi(x) = \int_{\mathcal{R}^m} F(x; y_1, \ldots, y_m) \mathrm{d}K_1(y_1) \ldots \mathrm{d}K_m(y_m)$, where $F(x; y_1, \ldots, y_m)$ is the uncertainty distribution of uncertain variable:

$$f(y_1, \ldots, y_m, \tau_1, \ldots, \tau_n)$$

for any real numbers $y_1, \ldots, y_m$.

Lemma 4. *(Liu [29]) Let $\kappa_1, \kappa_2, \ldots, \kappa_m$ be independent random variables with probability distributions $K_1, K_2, \ldots, K_m$, respectively, and let $\tau_1, \tau_2, \ldots, \tau_n$ be uncertain variables (not necessarily independent), then the uncertain random variable:*

$$\eta = f(\kappa_1, \ldots, \kappa_m, \tau_1, \ldots, \tau_n)$$

has an expected value:

$$\mathrm{E}[\eta] = \int_{\mathcal{R}^m} \mathrm{E}\left[f(y_1, \ldots, y_m, \tau_1, \ldots, \tau_n)\right] \mathrm{d}K_1(y_1) \ldots \mathrm{d}K_m(y_m),$$

where $E\left[f(y_1, \ldots, y_m, \tau_1, \ldots, \tau_n)\right]$ is the expected value of the uncertain variable $f(y_1, \ldots, y_m, \tau_1, \ldots, \tau_n)$ for any real numbers $y_1, \ldots, y_m$.

The model without prospect utility

In this section, the consumer search model without prospect utility is presented in a hybrid uncertain environment. Furthermore, the optimal stopping rule and optimal search strategy are proposed, and we discuss how the search cost, the probability of stockout, and the risky ordering of products affect the consumer's purchase threshold.

Consider a consumer who seeks sequentially to buy the best selection among a finite number $n \geq 2$ of differentiated but substitutable products indexed by $1, 2, \ldots, n$. The consumer must incur a search cost c_i in order to search product i and evaluate it, $\forall\, i \in \{1, 2, \ldots, n\}$. It is assumed that the consumer searches sequentially, such that a consumer is able to evaluate product i before deciding whether or not to search the next product. The subjective valuation of the consumer to product i is assumed to be an uncertain variable ξ_i with support $[a, b]$, where $0 \leq a < b < +\infty$. Denote by $\Phi_i(x)$ the uncertainty distribution of ξ_i and let $\phi_i(x) = \mathrm{d}\Phi_i(x)/\mathrm{d}x$. Search is also with limited recall. Having searched product i, he/she is free to return to buy its product at any later point only if product i is in stock with probability p. The consumer has two choices, one is to continue searching, i.e., he/she would compare the current product with the best previous product and choose a better one as a reserved product; alternatively, he/she can stop searching and buy the reserved product.

Let the collection $I = \{1, 2, \ldots, n\}$ be partitioned into the set Q of encountered products and its complement $\overline{Q}$ of not encountered products. That is, $Q \cup \overline{Q} = I$, and $Q \cap \overline{Q} = \varnothing$.

Let z be the valuation of the reserved product when the set of encountered products is Q, i.e.,

$$z = \max_{i \in Q} \left\{ \hat{\xi}_i \right\}, \tag{1}$$

where $\hat{\xi}_i$ is a realization of ξ_i. For ease of exposition, the product with valuation z is simply expressed as the reserved product z. Let $V(z, Q)$ represent the value function of the consumer, with the choice variable being a binary decision between stopping or continuing the search when the set of encountered products is Q. Define $\Pi(z, Q)$ as the consumer's benefit from choosing to stop now. When the consumer buys a product, we suppose that the benefit is equal to the subjective valuation of the product. That is to say,

$$\Pi(z, Q) = z. \tag{2}$$

The objective of the consumer is to find an optimal search strategy maximizing the benefit from the search. Such a problem can be formally described in a dynamic programming format [41]. Therefore, we can express the value function $V(z, Q)$ as follows:

$$V(z, Q) = \max \left\{ \Pi(z, Q), \max_{i \in \overline{Q}} \{ \beta \mathrm{E} \left[V\left(\eta_i, Q \cup \{i\} \right) \right] - c_i \} \right\}, \tag{3}$$

where $\eta_i = \max\{\xi_i, z\}$ is an uncertain random variable and β denotes a discount factor with $0 < \beta < 1$. That is, the reserved product z may be unpurchasable due to out of stock with probability $1 - p$, and the reserved product z may be purchasable with probability p.

According to Zhou et al. [26] and Lippman and McCall [42], the solution to the above multi-period dynamic programming problem is equivalent to solving a one-period static problem satisfying myopic property.

Starting from a point in time, we suppress the time dimension of the problem. To a consumer, the benefit from stopping the search is $\Pi_{\text{stop}}(z) = z$. If the consumer continues the search and the realized valuation of the next product is $\max\{\hat{\xi}_i, z\}$, the *ex post* benefit from continuing the search is:

$$\Pi_{\text{search}} \left(\hat{\xi}_i, z \right) = \beta \max \left\{ \hat{\xi}_i, z \right\} - c_i. \tag{4}$$

Therefore, the expected benefit from continuing the search becomes:

$$\begin{aligned} \Pi_{\text{search}}(z) &= \mathrm{E} \left[\Pi_{\text{search}}(\xi_i, z) \right] = \beta \mathrm{E} \left[\max\{\xi_i, z\} \right] - c_i \\ &= \beta p \mathrm{E} \left[\max\{\xi_i, z\} \right] + \beta(1-p) \mathrm{E} \left[\max\{\xi_i, 0\} \right] - c_i. \end{aligned} \tag{5}$$

By Lemma 3, Equation 5 can be rewritten as follows:

$$\Pi_{\text{search}}(z) = \beta b - \beta p z \Phi_i(z) - \beta \int_z^b \Phi_i(x) \mathrm{d}x - \beta(1-p) \int_a^z \Phi_i(x) \mathrm{d}x - c_i. \tag{6}$$

By the myopic property of the optimal search strategy, we define a difference function as follows:

$$\begin{aligned} D(z) &\triangleq \Pi_{\text{search}}(z) - \Pi_{\text{stop}}(z) \\ &= \beta b - z - \beta p z \Phi_i(z) - \beta \int_z^b \Phi_i(x) \mathrm{d}x - \beta(1-p) \int_a^z \Phi_i(x) \mathrm{d}x - c_i. \end{aligned} \tag{7}$$

Note that:

$$\frac{\mathrm{d}D(z)}{\mathrm{d}z} = -1 - \beta p z \phi_i(z) < 0. \tag{8}$$

Assume that $D(a) > 0$ and $D(b) < 0$. Since $D(z)$ is decreasing with z, the solution z_i^{LN} of the equation $D\left(z_i^{LN}\right) = 0$ exists uniquely. Thus, the critical number z_i^{LN} which satisfies:

$$\beta b - z_i^{LN} - \beta p z_i^{LN} \Phi_i(z_i^{LN}) - \beta \int_{z_i^{LN}}^{b} \Phi_i(x)\mathrm{d}x - \beta(1-p) \int_{a}^{z_i^{LN}} \Phi_i(x)\mathrm{d}x - c_i = 0 \quad (9)$$

is called the reservation value of product i.

Remark. If the consumer has historical data of the product value, then the valuation may be rationally characterized as random variable. A model similar to Equation 3 can be built to maximize the consumer's expected benefit. Suppose that the probability distribution function of the product i's valuation is $F_i(x)$. The other parameters and assumptions are similar to those of the model without prospect utility. However, the reservation value z_i^* satisfies:

$$\beta b - z_i^* - \beta \int_{a}^{b} F_i(x)\mathrm{d}x + \beta p \int_{a}^{z_i^*} F_i(x)\mathrm{d}x - c_i = 0.$$

□

According to the properties of reservation value, the optimal stopping rule and the optimal search strategy can be prescribed as follows:

Theorem 1. *(Optimal stopping rule) If $D(z) \le 0$, stop the search by buying the better product between the current product and the reserved product z, or else continue the search.*

The optimal search strategy specifies the order in which alternative reserved products should be evaluated in detail and at which point evaluations should stop and a product be purchased. This strategy is specified and proven to be optimal in Weitzman [43] and involves the following procedure:

Step 1. Rank the products according to their respective reservation value.
Step 2. Evaluate the products in this order.
Step 3. Buy the first product which obtained score after the detailed evaluation is greater than the reservation value of the next product to be evaluated.

To discuss more properties of the reservation value, we obtain the following proposition:

Proposition 1.

1) *The reservation value is decreasing with respect to the search cost.*
2) *The reservation value is increasing with respect to the probability of unpurchasable.*
3) *If product ξ_i uncertainly dominates product ξ_j in the first order and $c_i = c_j$, $\forall i, j \in I$, then the reservation value z_i^{LN} of firm i's product is larger than the reservation value z_j^{LN} of product j.*

Proof.

1) For convenience, define:

$$H_i(z) = \beta b - z - \beta p z \Phi_i(z) - \beta \int_{z}^{b} \Phi_i(x)\mathrm{d}x - \beta(1-p) \int_{a}^{z} \Phi_i(x)\mathrm{d}x. \quad (10)$$

By taking the derivative of two sides of the equation $H_i\left(z_i^{LN}\right) = c_i$ with respect to c_i, we have $-\frac{\partial z_i^{LN}}{\partial c_i} - \beta p z_i^{LN} \phi_i\left(z_i^{LN}\right) \frac{\partial z_i^{LN}}{\partial c_i} = 1$. Furthermore, $\frac{\partial z_i^{LN}}{\partial c_i} = \frac{1}{-1-\beta p z_i^{LN} \phi_i\left(z_i^{LN}\right)} < 0$. Therefore, the reservation value z_i^{LN} is decreasing with the search cost c_i.

2) By taking the partial derivative of the variable z_i^{LN} of equation $D\left(z_i^{LN}\right) = 0$ with respect to p, we have:

$$-\frac{\partial z_i^{LN}}{\partial p} - \beta z_i^{LN} \Phi_i\left(z_i^{LN}\right) - \beta p z_i^{LN} \phi_i\left(z_i^{LN}\right) \frac{\partial z_i^{LN}}{\partial p} + \beta \int_a^{z_i^{LN}} \Phi_i(x) \mathrm{d}x = 0.$$

Furthermore,

$$\left[1 + \beta p z_i^{LN} \phi_i\left(z_i^{LN}\right)\right] \frac{\partial z_i^{LN}}{\partial p} = -\beta z_i^{LN} \Phi_i\left(z_i^{LN}\right) + \beta \int_a^{z_i^{LN}} \Phi_i(x) \mathrm{d}x.$$

By using integration by parts,

$$\left[1 + \beta p z_i^{LN} \phi_i\left(z_i^{LN}\right)\right] \frac{\partial z_i^{LN}}{\partial p} = -\beta \int_a^{z_i^{LN}} x \phi_i(x) \mathrm{d}x.$$

Therefore,

$$\frac{\partial z_i^{LN}}{\partial p} = \frac{-\beta \int_a^{z_i^{LN}} x \phi_i(x) \mathrm{d}x}{1 + \beta p z_i^{LN} \phi_i\left(z_i^{LN}\right)} < 0.$$

Hence, the reservation value z_i^{LN} is increasing with the probability $1 - p$ of unpurchasable.

3) Since $\xi_i \succeq \xi_j$, it follows from Definition 1 that $\Phi_i(x) \leq \Phi_j(x), \forall i, j \in I$. Therefore, we can obtain $\int_a^z \Phi_i(x) \mathrm{d}x \leq \int_a^z \Phi_j(x) \mathrm{d}x$ and $\int_z^b \Phi_i(x) \mathrm{d}x \leq \int_z^b \Phi_j(x) \mathrm{d}x$. From Equation 10,

$$\begin{aligned} H_i(z) &= \beta b - z - \beta p z \Phi_i(z) - \beta \int_z^b \Phi_i(x) \mathrm{d}x - \beta(1-p) \int_a^z \Phi_i(x) \mathrm{d}x \\ &\geq \beta b - z - \beta p z \Phi_j(z) - \beta \int_z^b \Phi_j(x) \mathrm{d}x - \beta(1-p) \int_a^z \Phi_j(x) \mathrm{d}x = H_j(z). \end{aligned} \tag{11}$$

Suppose that $c_i = c_j$. By Equation 11, $H_i\left(z_i^{LN}\right) = c_i = c_j = H_j\left(z_j^{LN}\right) \geq H_j\left(z_i^{LN}\right)$. Since $H_i(z)$ is strictly decreasing in z, we have $z_i^{LN} \geq z_j^{LN}$.

□

The result 1) in Proposition 1 shows that an increase in the search cost decreases the reservation value. That is, if the search cost is very costly, the consumer's purchase threshold will be lowered. There is widespread evidence that the search cost often influences the consumer's purchase threshold negatively. Furthermore, we demonstrate that an increase in the probability of stockout will yield higher reservation value. It implies that greater probability of stockout leads to higher purchase threshold. Finally, the result implies that the product that has a good reputation will be preferred in the search process.

The model with prospect utility

In this section, the optimal consumer search model is investigated when consumers are loss averse for a prospect utility function. As discussed before, the optimal stopping rule and optimal search strategy are also proposed. In addition, we contrast the two reservation values in the models with and without prospect utility.

Prospect theory has attracted a good deal of attention as an alternative to the well-established expected utility theory. The basic idea of prospect theory is that alternatives are evaluated by a prospect utility function in terms of gains and losses with respect to a reference point. According to the descriptive foundations of value function proposed by Kahneman and Tversky [2,4], define the prospect utility function as follows:

$$U(h,k) = u(h) + R(u(h) - u(k)), \tag{12}$$

where $u(\cdot)$ represents the instantaneous utility function, and $R(\cdot)$ represents the gain-loss utility function that satisfies the following form:

$$R(u(h) - u(k)) = \begin{cases} (u(h) - u(k))^g, & \text{if } u(h) - u(k) \geq 0 \\ -\gamma\,(u(k) - u(h))^l, & \text{if } u(h) - u(k) < 0, \end{cases} \tag{13}$$

where the parameters $0 < g < 1$ and $0 < l < 1$ represent the coefficients of risk attitude for gains and losses, respectively. As g and l become higher, the consumer tends to take more risks. In addition, the parameter γ represents the coefficient of loss aversion. When $\gamma > 1$, the consumer is more sensitive to the loss. It means that people have a relative distaste for losses when comparing very small losses to very small gains. There is strong empirical evidence on loss aversion, in which the loss aversion index γ is estimated between 1.75 and 2.5 [44]. Therefore, the gain-loss utility function $R(\Delta u)$ is an S-shaped function of the reference utility gap $\Delta u = u(h) - u(k)$, with a concave kink at $\Delta u = 0$.

Following the above prospect utility function, we define the prospect utility function of optimal consumer search as follows:

$$U_{\text{search}}(\hat{\xi}_i, z) = \Pi_{\text{search}}(\hat{\xi}_i, z) + R(\Pi_{\text{search}}(\hat{\xi}_i, z) - \Pi_{\text{stop}}(z)), \tag{14}$$

where

$$\begin{aligned} &R(\Pi_{\text{search}}(\hat{\xi}_i, z) - \Pi_{\text{stop}}(z)) \\ &= \begin{cases} \left(\Pi_{\text{search}}(\hat{\xi}_i, z) - \Pi_{\text{stop}}(z)\right)^g, & \text{if } \Pi_{\text{search}}(\hat{\xi}_i, z) - \Pi_{\text{stop}}(z) \geq 0 \\ -\gamma\left(\Pi_{\text{stop}}(z) - \Pi_{\text{search}}(\hat{\xi}_i, z)\right)^l, & \text{if } \Pi_{\text{search}}(\hat{\xi}_i, z) - \Pi_{\text{stop}}(z) < 0. \end{cases} \end{aligned} \tag{15}$$

Before the consumer decides to make the investment to search, he/she is assumed to be able to anticipate his/her feelings of gains and losses based on the realized prospect utility. When the reserved product z is given, the expected utility from continuing the search can be written as follows:

$$\begin{aligned} U_{\text{search}}(z) &= \mathrm{E}\left[U_{\text{search}}(\xi_i, z)\right] \\ &= \mathrm{E}\left[\Pi_{\text{search}}(\xi_i, z)\right] + \mathrm{E}\left[R(\Pi_{\text{search}}(\xi_i, z) - \Pi_{\text{stop}}(z))\right] \\ &= \Pi_{\text{search}}(z) + \mathrm{E}\left[R(\Pi_{\text{search}}(\xi_i, z) - \Pi_{\text{stop}}(z))\right]. \end{aligned} \tag{16}$$

Since $R(\Pi_{\text{search}}(\xi_i, z) - \Pi_{\text{stop}}(z))$ is an uncertain random variable, it follows from Lemma 4 that:

$$\begin{aligned} T_R(z) &= \mathrm{E}\left[R(\Pi_{\text{search}}(\xi_i, z) - \Pi_{\text{stop}}(z))\right] \\ &= p\mathrm{E}\left[R\left(\beta \max\{\xi_i, z\} - c_i - z\right)\right] + (1-p)\mathrm{E}\left[R\left(\beta\xi_i - c_i - \xi_i\right)\right] \\ &= p\int_{\frac{z+c_i}{\beta}}^{b} (\beta x - c_i - z)^g\, \mathrm{d}\Phi_i(x) - p\gamma \int_z^{\frac{z+c_i}{\beta}} (z - \beta x + c_i)^l\, \mathrm{d}\Phi_i(x) \\ &\quad -p\gamma\int_a^z (z - \beta z + c_i)^l\, \mathrm{d}\Phi_i(x) - (1-p)\gamma\int_a^b (x - \beta x + c_i)^l\, \mathrm{d}\Phi_i(x). \end{aligned} \tag{17}$$

The first term on the right hand side of Equation 17 represents the expected gain utility from reserving the current product and continuing the search. The second term $-p\gamma \int_z^{\frac{z+c_i}{\beta}} (z-\beta x+c_i)^l \,\mathrm{d}\Phi_i(x)$ represents the expected loss utility from buying the current product and stopping the search. The third term $-p\gamma \int_a^z (z-\beta z+c_i)^l \,\mathrm{d}\Phi_i(x)$ represents the expected loss utility from buying the reserved product z and stopping the search if it is available. The last term $-(1-p)\gamma \int_a^b (x-\beta x+c_i)^l \,\mathrm{d}\Phi_i(x)$ is the expected loss utility from buying the current product and stopping the search if it is not available.

In a similar way without regard to the prospect utility, we consider that the expected utility from stopping the search is equivalent to the realized evaluation value z of the reserved product. Hence, we have:

$$\mathcal{U}_{\text{stop}}(z) = \Pi_{\text{stop}}(z) = z. \tag{18}$$

Similarly, we also define a difference function as follows:

$$\begin{aligned} D^p(z) &\triangleq \mathcal{U}_{\text{search}}(z) - \mathcal{U}_{\text{stop}}(z) \\ &= \Pi_{\text{search}}(z) + T_R(z) - \Pi_{\text{stop}}(z) \\ &= D(z) + T_R(z). \end{aligned} \tag{19}$$

Therefore, the optimal stopping rule with prospect utility can be prescribed as follows:

Theorem 2. *(Optimal stopping rule) If $D^p(z) \le 0$, stop the search by buying the better product between the current product and the reserved product z, or else continue the search.*

To derive the reservation value of optimal consumer search with prospect utility, it is easy to prove that:

$$\frac{\mathrm{d}D^p(z)}{\mathrm{d}z} < 0. \tag{20}$$

Note that $\frac{\mathrm{d}D(z)}{\mathrm{d}z} < 0$,

$$\frac{\partial\left(p\int_{\frac{z+c_i}{\beta}}^b (\beta x - c_i - z)^g \,\mathrm{d}\Phi_i(x)\right)}{\partial z} = -pg\int_{\frac{z+c_i}{\beta}}^b \frac{1}{(\beta x - c_i - z)^{1-g}}\mathrm{d}\Phi_i(x) < 0, \tag{21}$$

$$\begin{aligned} &\frac{\partial\left(-p\gamma\int_z^{\frac{z+c_i}{\beta}} (z-\beta x+c_i)^l \mathrm{d}\Phi_i(x)\right)}{\partial z} \\ &= -p\gamma l\int_z^{\frac{z+c_i}{\beta}} \frac{1}{(z-\beta x+c_i)^{1-l}}\mathrm{d}\Phi_i(x) + p\gamma(z-\beta z+c_i)^l\phi_i(z), \end{aligned}$$

$$\begin{aligned} &\frac{\partial\left(-p\gamma\int_a^z (z-\beta z+c_i)^l \mathrm{d}\Phi_i(x)\right)}{\partial z} \\ &= -p\gamma l(1-\beta)\int_a^z \frac{1}{(z-\beta z+c_i)^{1-l}}\mathrm{d}\Phi_i(x) - p\gamma(z-\beta z+c_i)^l\phi_i(z). \end{aligned}$$

Futhermore, we obtain:

$$\frac{\partial\left(-p\gamma\int_z^{\frac{z+c_i}{\beta}} (z-\beta x+c_i)^l \,\mathrm{d}\Phi_i(x)\right)}{\partial z} + \frac{\partial\left(-p\gamma\int_a^z (z-\beta z+c_i)^l \,\mathrm{d}\Phi_i(x)\right)}{\partial z} < 0. \tag{22}$$

Hence, it follows from Equations 8, 21, and 22 that $\frac{\mathrm{d}D^p(z)}{\mathrm{d}z} < 0$. It implies that an increase in the evaluation value z of reserved product will decrease the expected prospect utility. Furthermore, since $D^p(z)$ is decreasing with z, the solution z_i^{LA} of the equation $D^p\left(z_i^{LA}\right) = 0$ also uniquely exists. Thus, the critical number z_i^{LA} which satisfies $D\left(z_i^{LA}\right) + T_R\left(z_i^{LA}\right) = 0$ is also the reservation value of firm i's product under the prospect utility.

Proposition 2. *If $\beta b - c_i < z_i^{LN} < b$, the reservation value in the model with prospect utility will be lower than that in the model without prospect utility.*

Proof. Note that $D\left(z_i^{LN}\right) = 0$. By Equation 19, we have:

$$\begin{aligned} D^p\left(z_i^{LN}\right) &= D\left(z_i^{LN}\right) + T_R\left(z_i^{LN}\right) \\ &= T_R\left(z_i^{LN}\right) \\ &= p\int_a^b R\left(\beta \max\{x, z_i^{LN}\} - c_i - z_i^{LN}\right) \mathrm{d}\Phi_i(x) + (1-p)\int_a^b R(\beta x - c_i - x)\, \mathrm{d}\Phi_i(x) \\ &= p\int_a^b R\left(\beta \max\{x, z_i^{LN}\} - c_i - z_i^{LN}\right) \mathrm{d}\Phi_i(x) - (1-p)\gamma\int_a^b (x - \beta x + c_i)^l\, \mathrm{d}\Phi_i(x). \end{aligned}$$

By using the first mean value theorem of integral, we obtain that if there exists $\bar{x} \in [a, b]$, the above equation can be rewritten as follows:

$$D^p\left(z_i^{LN}\right) = pR\left(\beta \max\left\{\bar{x}, z_i^{LN}\right\} - c_i - z_i^{LN}\right) \phi_i(\bar{x})(b-a) - (1-p)\gamma\int_a^b (x - \beta x + c_i)^l\, \mathrm{d}\Phi_i(x).$$

If $\beta b - c_i < z_i^{LN} < b$, we have $\bar{x} \leq b < \frac{z_i^{LN} + c_i}{\beta}$. To prove the assertion, we consider the following two cases:

Case 1. When $z_i^{LN} \leq \bar{x} < \frac{z_i^{LN} + c_i}{\beta}$, we have:

$$\begin{aligned} D^p\left(z_i^{LN}\right) &= pR\left(\beta \bar{x} - c_i - z_i^{LN}\right) \phi_i(\bar{x})(b-a) - (1-p)\gamma\int_a^b (x - \beta x + c_i)^l\, \mathrm{d}\Phi_i(x) \\ &< -(1-p)\gamma\int_a^b (x - \beta x + c_i)^l\, \mathrm{d}\Phi_i(x) < 0. \end{aligned}$$

□

Case 2. When $\bar{x} < z_i^{LN}$, we also have:

$$\begin{aligned} D^p\left(z_i^{LN}\right) &= pR\left(\beta z_i^{LN} - c_i - z_i^{LN}\right) \phi_i(\bar{x})(b-a) - (1-p)\gamma\int_a^b (x - \beta x + c_i)^l\, \mathrm{d}\Phi_i(x) \\ &< -(1-p)\gamma\int_a^b (x - \beta x + c_i)^l\, \mathrm{d}\Phi_i(x) < 0. \end{aligned}$$

Hence, if $\beta b - c_i < z_i^{LN} < b$, we have $D^p\left(z_i^{LN}\right) < 0 = D^p\left(z_i^{LA}\right)$. Since $D^p(z)$ is decreasing with z, we have $z_i^{LA} < z_i^{LN}$. The theorem is thus proven. □

Proposition 2 indicates that the prospect utility leads to low reservation value. If the consumer is more loss averse, we conclude that the purchase threshold will be lower. The optimal search strategy and Proposition 1 also hold for the prospect utility case.

Comparative statics

In this section, we derive some comparative statics of how the reservation value under the prospect utility would respond to the changes in the parameters g, l, and γ. We first examine the effect of risk coefficient g for gains on reservation value z_i^{LA}.

Proposition 3. *Holding other parameters of the model constant, the reservation value of the firm's product under the prospect utility, which satisfies $a \leq z_i^{LA} < \beta b - c_i - 2$, is increasing with respect to the coefficient of risk attitude for gains. And that which satisfies $\beta b - c_i - 2 \leq z_i^{LA} \leq b$ is decreasing with respect to the coefficient of risk attitude for gains.*

Proof. From the previous discussion of $D^p(z)$, we can obtain that the function $D^p(z)$ is continuous, differentiable in z. By taking the partial derivative of the variable z_i^{LA} of equation $D^p\left(z_i^{LA}\right) = 0$ with respect to g, we have:

$$-\frac{\partial z_i^{LA}}{\partial g} - \beta pz\phi_i(z)\frac{\partial z_i^{LA}}{\partial g} + p\int_{\frac{z+c_i}{\beta}}^{b} (\beta x - c_i - z)^g \ln(\beta x - c_i - z)\phi_i(x)\mathrm{d}x$$
$$-pg\int_{\frac{z+c_i}{\beta}}^{b} (\beta x - c_i - z)^{g-1}\frac{\partial z_i^{LA}}{\partial g}\phi_i(x)\mathrm{d}x - p\gamma l\int_z^{\frac{z+c_i}{\beta}} (z-\beta x+c_i)^{l-1}\phi_i(x)\frac{\partial z_i^{LA}}{\partial g}\mathrm{d}x$$
$$-p\gamma l(1-\beta)\int_a^z (z-\beta z+c_i)^{l-1}\phi_i(x)\frac{\partial z_i^{LA}}{\partial g}\mathrm{d}x = 0$$

by using the equation:

$$\frac{\partial\,(\beta x - c_i - z)^g}{\partial g} = (\beta x - c_i - z)^g \ln(\beta x - c_i - z) - g\,(\beta x - c_i - z)^{g-1}\,\frac{\partial z}{\partial g}.$$

Furthermore,

$$\frac{\partial z_i^{LA}}{\partial g} = \frac{p\int_{\frac{z+c_i}{\beta}}^{b} (\beta x - c_i - z)^g \ln(\beta x - c_i - z)\phi_i(x)\mathrm{d}x}{N},$$

where:

$$N = 1 + \beta pz\phi_i(z) + pg\int_{\frac{z+c_i}{\beta}}^{b} (\beta x - c_i - z)^{g-1}\,\phi_i(x)\mathrm{d}x$$
$$+p\gamma l\int_z^{\frac{z+c_i}{\beta}} (z - \beta x + c_i)^{l-1}\,\phi_i(x)\mathrm{d}x + p\gamma l(1-\beta)\int_a^z (z - \beta z + c_i)^{l-1}\,\phi_i(x)\mathrm{d}x.$$

Since $\ln(\beta x - c_i - z)$ could be positive or negative, it holds that $\frac{\partial z_i^{LA}}{\partial g} > 0$ if $a \le z_i^{LA} < \beta b - c_i - 2$, and $\frac{\partial z_i^{LA}}{\partial g} < 0$ if $\beta b - c_i - 2 \le z_i^{LA} \le b$. □

Proposition 3 indicates that as the risk coefficient g for gains becomes higher, the consumer raises his/her reservation value of low-value product and reduces his/her reservation value of high-value product. This implies that the consumer's purchase threshold for low-value product will be increased with the coefficient of risk attitude for gains, and the consumer's purchase threshold for high-value product will be decreased with the coefficient of risk attitude for gains.

Next, we examine the effect of risk coefficient l for losses on reservation value z_i^{LA}.

Proposition 4. *Holding other parameters of the model constant, if* $a \le z_i^{LA} < \frac{1-c_i}{1-\beta}$ *and* $\frac{a+b}{2} \le \frac{1-c_i}{1-\beta}$, *the reservation value under the prospect utility is increasing with respect to the coefficient of risk attitude for losses. If* $\frac{2-c_i}{1-\beta} < z_i^{LA} \le b$ *and* $\frac{1-c_i}{1-\beta} < \frac{a+b}{2}$, *the reservation value under the prospect utility is decreasing with respect to the coefficient of risk attitude for losses.*

Proof. By taking the partial derivative of the variable z_i^{LA} of equation $D^p\left(z_i^{LA}\right) = 0$ with respect to l, we have:

$$-\frac{\partial z_i^{LA}}{\partial l} - \beta pz\phi_i(z)\frac{\partial z_i^{LA}}{\partial l} - pg\int_{\frac{z+c_i}{\beta}}^{b} (\beta x - c_i - z)^{g-1}\,\phi_i(x)\frac{\partial z_i^{LA}}{\partial l}\mathrm{d}x$$
$$-p\gamma\int_z^{\frac{z+c_i}{\beta}}(z-\beta x+c_i)^l\ln(z-\beta x+c_i)\phi_i(x)\mathrm{d}x - p\gamma l\int_z^{\frac{z+c_i}{\beta}}(z-\beta x+c_i)^{l-1}\phi_i(x)\frac{\partial z_i^{LA}}{\partial l}\mathrm{d}x$$
$$-p\gamma\int_a^z(z-\beta z+c_i)^l\ln(z-\beta z+c_i)\phi_i(x)\mathrm{d}x - p\gamma l(1-\beta)\int_a^z(z-\beta z+c_i)^{l-1}\,\phi_i(x)\frac{\partial z_i^{LA}}{\partial l}\mathrm{d}x$$
$$-(1-p)\gamma\int_a^b(x-\beta x+c_i)^l\ln(x-\beta x+c_i)\phi_i(x)\mathrm{d}x = 0.$$

Furthermore,

$$\frac{\partial z_i^{LA}}{\partial l} = \frac{-p\gamma \int_z^{\frac{z+c_i}{\beta}} S(x)\phi_i(x)\mathrm{d}x - p\gamma \int_a^z S(z)\phi_i(x)\mathrm{d}x - (1-p)\gamma \int_a^b S^*(x)\phi_i(x)\mathrm{d}x}{N},$$

where:

$$N = 1 + \beta pz\phi_i(z) + pg \int_{\frac{z+c_i}{\beta}}^b (\beta x - c_i - z)^{g-1} \phi_i(x)\mathrm{d}x$$
$$+ p\gamma l \int_z^{\frac{z+c_i}{\beta}} (z - \beta x + c_i)^{l-1} \phi_i(x)\mathrm{d}x + p\gamma l(1-\beta)\int_a^z (z - \beta z + c_i)^{l-1} \phi_i(x)\mathrm{d}x,$$

$S(x) = (z - \beta x + c_i)^l \ln(z - \beta x + c_i)$ and $S^*(x) = (x - \beta x + c_i)^l \ln(x - \beta x + c_i)$. Since $a \le z_i^{LA} < \frac{1-c_i}{1-\beta} < \frac{2-c_i}{1-\beta}$, it holds that $-p\gamma \int_z^{\frac{z+c_i}{\beta}} S(x)\phi_i(x)\mathrm{d}x > 0$. Furthermore, it follows from $z_i^{LA} < \frac{1-c_i}{1-\beta}$ that $-p\gamma \int_a^z S(z)\phi_i(x)\mathrm{d}x > 0$. When $\frac{a+b}{2} \le \frac{1-c_i}{1-\beta}$, we have $-(1-p)\gamma \int_a^b S^*(x)\phi_i(x)\mathrm{d}x > 0$. Hence, if $a \le z_i^{LA} < \frac{1-c_i}{1-\beta}$ and $\frac{a+b}{2} \le \frac{1-c_i}{1-\beta}$, $\frac{\partial z_i^{LA}}{\partial l} > 0$.

If $\frac{2-c_i}{1-\beta} < z_i^{LA} \le b$, it holds that $-p\gamma \int_z^{\frac{z+c_i}{\beta}} S(x)\phi_i(x)\mathrm{d}x < 0$. Moreover, it follows from $\frac{1-c_i}{1-\beta} < \frac{2-c_i}{1-\beta} < z_i^{LA}$ that $-p\gamma \int_a^z S(z)\phi_i(x)\mathrm{d}x < 0$. When $\frac{1-c_i}{1-\beta} < \frac{a+b}{2}$, we have $-(1-p)\gamma \int_a^b S^*(x)\phi_i(x)\mathrm{d}x < 0$. Consequently, if $\frac{2-c_i}{1-\beta} < z_i^{LA} \le b$ and $\frac{1-c_i}{1-\beta} < \frac{a+b}{2}$, we have $\frac{\partial z_i^{LA}}{\partial l} < 0$. □

Proposition 4 indicates that an increase in the the risk coefficient l for losses raises the reservation value of low-value product and reduces the reservation value of high-value product. This means that when the coefficient of risk attitude for losses is very high, the consumer would raise the purchase threshold for low-value product and reduce the purchase threshold for high-value product.

Finally, we can investigate the effect of coefficient of loss aversion on the reservation value z_i^{LA}.

Proposition 5. *Holding other parameters of the model constant, the reservation value of the firm's product under the prospect utility is decreasing with respect to the coefficient of loss aversion.*

Proof. By taking the partial derivative of the variable z_i^{LA} of equation $D^p\left(z_i^{LA}\right) = 0$ with respect to γ, we obtain:

$$-\frac{\partial z_i^{LA}}{\partial \gamma} - \beta pz\phi_i(z)\frac{\partial z_i^{LA}}{\partial \gamma} - pg \int_{\frac{z+c_i}{\beta}}^b (\beta x - c_i - z)^{g-1} \phi_i(x)\frac{\partial z_i^{LA}}{\partial \gamma}\mathrm{d}x$$
$$-p \int_z^{\frac{z+c_i}{\beta}} (z-\beta x+c_i)^l\phi_i(x)\mathrm{d}x - p\gamma l \int_z^{\frac{z+c_i}{\beta}} (z-\beta x+c_i)^{l-1}\phi_i(x)\frac{\partial z_i^{LA}}{\partial \gamma}\mathrm{d}x$$
$$-p\int_a^z (z-\beta z+c_i)^l\phi_i(x)\mathrm{d}x - p\gamma l(1-\beta)\int_a^z (z - \beta z + c_i)^{l-1} \phi_i(x)\frac{\partial z_i^{LA}}{\partial \gamma}\mathrm{d}x$$
$$-(1-p)\int_a^b (x - \beta x + c_i)^l \phi_i(x)\mathrm{d}x = 0.$$

Moreover, we have:

$$\frac{\partial z_i^{LA}}{\partial \gamma} = \frac{-p \int_z^{\frac{z+c_i}{\beta}} W_1\phi_i(x)\mathrm{d}x - p\int_a^z W_2\phi_i(x)\mathrm{d}x - (1-p)\int_a^b W_3\phi_i(x)\mathrm{d}x}{N} < 0,$$

where:

$$N = 1 + \beta pz\phi_i(z) + pg \int_{\frac{z+c_i}{\beta}}^b (\beta x - c_i - z)^{g-1} \phi_i(x)\mathrm{d}x$$
$$+p\gamma l \int_z^{\frac{z+c_i}{\beta}} (z - \beta x + c_i)^{l-1} \phi_i(x)\mathrm{d}x + p\gamma l(1-\beta)\int_a^z (z - \beta z + c_i)^{l-1} \phi_i(x)\mathrm{d}x$$

and $W_1 = (z - \beta x + c_i)^l$, $W_2 = (z - \beta z + c_i)^l$, $W_3 = (x - \beta x + c_i)^l$. □

From Proposition 5, it shows that the reservation value of the firm's product is decreasing with the coefficient of loss aversion, that is to say, the consumer who has more sensitive to the loss will search for much less products. Hence, the above propositions make it clear that risk attitude for searching high-value product and loss aversion would indeed lead to 'too little search.'

Numerical examples

In this section, some numerical examples are given to illustrate the reservation values under the models without and with prospect utility and examine the effects of risk attitude for gains, risk attitude for losses, and loss aversion on the optimal search strategy.

Suppose a consumer must decide to buy one among three products. Because the three products are differentiated but substitutable, the consumer only knows different uncertainty distributions about the three products. Without loss of generality, suppose the consumer's valuations for the three products are linear uncertain variables $\mathcal{L}(1,9)$, $\mathcal{Z}(1,4,9)$, and $\mathcal{Z}(1,7,9)$, respectively. Let $\beta = 0.9$, $p = 0.7$, $c_1 = 0.5$, $c_2 = 1$, and $c_3 = 1.5$. By Equation 9, the reservation values of products can be calculated as $z_1^{LN} = 3.5$, $z_2^{LN} = 2.4$, and $z_3^{LN} = 3.6$. Therefore, the consumer can rank the products according to their respective reservation value so that $z_3^{LN} > z_1^{LN} > z_2^{LN}$. Then, the consumer can evaluate three products in accordance with the order of the reservation value and make a purchase decision through the optimal search strategy without prospect utility. Similarly, we can also obtain the prospect utility case by the equation $D^p\left(z_i^{LA}\right) = 0$.

To illustrate the reservation values under the models without and with prospect utility numerically, suppose the parameters $g = 0.5$, $l = 0.5$, and $\gamma = 2$. Without loss of generality, we only consider the first product when the parameters $\beta = 0.9$, $p = 0.7$, and $c_1 = 0.5$ remain unchanged. From Figure 1, we observe that $z_1^{LA} < z_1^{LN}$. However, the sufficient condition $7.6 < z_1^{LN} < 9$ of Proposition 2 cannot hold. It implies that the reservation value z_i^{LN} which does not satisfy the sufficient condition may give rise to the result of Proposition 2.

We find our key results continue to hold in our numerical studies. When the risk coefficient g for gains can be chosen in the set $\{0.5, 0.7, 0.9\}$, Figure 2 reports our findings for $\beta = 0.9, p = 0.7, c_1 = 0.5, l = 0.5$, and $\gamma = 2$, which is representative, and other parameter combinations considered in our studies also yield similar results. As expected, Figure 2 shows the reservation value of low-value product which satisfies $1 \leq z_1^{LA} < 5.6$ will be increased with the coefficient of risk attitude for gains. When the risk coefficient l for losses can be chosen in the set $\{0.5, 0.7, 0.9\}$ and the other parameters $\beta = 0.9, p = 0.7$, $c_1 = 0.5$, $g = 0.5$, and $\gamma = 2$ remain unchanged, Figure 3 shows that the reservation value of low-value product which satisfies $1 \leq z_1^{LA} < 5$ is increasing with respect to the coefficient of risk attitude for losses.

Finally, the effect of loss aversion on the reservation value can be examined. Without loss of generality, suppose the coefficient γ of loss aversion can be chosen in the set $\{1.5, 2, 2.5\}$ and the other parameters $\beta = 0.9, p = 0.7, c_1 = 0.5, g = 0.5$, and $l = 0.5$ remain unchanged. From Figure 4, it is apparent that the reservation value of the firm's product is decreasing with the coefficient of loss aversion.

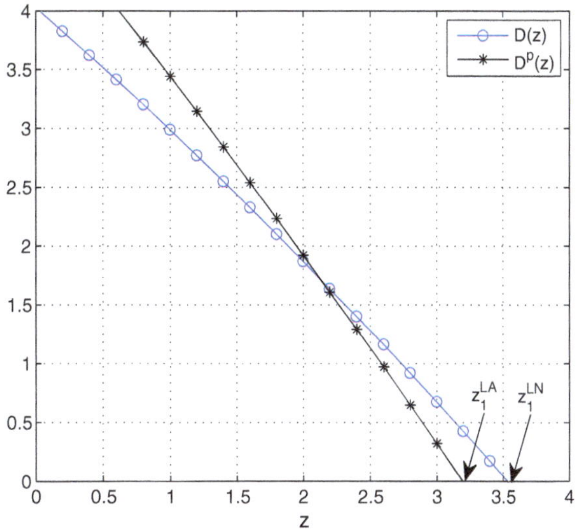

Figure 1 A comparison between reservation values.

Conclusion

This paper presented a consumer search model with prospect utility, in which the subjective valuations of the consumer to every firm's product were assumed to be independent but not identically distributed uncertain variables, and the valuation of a reserved product with probabilistic stockout was consequently characterized as an uncertain random variable. An optimal stopping rule and an optimal search strategy were proposed to maximize the benefit of the consumer. Moreover, we studied the consumer's purchase threshold in the model without and with prospect utility. The results demonstrated that the prospect utility will lead to the low purchase threshold. We found that the consumer's purchase threshold is decreasing with search cost and is increasing with the probability of stockout. Furthermore, we also found that the product that has a good reputation will become the preferred product in the search process. In addition, the risk

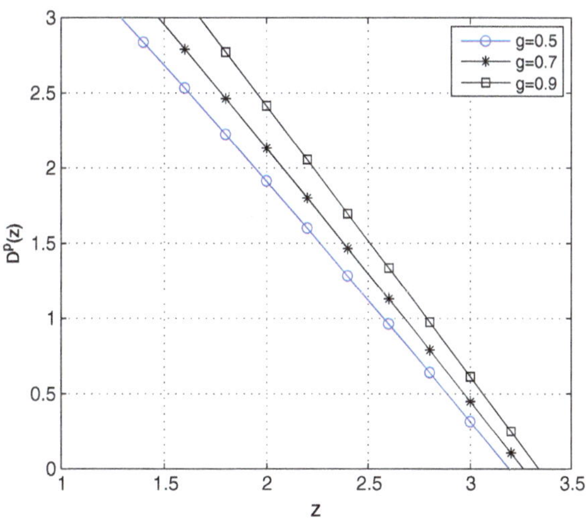

Figure 2 Effect of risk attitude for gains on the reservation value of low-value product.

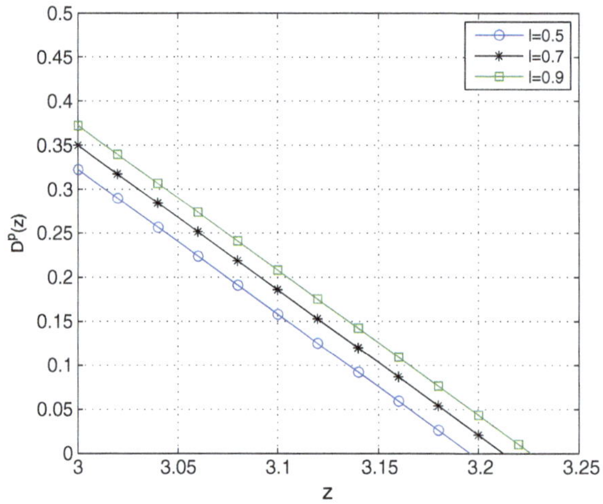

Figure 3 Effect of risk attitude for losses on the reservation value of low-value product.

coefficients for gains and losses will raise the consumer's purchase threshold for low-value product, while they will reduce the purchase threshold for high-value product. To provide a simple explanation for why consumers generally stop searching too early, we showed that the loss aversion will reduce the consumer's purchase threshold. Finally, some numerical examples were given to illustrate the effectiveness of the proposed models and conclusions.

In the future work, we can incorporate advertisement into consumer search in the presence of loss aversion. For instance, firms will post-advertise to improve their goodwill and induce consumers to buy their products. In addition, it is also interesting to study consumer inertia in the dynamic search and examine the effects of prospect utility on the consumer inertia.

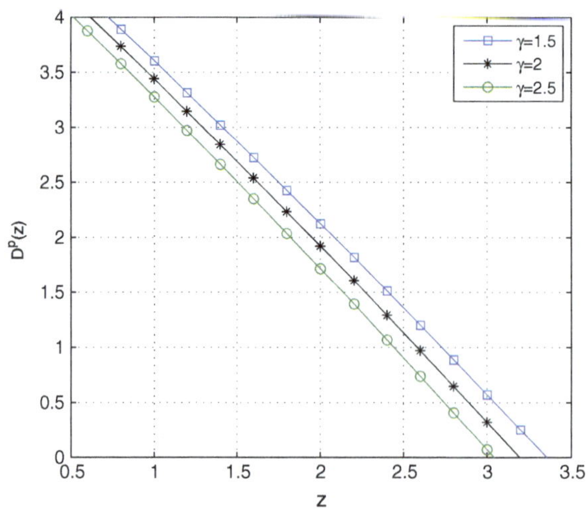

Figure 4 Effect of loss aversion on the reservation value.

Endnotes

[a]This statistical result is publicly available at http://hbr.org/2004/05/stock-outs-cause-walkouts.

[b]This means the product which obtained low score. Similarly, 'high-value product' means the product which obtained high score.

Competing interests

The authors declare that they have no competing interests.

Acknowledgements

This work is supported by the National Natural Science Foundation of China (Nos. 71071106 and 71371133) and supported partially by the Specialized Research Fund for the Doctoral Program of Higher Education of China (No. 20120032110071) and the Program for New Century Excellent Talents in Universities of China.

References

1. Schunk, D: Behavioral heterogeneity in dynamic search situations: theory and experimental evidence. J. Econ. Dyn. Control. **33**(9), 1719–1738 (2009)
2. Tversky, A: Prospect theory: an analysis of decision under risk. Econometrica: J. Econometric Soc. **47**(2), 263–291 (1979)
3. Tversky, A, Kahneman, D: Loss aversion in riskless choice: a reference-dependent model. Q. J. Econ. **106**(4), 1039–1061 (1991)
4. Tversky, A, Kahneman, D: Advances in prospect theory: cumulative representation of uncertainty. J. Risk Uncertainty. **5**(4), 297–323 (1992)
5. Liu, B: Uncertainty Theory. 2nd edn. Springer, Berlin (2007)
6. Liu, Y: Uncertain random variables: a mixture of uncertainty and randomness. Soft Comput. **17**(4), 625–634 (2013)
7. Branco, F, Sun, M, Villas-Boas, JM: Optimal search for product information. Manage. Sci. **58**(11), 2037–2056 (2012)
8. Schunk, D, Winter, J: The relationship between risk attitudes and heuristics in search tasks: a laboratory experiment. J. Econ. Behavior Organization. **71**(2), 347–360 (2009)
9. Stahl, DO: Oligopolistic pricing with sequential consumer search. Am. Econ. Rev. **79**(4), 700–712 (1989)
10. Cachon, GP, Terwiesch, C, Xu, Y: On the effects of consumer search and firm entry in a multiproduct competitive market. Marketing Sci. **27**(3), 461–473 (2008)
11. Arbatskaya, M: Ordered search. RAND J. Econ. **38**(1), 119–126 (2007)
12. Wilson, CM: Ordered search and equilibrium obfuscation. Int. J. Ind. Organization. **28**(5), 496–506 (2010)
13. Zhou, J: Ordered search in differentiated markets. Int. J. Ind. Organization. **29**(2), 253–262 (2011)
14. Chen, Y, Zhang, T: Equilibrium price dispersion with heterogeneous searchers. Int. J. Ind. Organization. **29**(6), 645–654 (2011)
15. Haan, MA, Moraga-González, JL: Advertising for attention in a consumer search model. Econ. J. **121**(552), 552–579 (2011)
16. Chen, Y, He, C: Paid placement: advertising and search on the internet. Econ. J. **121**(556), 309–328 (2011)
17. Hardie, BG, Johnson, EJ, Fader, PS: Modeling loss aversion and reference dependence effects on brand choice. Marketing Sci. **12**(4), 378–394 (1993)
18. Weng, Z: Consumer search and firm-worker reciprocity: a behavioral approach. PhD thesis, The Ohio State University (2010)
19. Zhou, J: Reference dependence and market competition. J. Econ. Manage. Strategy. **20**(4), 1073–1097 (2011)
20. Kuksov, D, Wang, K: The bright side of loss aversion in dynamic and competitive markets. Marketing Sci. **33**(5), 693–711 (2014)
21. Bowman, D, Minehart, D, Rabin, M: Loss aversion in a consumption-savings model. J. Econ. Behavior Organization. **38**(2), 155–178 (1999)
22. Popescu, I, Wu, Y: Dynamic pricing strategies with reference effects. Operations Res. **55**(3), 413–429 (2007)
23. Foellmi, R, Rosenblatt-Wisch, R, Schenk-Hoppé, KR: Consumption paths under prospect utility in an optimal growth model. J. Econ. Dyn. Control. **35**(3), 273–281 (2011)
24. Mu, R, Lan, Y, Tang, W: An uncertain contract model for rural migrant worker's employment problems. Fuzzy Optimization Decision Making. **12**(1), 29–39 (2013)
25. Lan, Y, Du, H: A warranty contract model with uncertain demands. Inf.-An Int. Interdisplinary J. **16**(2), 1037–1045 (2013)
26. Zhou, C, Tang, W, Zhao, R: An uncertain search model for recruitment problem with enterprise performance. J. Intell. Manuf. (2014). doi:10.1007/s10845-014-0997-1
27. Wang, X, Lan, Y, Wang, J: An uncertain wage contract model with adverse selection and moral hazard. J. Appl. Mathematics. **2014**(Article ID 282867), 1–9 (2014)
28. Wu, X, Zhao, R, Tang, W: Optimal contracts for the agency problem with multiple uncertain information. Knowledge-Based Syst. **59**, 161–172 (2014)
29. Liu, Y: Uncertain random programming with applications. Fuzzy Optimization Decision Making. **12**(2), 153–169 (2013)
30. Gao, J, Yao, K: Some concepts and theorems of uncertain random process. Int. J. Intell. Syst. to be published (2014)
31. Liu, B: Uncertain random graph and uncertain random network. J. Uncertain Syst. **8**(1), 3–12 (2014)
32. Sheng, Y, Yao, K: Some formulas of variance of uncertain random variable. J. Uncertainty Anal. Appl. **2**(1), 12 (2014)
33. Zhou, J, Yang, F, Wang, K: Multi-objective optimization in uncertain random environments. Fuzzy Optimization Decision Making. **13**(4), 397–413 (2014)

34. Liu, B: Some research problems in uncertainty theory. J. Uncertain Syst. **3**(1), 3–10 (2009)
35. Peng, Z, Iwamura, K: A sufficient and necessary condition of uncertainty distribution. J. Interdisciplinary Mathematics. **13**(3), 277–285 (2010)
36. Liu, B: Uncertainty Theory: A Branch of Mathematics for Modeling Human Uncertainty. Springer, Berlin (2010)
37. Liu, B: Toward uncertain finance theory. J. Uncertainty Anal. Appl. **1**, 1 (2013)
38. Liu, Y, Ha, M: Expected value of function of uncertain variables. J. Uncertain Syst. **4**(3), 181–186 (2010)
39. Gao, Z, Guo, H: Some properties of uncertain dominance. Int. Inf. Inst. (Tokyo). Inf. **15**(12), 168–173 (2013)
40. Hou, Y: Subadditivity of chance measure. Journal of Uncertainty Analysis and Applications. **2**, 14 (2014)
41. DeGroot, MH: Optimal Statistical Decisions. (John Wiley & Sons, Inc., Hoboken (2004)
42. Lippman, SA, McCall, JJ: The economics of job search: a survey. Econ. Inquiry. **14**(2), 155–189 (1976)
43. Weitzman, ML: Optimal search for the best alternative. Econometrica: J. Econometric Soc. **47**(3), 641–654 (1979)
44. Abdellaoui, M, Bleichrodt, H, Paraschiv, C: Loss aversion under prospect theory: a parameter-free measurement. Manage. Sci. **53**(10), 1659–1674 (2007)

Permissions

All chapters in this book were first published in JUAA, by Springer; hereby published with permission under the Creative Commons Attribution License or equivalent. Every chapter published in this book has been scrutinized by our experts. Their significance has been extensively debated. The topics covered herein carry significant findings which will fuel the growth of the discipline. They may even be implemented as practical applications or may be referred to as a beginning point for another development.

The contributors of this book come from diverse backgrounds, making this book a truly international effort. This book will bring forth new frontiers with its revolutionizing research information and detailed analysis of the nascent developments around the world.

We would like to thank all the contributing authors for lending their expertise to make the book truly unique. They have played a crucial role in the development of this book. Without their invaluable contributions this book wouldn't have been possible. They have made vital efforts to compile up to date information on the varied aspects of this subject to make this book a valuable addition to the collection of many professionals and students.

This book was conceptualized with the vision of imparting up-to-date information and advanced data in this field. To ensure the same, a matchless editorial board was set up. Every individual on the board went through rigorous rounds of assessment to prove their worth. After which they invested a large part of their time researching and compiling the most relevant data for our readers.

The editorial board has been involved in producing this book since its inception. They have spent rigorous hours researching and exploring the diverse topics which have resulted in the successful publishing of this book. They have passed on their knowledge of decades through this book. To expedite this challenging task, the publisher supported the team at every step. A small team of assistant editors was also appointed to further simplify the editing procedure and attain best results for the readers.

Apart from the editorial board, the designing team has also invested a significant amount of their time in understanding the subject and creating the most relevant covers. They scrutinized every image to scout for the most suitable representation of the subject and create an appropriate cover for the book.

The publishing team has been an ardent support to the editorial, designing and production team. Their endless efforts to recruit the best for this project, has resulted in the accomplishment of this book. They are a veteran in the field of academics and their pool of knowledge is as vast as their experience in printing. Their expertise and guidance has proved useful at every step. Their uncompromising quality standards have made this book an exceptional effort. Their encouragement from time to time has been an inspiration for everyone.

The publisher and the editorial board hope that this book will prove to be a valuable piece of knowledge for researchers, students, practitioners and scholars across the globe.

List of Contributors

Dipak Kumar Jana
Department of Engineering Science, Haldia Institute of Technology, Haldia Purba Midna Pur 721657, West Bengal, India

Kalipada Maity
Department of Mathematics, Mugberia Gangadhar Mahavidyalaya, Mugberia Purba Medinipur 721425, West Bengal, India

Tapan Kumar Roy
Department of Mathematics, Bengal Engineering and Science University, Shibpur Howrah 711103, West Bengal, India

Xiaosheng Wang
School of Science, Hebei University of Engineering, Handan, 056038, China

Zixiong Peng
School of Applied Mathematics, Central University of Finance and Economics, Beijing 100081, China

Hua Ke
School of Economics and Management, Tongji University, Shanghai 200092, China

Praveen Kumar Shukla
Department of Information Technology, Babu Banarasi Das Northern India Institute of Technology, Lucknow 226028, India

Surya Prakash Tripathi
Department of Computer Science & Engineering, Institute of Engineering & Technology, Lucknow 226021, India

Arthur Seibel
Workgroup on System Technologies and Engineering Design Methodology, Hamburg University of Technology, 21073 Hamburg, Germany

Andreas Japing
Workgroup on System Technologies and Engineering Design Methodology, Hamburg University of Technology, 21073 Hamburg, Germany

Josef Schlattmann
Workgroup on System Technologies and Engineering Design Methodology, Hamburg University of Technology, 21073 Hamburg, Germany

Xiaowei Chen
Department of Risk Management and Insurance, Nankai University, Tianjin 300071, China

Yuhong Sheng
Department of Mathematical Sciences, Tsinghua University, Beijing 100084, China

Kai Yao
School of Management, University of Chinese Academy of Sciences, Beijing 100190, China

Samar Hazari
Department of Mathematics, National Institute of Technology, Durgapur, West Bengal 713209, India

Kalipada Maity
Department of Mathematics, Mugberia Gangadhar Mahavidyalaya, Bhupatinagar Purba Medinipur, West Bengal 721425, India

Jayanta Kumar Dey
Department of Mathematics, Mahisadal Raj College, Purba-Medinipur, West Bengal 721628, India

Samarjit Kar
Department of Mathematics, National Institute of Technology, Durgapur, West Bengal 713209, India

Jin Peng
Institute of Uncertain Systems, Huanggang Normal University, Huanggang, Hubei 438000, China

Bo Zhang
School of Statistics and Mathematics, Zhongnan University of Economics and Law, Wuhan, Hubei 430073, China

Shengguo Li
Institute of Uncertain Systems, Huanggang Normal University, Huanggang, Hubei 438000, China

Jinwu Gao
School of Information, Renmin University of China, Beijing 100872, China

Haiying Guo
School of Sciences, Hebei University of Engineering, Handan 056038, China

Xiang Li
School of Economics and Management, Beijing University of Chemical Technology, No. 15, Beisanhuan East Road, 100029 Beijing, China

Yuan Gao
State Key Laboratory of Rail Traffic Control and Safety, Beijing Jiaotong University, Beijing 100044, China

Rong Gao
Department of Mathematical Sciences, Tsinghua University, Beijing 100084, China

Lixing Yang
State Key Laboratory of Rail Traffic Control and Safety, Beijing Jiaotong University, Beijing 100044, China

Agustín Alvarez
Instituto de Calculo, FCEN, University of Buenos Aires, Intendente Guiraldes 2160, Ciudad Universitaria Pab II (1428), Buenos Aires, Argentina

Juan P Pinasco
IMAS (UBA-CONICET) and Departamento de Matematica, FCEN, University of Buenos Aires, Intendente Guiraldes 2160, Ciudad Universitaria Pab I (1428), Buenos Aires, Argentina

Yan Zhang
State Key Laboratory of Rail Traffic Control and Safety, Beijing Jiaotong University, Beijing 100044, China

Pei Liu
State Key Laboratory of Rail Traffic Control and Safety, Beijing Jiaotong University, Beijing 100044, China

Ling Chen
School of Science, Nanjing University of Science and Technology, Nanjing 210094, China

Yun Sun
School of Science, Nanjing University of Science and Technology, Nanjing 210094, China

Yuanguo Zhu
School of Science, Nanjing University of Science and Technology, Nanjing 210094, China

Chi Zhou
Institute of Systems Engineering, Tianjin University, Tianjin 300072, China

Wansheng Tang
Institute of Systems Engineering, Tianjin University, Tianjin 300072, China

Ruiqing Zhao
Institute of Systems Engineering, Tianjin University, Tianjin 300072, China